AF687218

BAR INTERNATIONAL SERIES 3035 | 2021

Sounds of Laughter, Shades of Life

Pleistocene to modern hominin occupations of the Bau de l'Aubesier rock shelter, Vaucluse, France

LUCY WILSON

BAR PUBLISHING

Published in 2021 by
BAR Publishing, Oxford

BAR International Series 3035

Sounds of Laughter, Shades of Life

ISBN 978 1 4073 5838 3 paperback
ISBN 978 1 4073 5839 0 e-format

DOI https://doi.org/10.30861/9781407358383

A catalogue record for this book is available from the British Library

© Lucy Wilson 2021

COVER IMAGE *The view from the Bau de l'Aubesier, looking north into the gorge of the Nesque river (photo L. Wilson). Inset pictures, left:* Equus mosbachensis *dental rows (photo P. Fernandez); centre: a modern horse, Storm, to serve as an example of what hominins hunted so much (photo K. Sterritt); right:* Bos primigenius primigenius *dental row (photo P. Fernandez).*
PHOTOGRAPHIES DE COUVERTURE: *Vue du Bau au nord vers les gorges de la Nesque (cliché L. Wilson). Images encadrées, à gauche: rangées dentaires d'*Equus mosbachensis *(cliché P. Fernandez); au centre: un cheval actuel, Storm, à titre d'exemple de cette proie si convoitée des hominidés (cliché K. Sterritt); à droite: rangée dentaire de* Bos primigenius primigenius *(cliché P. Fernandez).*

The Author's moral rights under the 1988 UK Copyright,
Designs and Patents Act, are hereby expressly asserted.

All rights reserved. No part of this work may be copied, reproduced, stored, sold, distributed, scanned, saved in any form of digital format or transmitted in any form digitally, without the written permission of the Publisher.

Links to third party websites are provided by BAR Publishing in good faith and for information only. BAR Publishing disclaims any responsibility for the materials contained in any third party website referenced in this work.

BAR
PUBLISHING

BAR titles are available from:

BAR Publishing
122 Banbury Rd, Oxford, OX2 7BP, UK
EMAIL info@barpublishing.com
PHONE +44 (0)1865 310431
FAX +44 (0)1865 316916
www.barpublishing.com

Of Related Interest

I coltelli degli ultimi Neandertal
Strategie tecnologiche e comportamentali alla fine del Paleolitico Medio
Davide Delpiano

Oxford, BAR Publishing, 2021 BAR International Series **3025**

Competition Between Humans and Large Carnivores
Case studies from the Late Middle and Upper Palaeolithic of the Central Balkans
Stefan Milošević

Oxford, BAR Publishing, 2020 BAR International Series **2963**

Crânes trophées, crânes d'ancêtres et autres pratiques autour de la tête : problèmes d'interprétation en archéologie
Actes de la table ronde pluridisciplinaire, musée national de Préhistoire, Les Eyzies-de-Tayac (Dordogne, France), 14-16 octobre 2010
Edited by Bruno Boulestin and Dominique Henry Gambier

Oxford, BAR Publishing, 2012 BAR International Series **2415**

Unravelling the Palaeolithic
Ten years of research at the Centre for the Archaeology of Human Origins (CAHO, University of Southampton)
Edited by Karen Ruebens, Iza Romanowska and Rachel Bynoe

Oxford, BAR Publishing, 2012 BAR International Series **2400**

For more information, or to purchase these titles, please visit **www.barpublishing.com**

Acknowledgements

This book benefitted greatly from reviews by Philippe Fernandez, whose knowledge of the site, of analytical techniques and of the relevant palaeontological literature are much appreciated by the author, and by Peter Mears, who helped me clarify certain explanations and did some statistical analyses. Philippe also helped to improve my French. Merci, Philippe! Thanks, Peter! Three anonymous reviewers also provided very useful feedback, and I thank them for their efforts. Any errors that remain are of course my own.

All photographs are either by the author or used with permission. The photo in Figure 2.6 was taken with my camera in 1989 but since I am in it, I did not take it; I do not know who did, and consider it “mine”. All maps, and Figure 1.11, were made by C. Browne and L. Wilson. Figure 1.12 was made by L. Wilson, C. Browne, and redrawn by P. Mears; all other figures are by the author. My thanks to Philippe Fernandez for pictures of fauna, Erik Trinkaus for pictures of the hominin teeth, and to Michel Olive and Xavier Delestre for permission to use the photograph of the mandible in chapter 4:

Photographie de la mandibule humaine du Bau de l’Aubesier :
©MC - DRAC/SRA PACA - Michel OLIVE.

The title, and some words in chapter 12, are from the song “Across The Universe” by the Beatles, written by John Lennon and Paul McCartney:

Across The Universe
Words and Music by John Lennon and Paul McCartney
Copyright (c) 1968, 1970 Sony Music Publishing LLC
Copyright Renewed
All Rights Administered by Sony Music Publishing LLC, 424 Church Street, Suite 1200, Nashville, TN 37219
International Copyright Secured All Rights Reserved
Reprinted by Permission of Hal Leonard LLC

C’est surtout mes amis Georges et Marie-Josée Ughetto, et Muriel Bertrand, qui m’ont poussée à écrire ce livre. Je trouve que c’était une très bonne idée, et je les remercie et dédie ce *magnum opus* à eux.

Contents

List of Figures

List of Tables

Abstract

The Bau de l'Aubesier is a large rock shelter in the gorges de la Nesque, Vaucluse, France. First excavated in the early 1900s, it was immediately recognised as containing abundant and important Middle Palaeolithic remains, but received little attention until a Canadian, Serge Lebel, led an international team to study it from 1987 until the early 2000s. These excavations showed that the site is deeper and more extensive than had previously been recognised, and unearthed very abundant lithic and faunal remains, as well as hominin remains (teeth and a pathological half-mandible) from several pre-Neandertal and Neandertal individuals. The nearly 13 m deep sediments consist primarily of cryoclastic debris, much of which has been strongly cemented into a dense breccia, which has protected and preserved the archaeological levels – and made them difficult to excavate! There are at least ten layers, with the main ones dating from the later Middle Pleistocene (about 250,000 years old) to the Upper Pleistocene, about 100,000 years old. The lowest layers, K2 through I, are rich both in lithics, including some of the earliest blade production in western Europe, and in faunal material, particularly from aurochs and horses. The site is one of the best archives of *Equus* species known from this time period in southern France. Layer I yielded the half-mandible, and also a hominin tooth with a toothpick groove, while an even older tooth was found in layer K. Towards the middle of the stratigraphy there is a highly burned layer, H1, which preserves large quantities of burned bones and tools, along with repeated instances of fire use, forming a thick, black, greasy layer. It also yielded a large fragmented rock, which served as an anvil so that bones could be broken open for their marrow. In the upper part of the deposits are two trenches: the Tranchée Moulin along the western wall, which was the scene of the earlier excavations, and the tranchée L, at the centre of the site. The tranchée L is well known for its layer 4 (the "couche 4"), which is intensely burned, and extends over more than 40 m^2. This layer provided the great majority of lithic remains in the site: more than 40,000 pieces, from a Typical Mousterian industry, rich in sidescrapers. It also dominates the faunal assemblage from the upper deposits, and provided 6 Neandertal teeth. While both of the two richest layers in the site, layers 4 and H1, are intensely burned, the extensive deposits in layer 4 are featureless (at least visibly), while the hearth area is more constrained in the older layer, H1. The uppermost metre or so of sediments yielded undiagnostic lithic and faunal remains which may be Middle Palaeolithic or even Upper Palaeolithic. In this synthesis volume, these results are combined, along with a detailed study of the raw material procurement behaviours for each layer, the evidence for different hunting strategies used, and the environmental indicators, to give an overall view of how hominins used the site and lived their lives throughout this important period of human evolution.

Résumé

Le Bau de l'Aubesier est un grand abri-sous-roche situé dans les gorges de la Nesque, Vaucluse, France. Fouillé pour la première fois au début du 20ème siècle, il a été reconnu très tôt pour ses restes abondants et importants du Paléolithique moyen, mais n'a été fouillé de manière scientifique qu'à partir de 1987, jusqu'au début des années 2000, par une équipe internationale dirigée par un canadien, Serge Lebel. Ces récentes fouilles ont montré que les dépôts étaient plus profonds et plus étendus que ceux dégagés auparavant, et ont livré des vestiges lithiques et fauniques très abondants ainsi que des restes humains de Néandertaliens et pré-Néandertaliens (des dents et une hémi-mandibule pathologique). Les sédiments, épais de presque 13 m, comprennent pour la plupart des fragments cryoclastiques, souvent cimentés qui forment une brèche dense qui a protégé et préservé les vestiges – et rendu leur fouille difficile! Il y a au moins 10 couches archéologiques datant essentiellement du Pléistocène moyen supérieur (à partir de 250 000 ans) et du Pléistocène supérieur. La partie supérieure des dépôts comprend environ un mètre de sédiments livrant des vestiges non-diagnostiques qui pourraient provenir du Paléolithique moyen ou peut-être supérieur. Les deux couches archéologiques les plus riches, 4 et H1, livrent des restes intensément brûlés dans des sédiments noirs. La couche 4 est épaisse et s'étend sur une quarantaine de mètres carrés sans particularité évidente, tandis que la couche H1, plus vieille, a livré une aire de combustion de même épaisseur mais plus circonscrite dans l'espace. Au Bau on trouve également une industrie laminaire parmi les plus vieilles en Europe de l'ouest ainsi qu'une faune du Pléistocène moyen et supérieur exceptionnellement riche avec en particulier beaucoup de chevaux. Cet ouvrage de synthèse combine pour chacune des couches les résultats obtenus concernant les comportements d'approvisionnement en matières premières lithiques, les différentes stratégies de chasse utilisées par les groupes humains ainsi que des informations sur leur cadre paléoenvironnemental. L'ensemble permet d'aborder la façon dont les hominidés ont fréquenté, utilisé et surtout vécu au Bau de l'Aubesier tout au long de cette période importante de l'évolution humaine.

1

Introduction and Context

1.1. Introduction

This book is about a rock shelter in the south of France, called the Bau de l'Aubesier (hereinafter the Bau) (Figure 1.1). It is about the place, its geographic context, what it is, how it evolved through time, and how people used it from time to time over a period of about 300,000 years. It is not about the excavations of the site, but of course almost all of what I know about the Bau stems from excavations, particularly those started in 1987, in which I participated. I sometimes want to write stories or relate details of those excavations. If the reason for that is to explain how our techniques influence the knowledge I have of the site, I include that material in the main body of the text. If it is more that I feel the need to get something off my chest, then those stories are asides, called "From The Chest", which do not need to fit in with the flow of the main text.

The book is primarily written in English, but figure and table captions are in both English and French, and excerpts from site reports, my own excavation notes, etc., have generally been left in French. The different parts of the site are often referred to using their French names, *e.g.* the tranchée L (tranchée means trench) and the couche 4 (couche means layer). Each chapter has a "Sommaire du chapitre" section at the end, for French-only readers.

Often, books like this about an archaeological site (or project) have a series of individual papers by various experts, and one chapter synthesising the collection. This book is not like that. Many people have contributed their expertise to the analysis of the finds and the site. However, years have passed since we stopped excavating and people have moved on. I cite their reports, use their results and appreciate their contributions, but my goal is to produce a synthesis of all available data, not a compartmentalised series of summaries of different aspects. The Bau *is* a very special place. People used it, over and over. They left so much behind that we can learn a lot about what they did and why they did it. That is what I want to show in this book.

One of my main reasons for doing prehistoric archaeology is that I wonder what people in the past were like. I use the word "people" because I have to call them something, but what I mean are pre-modern hominins. I am not particularly interested in the archaeology of modern humans, because they are modern: they have all modern capabilities. Humans are, of course, intrinsically interesting, even modern ones, but what I am *particularly* interested in is figuring out the capacities of pre-modern humans/hominins. No such people exist today, and we can not understand them just by analogy with modern humans or with other primates such as chimpanzees. Pre-modern people were not modern: they did not have our capabilities and behaviour patterns, or at least not all of them. But which did they have and when did they get them? And what did they do or could they do which was different from us? They also were not apes or other primates. Although there are commonalities between all primates, and between apes and hominins especially, there is no simple line from monkeys-to-apes-to-hominins-to-humans, with upgrades added at each step along the way. We are all unique, with some shared characteristics and some entirely our own. Thus, although I recognise that ethnographic and primatological information provide very useful contexts for our interpretations, what I have always tried to work on is figuring out how to find out what prehistoric people actually did, in order to see if I can figure out how and why they did it. What were *their* capabilities, whether or not some of them were like ours? The Bau is such a rich site that it contributes real data towards solving those questions, which is another reason that it is a special place.

Mostly, though, it is a special place because of the feeling that people get when they are there: it is majestic and welcoming. It has a spirit. It has been a refuge and a

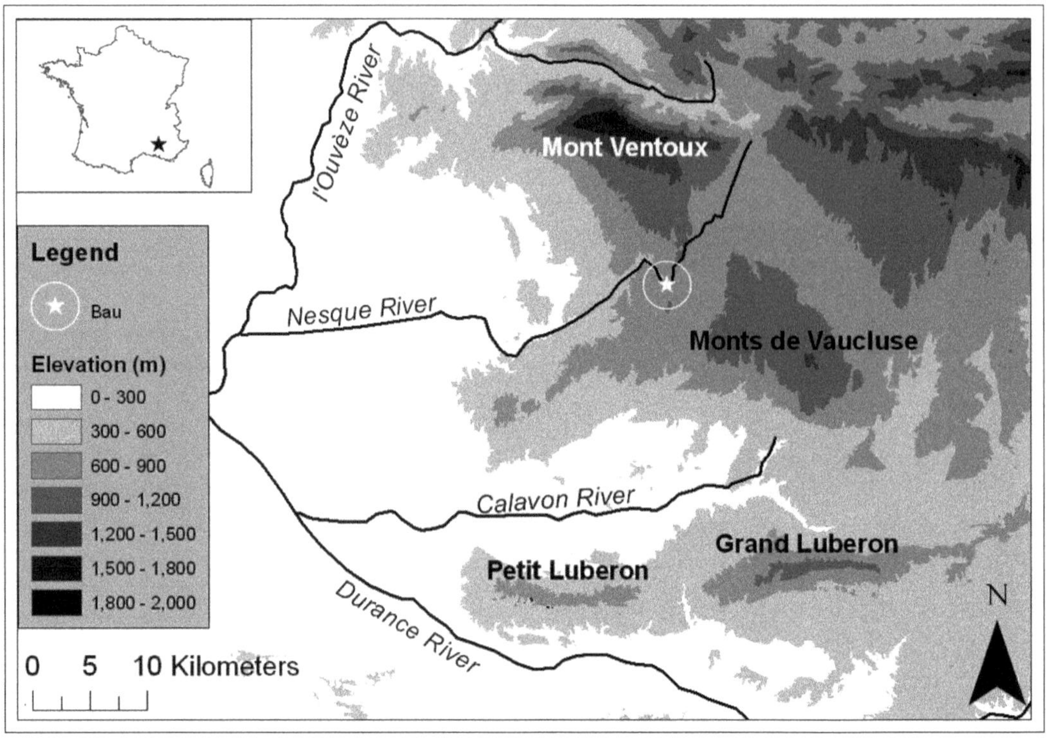

Figure 1.1. Map of the study area, showing the location of the Bau de l'Aubesier, and other important regional geographic features. ***Carte de la région d'étude, avec la localisation du Bau de l'Aubesier, et la géographie régionale.***

home, even if only a temporary one, for hundreds, even thousands of generations of hominins, including modern archaeologists. I do not know if I can explain that well enough for readers to feel it: maybe you have to go there. But I do try.

I start by describing the rock shelter and its location. Then I begin at the base, and work my way up through the stratigraphy, describing the sediments (and how they accumulated), the environment at the time of each level, human occupation and what they left behind (fauna, tools, in some cases hearths), and what these tell us. My own research concerns stone tool raw materials, so I include not only what types of tools there are, but also what they are made of, where that came from, and what the use of raw materials and fauna can tell us about use of the landscape: what strategies each group or groups (since most layers are palimpsests created by an unknown number of people over an uncertain length of time) used to live their lives. The book includes plenty of photographs, maps and other figures to help people envisage the site, its environs, and its contents. Some photographs were taken under field conditions and are not very clear; I apologise for this. They are all I have. I also apologise for not having any scientific drawings of the stone tools, although they would add clarity to the presentation of the tools. Given the recent circumstances, I have not been able to go to France since 2019, and have had to work with what I already had.

1.2. Why is the Bau a special place? What IS the Bau?

The Bau is a large rock shelter, approximately 25 to 30 metres wide, with deposits extending approximately 26 or more metres from the back wall to the front opening, although not all of that area is currently covered by the rock overhang. Part of the current surface of the deposits is covered by huge boulders, chunks of the rock shelter roof, which fell down during either late glacial or post-glacial times. They have had time to become partly buried in sediments, but the archaeological layers are safely underneath them. Within the site there are sediments which accumulated in roughly horizontal layers; these are known to be at least 13 metres deep, but the actual bedrock base of the site has not been attained. One striking aspect of these sediments is that they are, in many places, very strongly cemented, forming a tough breccia which both preserves the archaeological material, and makes it very difficult to dig that material up!

One apparently odd thing about the Bau, in terms of it being a place that people would choose to live at, over and over again, is that it faces north. It gets very little sunlight even in summer today. In the past, when the overhang was more extensive, no sunlight would have reached even the front of the deposits. This might be an advantage during a heatwave in an interglacial, but during glacial periods? It does seem strange that people would choose to live in the shade when it was cold, but they did.

The front of the deposits is presently sloped, the exposed parts of the walls of the rock shelter are irregular, and there may well be bedrock steps underneath parts of the site, so the total content is impossible to calculate. Nonetheless, it must contain hundreds of cubic metres of sediments. Even though all excavations combined have probably removed (as a rough but conservative estimate) over 160 m^3 of sediments from the site, there is still a lot of sediment there. Within that sediment, we have identified a number of archaeological layers, as will be described below. Dates for the deposits will also be described below, but the sediments cover the period running from the later part of the Middle Pleistocene through the Upper Pleistocene; the oldest known archaeological layer is probably around 250,000 years old, and the youngest main archaeological layer is probably younger than 80,000 years old (and may be considerably younger than that, but still Pleistocene). That is excluding hearsay accounts of occasional Neolithic potsherds found on top of the deposits by early excavators (corroborated by personal observation of some old collections).

The Bau is tucked away in the corner of a huge semi-circular escarpment (Figure 1.2), with a vertical cliff of about 50 m height above it, and a steep wooded slope, also about 50 m in height, below it. It is in the gorge of the Nesque river (Figure 1.1), near Monieux, in the northeastern part of the Vaucluse. The Nesque is presently an intermittent river, with most of its course through the gorge containing water only in periods of sufficient rainfall or snowmelt. Some areas do retain water longer, but turn into more or less deep stagnant puddles in the summer, or dry up completely. The intermittent nature of the river has three causes: the fact that a lot of the water is captured upstream to supply local villages and farmers; a relatively warm, dry Mediterranean-type climate with seasonal precipitation; and the influence of karst phenomena. The gorge carves through the Lower Cretaceous (Bedoulian) limestone which underlies much of the region. Over time limestone dissolves in water, giving characteristic karst topography including sinkholes, disappearing streams, springs, and so on. In such areas, the bedrock is riddled with solution channels which take surface water underground, and lead it through intricate pathways until it finds an outlet at some lower elevation. In this case, that outlet is partly about 18 km away at Venasque, where the Nesque reappears, and partly at the Fontaine de Vaucluse, about 28 km from the Bau in a straight line, but much farther by the underground channels (du Merle & Guende 1978).

Figure 1.2. Photo of the Bau in its geographic context, in the gorges of the Nesque. The arrow is pointing to the position of the rock shelter. ***Photographie du Bau dans son cadre géographique, dans les gorges de la Nesque. La flèche montre la position de l'abri sous roche.***

The action of water flowing through the gorge, and the dissolution of the limestone by the water, are, however, only contributors to the current size and shape of the gorges, as is mass wasting of the steep rocky walls. The origin of the structure is a zig-zagging series of faults that ripped the land open, as part of the tectonic activity that created the Mont Ventoux and the Monts de Vaucluse.

1.3. Geologic history of the region

Some time before about 200 million years ago the single continent, Pangaea, started separating into two large pieces, Laurasia and Gondwana. Even when these were still connected, a branch of the ocean formed a new sea, the Tethys sea, between them. The Massif Central of present-day France represented the continental edge of Laurasia at that time, and what is now Provence, in the south of France, was a shallow marine basin (Truc 2012, Ughetto & Ughetto 2007). The area was located about 1000 km farther south than it is now, and the climate was hot and arid. During the last 30 million years of the Triassic, a repetitive cycle of climatic variation, from more to less arid, resulted in the accumulation of 600 m of sediments in this basin. These sediments consist of layers of salt (deposited in times of extreme aridity), layers of gypsum and limestone (when it was somewhat less arid), and layers of clay and fine sands deposited during wetter periods, when there was enough rainfall to support vegetation and rivers on land, so that sediments were carried out to sea (Truc 2012).

In the Jurassic, in a humid tropical climate, well over 1000 m (possibly more than twice that amount) of clays and calcite accumulated in the subsiding basin (Truc 2012), resulting in the marls now known as the *terres noires*. In the later Jurassic these passed progressively into bedded limestones, as increased vegetation on land reduced the input of terrigenous sediments. Moving into the Cretaceous, a subsiding chunk of crust under this area resulted in what is known as the *bassin vocontien* or *fosse vocontienne*, which accumulated hundreds of meters of limestones, at first mixed with marls, but through much of the Lower and mid-Cretaceous purely marine and very fossiliferous (Masse 1993, Truc 2012). These are the rocks within which we find the gorges of the Nesque.

These Lower Cretaceous (especially Bedoulian) limestones are important to the story of the Bau de l'Aubesier not only because the rock shelter formed (much later) as a cavity within them, but also because they contain abundant flint nodules. As sediments accumulate and get buried deeper and deeper, they get subjected to higher and higher pressures and temperatures. This compacts them and turns them into rocks, and it also changes the solubility characteristics of the minerals in them, as well as of the ions dissolved into the interstitial liquids that circulate through the deposits. A complex series of diagenetic changes occurs, whereby minerals change their structures and compositions, or dissolve and are replaced by new minerals, often more than once. In the process that interests us, the circulating liquids have silica dissolved in them. This may be of terrestrial or biogenic origin, derived either directly from weathering of terrigenous minerals, or from the dissolution of siliceous body parts such as sponge spicules. Under the right conditions, the silica precipitates, replacing calcite wholly or in part in certain areas, and thus forming flint nodules. The flint can contain textures and structures, such as fossils, that were part of the original limestone; calcareous fossils can still be in their original mineral form, can be partly replaced or infilled, or can have been completely converted to silica, depending on the circumstances (Wilson et al. 2010).

By the late Cretaceous subsidence stopped and the sea became shallower. Terrigenous inputs plus strong currents led to the formation of undersea sand dunes (Truc 2012), and a mineral called glauconite formed from alteration of silicate minerals in the warm shallow sea. These sands eventually became glauconitic sandstones (sometimes known as greensands), which later were exposed and highly weathered, causing the glauconite to break down and leave behind kaolinite (a very pure clay) and iron oxides and hydroxides: ochre deposits (Figure 1.3). Small deposits of these exist today scattered throughout the region, with major deposits in Mormoiron (about 11 km west of the Bau), Roussillon (18 km southwest) and Rustrel (about 21 km southeast of the Bau), among other localities.

Tectonism came back into play in the Eocene. The situation in Provence was inherently unstable: hundreds of meters of limestone beds on top of hundreds to thousands of meters of marls and evaporites. Limestone is rigid. Marl and evaporites are soft and flexible. As the African plate moved northward again and affected the European plate (the *plissement pyrénéo-provençal*), the rocks of the Vaucluse folded into asymmetrical to overturned anticlines – but since the softer rocks underneath were better at folding than the more rigid rocks on top, the folds sometimes broke into what are more like monoclines. That is the case for the Mont Ventoux: it is a huge slab, corresponding to the more gently-sloped southern limb of the anticline (Figure 1.4). On its northern side we can see the series of strata that make up the limb, with the oldest rocks at the bottom. The original crest of the fold has been eroded away, and the remnants of the overturned, steeper limb are on the other side of the valley.

This compressive phase was followed in turn by a phase of extension, where the crust became stretched instead of squeezed. Intersecting sets of faults running either NE-SW

Figure 1.3. Photo of kaolinite and ochre deposits in Mormoiron, about 11 km west of the Bau. Photo by P. Mears.
Photographie des kaolinites et ocres de Mormoiron, à environ 11 km à l'ouest du Bau. Cliché P. Mears.

Figure 1.4. Photo of the Mont Ventoux from the eastern side, showing its asymmetrical slopes. ***Photographie du Mont Ventoux de l'est, avec ses versants asymétriques.***

or NW-SE were created or reactivated, and slabs of crust split apart or sank. The gorges of the Nesque follow some of these faults (see Figure 1.1).

In the Oligocene, the area became a shifting series of lakes and lagoons, sometimes freshwater, sometime brackish, sometimes hypersaline. More limestones were laid down in these, generally in thin layers, and more flint formed within these limestones. The Oligocene beds are preserved now in downfaulted blocks which protected them from later erosion, so in many cases they are at the same level, or lower, than the Cretaceous rocks.

During the Miocene the sea re-invaded southern France, and the Mont Ventoux became an island for a few million years. At the end of the Miocene, however, the compressive forces started again, in what was finally going to cause the creation of the Alps (the Mont Ventoux having been a solitary mountain for many years by then). The rest of southern France also rose as the Alps did, and Provence became dry land. Rivers started to flow, taking advantage of cracks caused by the faults, where available. As tectonic plates moved around, slabs of them changed their orientations relative to each other, and, for a few hundred thousand years (approximately 5.9 to 5.3 million years ago) the Mediterranean Sea periodically became cut off from the Atlantic Ocean (Triat 2015). During this time, known as the Messinian Salinity Crisis, the sea partially evaporated away. Sea level declined by 3 to 5 km below present-day levels. The rivers on the continents suddenly had to try to flow down to this drastically lower sea level. They did so by carving deep valleys. Many of those valleys are now obscured by later sediment deposits, or are under the sea, but the Nesque's deep gorge, up to 350 m deep at the Rocher du Cire (Figure 1.5), still stands as testament to the energy of a down-cutting stream.

The landscape as we know it now was pretty much in place by the end of the Miocene or the beginning of the Pliocene: the Mont Ventoux standing in solitary splendour, dominating the surroundings hills and plateaus, the deep gorges of the Nesque. . . all this in time for the Pleistocene ice ages, just after 2 million years ago. This area was never glaciated (Legal 2007), but glaciers extended down from the Alps to within a few tens of km of the Bau, at times, in the valley of the Durance river. Even without actual glaciers, during the cold periods the action of periglacial processes, such as erosion and cracking of rocks by frost, was very important. We see this in the deposits in the Bau, as well as even now on the slopes of the Mont Ventoux.

Figure 1.5. Photo of the Rocher du Cire, near the Bau in the gorges de la Nesque. ***Photographie du Rocher du Cire, près du Bau dans les gorges de la Nesque.***

This, then, is the landscape within which the people who occupied the Bau lived: hilly, with elevations ranging from 300 m above sea level (asl) on the lowest plains, to 650 m asl at the Bau, to around 1000 to 1100 m asl on the high plateaus, and culminating at the summit of the Mont Ventoux, 1909 m asl. The landscape is dissected by gorges and steep valleys, making the terrain difficult to hike across in many areas. The limestone bedrock would have always made it difficult to retain water on the surface, although during wetter climatic periods it is likely that the Nesque was a perennial stream. The elevations would have meant that winters were cold and windy; even today, snow occasionally falls in this area, and during a glacial period the mountains and plateaus must have been snow-covered for months at a time.

Although pre-modern hominins surely did not care how old the rocks were, the variety of geologic formations in this region must have mattered to them. For one thing, some layers contain flint; others do not. Also, some layers are less permeable than the karst-affected limestone, and surface water can flow across them, rather than sinking in. This explains why the Nesque does exist as a channel of surface water for several km, flowing from its source near Aurel, across the valley of Sault and Monieux, until it meets the limestone at the beginning of the gorges. The mineralogical nature and the permeability characteristics of the rocks also affect what plants can live in which places, which in turn affects which animals can live in which places. And finally, in limestone we can find caves.

The Bau was created by a combination of several forces: dissolution by groundwater, mechanical erosion by flowing water, and frost-cracking. Even before the gorge had been created and lifted to its current height, water must have flowed through the limestone, dissolving channels both downward and horizontally: water flows down until it reaches the level of groundwater, then it flows nearly horizontally until it finds an outlet. As the land rose, the groundwater level sank, so the higher horizontal channels would be abandoned. Tectonic action split the land open and exposed those old channels as caves and rock shelters. It is also possible that streams flowed in the channels, carrying rocks with them and wearing away the rock walls even further. Finally, when the Bau had been created and was exposed to the open air, water percolating through the roof and walls would freeze during cold periods. The limestone acts like a sponge, absorbing then giving up rainwater. Water expands when it converts to ice, so the rocks cracked, and smaller or larger chunks fell off.

Over time, the ceiling and floor of the rock shelter both moved upwards, as chunks of ceiling rained down and accumulated. When we excavated there, we knew that on rainy days we were safely sheltered and could stay dry – but four days later water would start dripping out of the ceiling. Luckily for us, in the summer there is no chance of that water freezing and knocking rocks off!

1.4. Excavations at the Bau

The Bau was identified as an archaeological site at least as early as the beginning of the 20th century. It was first dug from about 1901 to 1904 by Franki Moulin, who uncovered a rich lithic industry, faunal remains, and one Neandertal tooth (a deciduous second molar). Moulin published his findings in the first issue of what would become the *Bulletin de la Société préhistorique française* (Moulin 1904). Moulin worked only in the upper part of the deposits, next to the western wall of the rock shelter, leaving a trench that by the 1980s was about 2 m deep, 2 m wide, and 10 m long (Figure 1.6).

Even at that early date, it was evident that the site was exceptionally interesting. Moulin stated that the finds came from a single, pure layer, protected between two stalagmitic layers, but his description of the lithic industry made it clear that although it included abundant characteristic Mousterian pieces (such as Mousterian points), there was also a large proportion of elongated forms, including blades. Since blades were thought to be an Upper Palaeolithic tool type, other commentators suggested that either the finds came from a mixture of layers, or they were due to a transitional phase between the Mousterian and the Solutrean (de Mortillet in Moulin 1904, p. 20).

Despite this evident promise, the site received only occasional attention in subsequent decades. A group of amateurs spent some time in the 1950s sifting through Moulin's debris piles, and in the 1960s B. Mary both excavated and sifted more of the others' debris. During this time a second Neandertal tooth was found, subsequently studied and published by M.A. de Lumley (de Lumley 1973). Henry de Lumley also studied the site very briefly in the 1960s, publishing a short description and a stratigraphic section in his monograph concerning the Palaeolithic in southern France (de Lumley 1971). He described the lithic industry as an "evolved" Mousterian, with precursors of Upper Palaeolithic industries, and named it the "complèxe du Bau de l'Aubesier".

Large-scale, serious, multi-disciplinary scientific investigation of the Bau started in 1987, directed by Dr. Serge Lebel, then of the Université du Québec à

Figure 1.6. Photo of the tranchée Moulin. *Photographie de la tranchée Moulin.*

Montréal (UQAM). This was a major undertaking, and Serge deserves all credit for those excavations. Serge is a Canadian, and did his doctorate in France, defending his dissertation in November 1984. By then he had decided that he wanted to excavate a site of his own, and after checking out several possibilities, settled on the Bau. I am also a Canadian, and was a fellow student of Serge's, but several years behind him (I defended in June 1986). Serge asked if I would like to participate in his excavation, and I agreed enthusiastically. Serge moved back to Canada but nonetheless managed to convince the authorities in France to give a Canadian permission to excavate a French site (co-led by Robert Brandi of the Direction régionale des Antiquités in Aix-en-Provence for the first two years). Serge obtained all of the necessary permissions, got together the funding, found willing scientists to study aspects such as the faunal remains, recruited students to serve as excavation teams (and for many years ran the dig as a field school for UQAM), found affordable lodging near the site during the excavations (no small feat in an area very popular with tourists), and arranged to get all of the necessary equipment, from jackhammers to brushes and plastic bags.

Serge also was the visionary who directed the day to day and year to year planning of how to extract information from the site. None of this was easy. The Bau is a difficult site: the deposits are mainly highly cemented, requiring heavy equipment, and the rock shelter is halfway down a one-hundred metre high slope/cliff. Access to the site involves a steep, narrow path into and along the gorge, which makes it difficult to bring in equipment. To overcome that, after our first excavation season Serge designed a hoist system, which we installed on the plateau above the site in 1988. When that system turned out to be dangerous (he received a serious blow to the head), he designed a much better system which we installed in 1989 (Figures 1.7 and 1.8), and which we used for every subsequent field season. Serge also arranged for support in renting pneumatic jackhammer equipment, including an air generator which we left on the plateau, 80 m of tubing to reach down to the site, and jackhammers of various sizes, from 2 kg to 30 kg. For several seasons, we started with a small team undertaking a couple of weeks of jackhammer work in order to get through the breccia and reach archaeological layers below (Figure 1.9). We also had an electric jackhammer for incidental use throughout the rest of each season. Naturally, these measures resulted in large quantities of sediments. We installed tubes to transport the sterile sediments outside of the excavation area, and even then had to periodically work hard getting the piles out of the way of the ends of the tubes (Figure 1.10).

The site also turned out to be much larger, deeper and more complex than expected, extending far beyond Moulin's trench. It is also extremely rich: we excavated tens of thousands of lithics and hundreds of thousands of bones, from multiple layers and areas. Serge led his teams through many seasons of very hard work, but always with an exemplary attention to detail.

Figure 1.7. The hoist system on the plateau above the site, installed in 1989. ***Le treuil sur le plateau en haut du site, installé en 1989.***

Figure 1.8. A box containing an electric jackhammer being lowered down to the site. ***Une caisse contenant un marteau-piqueur électrique descendant vers le site.***

Figure 1.9. The south wall of A15-B15-C15, showing archaeological layer H1 (dark) below and breccia above. The white lines on the breccia are marks of the jackhammer bit, demonstrating how difficult it was to get through this highly cemented deposit. ***La coupe sud de A15-B15-C15, avec la couche archéologique H1 (couleur noire) en bas, et la brèche au-dessus. Les marques blanches sur la brèche sont des traces du marteau-piqueur, montrant la difficulté pour arriver à cette couche fortement cimentée.***

At the beginning of our first season, in 1987, using a theodolite, we marked lines of equal height on three walls of the rock shelter; the top of each line constitutes our datum level. All depths that I give throughout the rest of this book are measured in cm or metres vis-à-vis that level (our z zero, or z Ø), expressed as negative numbers because they are below datum. We also divided the site into metre-grids (Figure 1.11). The site is marked off from back to front by numbers, starting with 1 inside the back wall and increasing towards the gorge, and from side to side by letters, with A being an incomplete metre that starts inside the western wall, followed by B, C and so on eastwards. All locations that I give below thus refer to this letter-number combination. Note that photos of the site may also show a different series of numbers painted on the walls; these increase towards the back of the cave. I have been told that those were painted on by Henry de Lumley in the 1960s. We never used them and this book will never again refer to them.

In addition, we labelled the layers within the deposits using two different systems (Figure 1.12). The western part of the site, including the tranchée Moulin and the area in front of and below it, the pente du remplissage (which means "slope of the deposits"), has the layers labelled in alphabetical order, starting with A at the top. Some of these are subdivided depending on sedimentological or archaeological criteria, using numbers or letters (e.g. E, EA, EB; or H1, H2). The lower layers in this part of the site, starting at the G/H boundary, are defined as being one metre thick: H extends from 605 to 705 cm below z Ø, I runs from 705 to 805 cm, J from 805 to 905 cm, and K is below 905 cm. No archaeological material was found below the K metre, although we did dig a test pit down to below 1200 cm. In the central part of the site, we dug a new trench and named it the tranchée L. Since we did not know how the layers in that trench lined up (or did not) with the tranchée Moulin, those layers are labelled with numbers, starting with 1 at the top. These, too, are often subdivided, using supplemental letters (e.g. 3, 3A, 3B and so on). These will all be explained as necessary throughout the rest of this book.

These excavations showed that the deposits are more than 12 metres thick, well stratified, and contain several well-

Figure 1.10. Clearing away sediments at the bottom of a disposal tube in 1993. ***Déblaiement des sédiments en bas d'un tube, en 1993.***

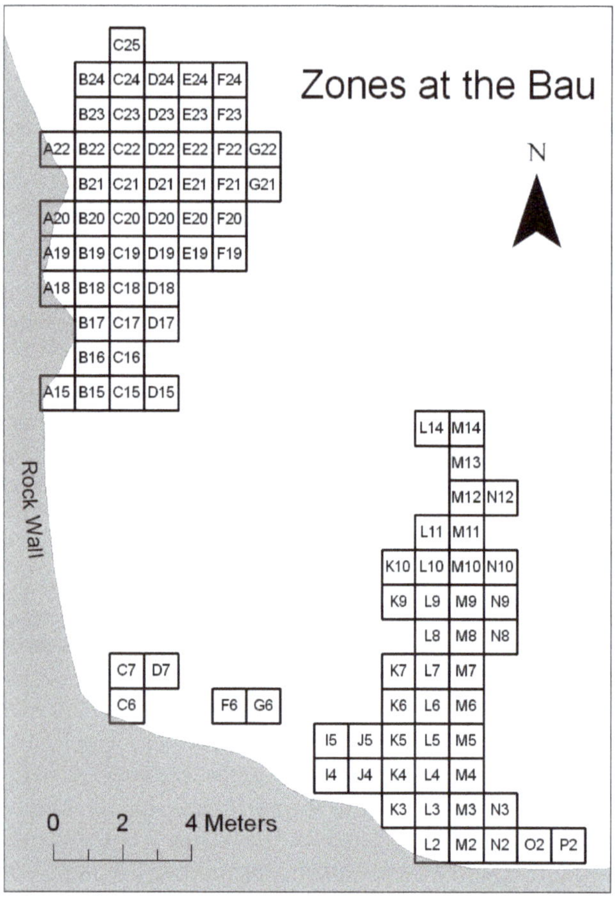

Figure 1.11. Map of the site showing the metre grid system and most important excavated areas. ***Carte du site avec le système de quadrillage par mètre carré, et les zones fouillées les plus importantes.***

differentiated archaeological layers, some highly burned, with the ensemble probably dating to oxygen isotope stages 7 to 5. Some layers are exceptionally rich in faunal remains (Fernandez 2001, Fernandez et al. 1998), others in lithics. Our excavations also turned up some more human remains, Neandertal and older, including teeth and part of a jaw (Trinkaus et al. 2000, Lebel et al. 2001, Lebel & Trinkaus 2002).

Excavations were conducted in the summers of 1987, 1988, 1989, 1990, 1992, 1993, 1994, 1995, 1998, 1999 and 2000, with a short scientific mission in 2006. Serge had an assistant every year, and these were:

Danielle Fournier	1987–1990
André Delisle	1992–1993
Dominique Paquin	1994–1995, 1998–1999
Jennifer Lehoux	2000
Eric Perreault	2006

The excavation teams included many people over the years. I am not going to list their names, because I am sure to miss somebody. Many of them worked very well, and we often had a lot of fun, too. I hope they all look back on it as an important learning experience.

This book is based partly on information from Franki Moulin's description of his excavation (Moulin 1904), and Henry de Lumley's site description and stratigraphy from the mid-60s (de Lumley 1971), but it is mainly based on materials from our excavations, including site reports

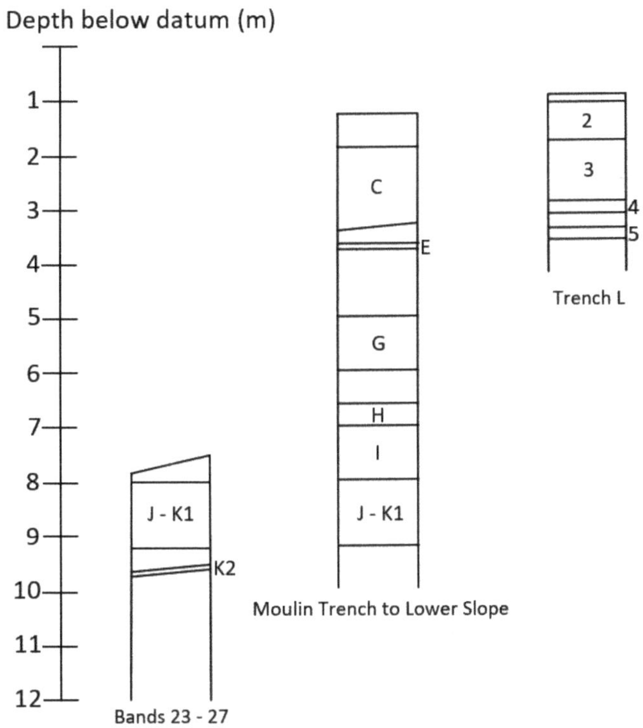

Figure 1.12. Summary site stratigraphy, mentioning major archaeological layers. ***Schéma stratigraphique sommaire, avec niveaux archéologiques majeurs.***

submitted to the French government, a variety of papers published about the human teeth and jaw, my own notes from the excavation seasons, a few theses written by students, and occasional analytical reports by myself and others (beyond those included in the site reports). These will be cited and listed in the bibliography. I have extensive data about the stone tools, both their raw materials (which I recorded as we excavated) and their typology (analysed in a basic way by myself and three students during the summers of 2013 and 2014, using the collections in storage at the archaeological depository in Vaison-la-Romaine). I will use this large collection of detailed typological data, rather than the summary data occasionally included in site reports by Serge Lebel, when I discuss typology throughout this book.

As mentioned above, the goal of this book is to present a synthesis of what went on at the site during prehistory. This is not a detailed site report, with lists of finds per layer. That is not really possible, anyway: a lot of the raw data are missing. For instance, although we recorded the exact location of many thousands of stone tools while we excavated them, all I currently have is information on which square meter and layer the tools came from. The reason for this lack is explained in section 1.5, "From The Chest". However, I choose to see this lack of detail as enforcing a "big-picture" view of the results, which is what I want to give in any case.

1.5. From The Chest: Serge

The Bau was excavated starting in 1987, as a project led by Dr. Serge Lebel, who at the time was employed by the Université du Québec à Montréal (UQAM). We excavated for about two months per summer in 1987-1990, 1992-1995, and 1998 and 1999. I was there for all of those. Serge also led an excavation in 2000, and I was present for part of that season. Serge did a short campaign in 2006, mainly testing and taking samples, with a very small team. I was not present for that, but I do have the report he wrote about it. During the 2000s, Serge withdrew more and more from contact with colleagues, and some time after 2006 left his employment at UQAM. I heard from him only infrequently during the period from 2002 to 2009. I last saw him in the summer of 2005, and last had any written word from him in 2009. Other friends and colleagues in France last heard from him even earlier than that. Serge had and may still have the excavation forms, sheets, notebooks, cross-sections, lists of samples, and so on. He has not responded to any attempts to contact him, directly or indirectly, by me or anyone I know of, since 2009. He has not published the results of the excavations. There is no sign of him participating any further in the analysis of the results. He does not, for instance, react when papers on the site are published. Since I am the person who knows second-best what went on during our excavations, and since I have continued to study the raw material geology of the Vaucluse (for the Bau and for other sites) ever since 1987, if anyone is going to write up an overall view of this site, it has to be me. Unfortunately, some potentially very useful information is just not available anymore.

1.6. The study of the lithic raw materials

My role in the Bau project was as on-site geologist and as archaeopetrographer – that is, studying the types of raw material the stone tools were made of. I was never an official assistant or co-director: I was in between, somehow, and in addition to doing my own specialist work, and a fair amount of prospecting throughout the region, I participated in the setting up and taking down of the dig equipment every year until 1999 inclusive, as well as in the excavations, the determination of stratigraphic questions and names for layers, teaching field-school participants and others about how to dig, and so on. After excavations were over, I continued studying the rock types.

As I studied the raw materials during the excavations, I classified them according to my série pétrographique (petrographic series): different varieties of rock (almost all flint), described and listed, and assigned a code letter. Table 1.1 gives the série, which ultimately came to 46 types, labelled in alphabetical order from A to AT. Starting in 1987, and continuing well into the 2000s, I undertook a very extensive prospecting campaign in the region, and identified more than 350 potential raw material sources, both outcrops and secondary sources such as colluvial and alluvial deposits (Figure 1.13). I then used a variety of methods (mainly petrographic, but also geochemical) to match the rock types from the série pétrographique to the source or sources where they could have been obtained (Wilson 2007, Wilson et al. 2010). Since many sources (e.g., separate outcrops of the same rock formation) give

Table 1.1. Petrographic series. ***Série pétrographique***

Type	Description
A	silex patiné blanc et/ou beige
B	silex rose tacheté
C	silex mauve grisâtre ou gris
D	silex patiné jaunâtre
E	silex mauve grisâtre clair tacheté de blanc
F	silex gris foncé legèrement translucide
G	silex blanc ou beige à cortex roux
H	silex blanc et brun zoné
I	silex marron foncé
J	silex gris clair
K	silex jaune-orange
L	silex blanc et beige rayé
M	silex brun clair et gris translucide
N	silex brun-mauve à cortex roux
O	silex gris foncé ou noir
P	silex beige à bandes ou zones sous-corticales gris foncées
Q	silex beige à bandes sous-corticales marrons
R	silex brun-jaunâtre clair tacheté de blanc
S	silex brun-grisâtre translucide
T	silex argileux gris à cortex épais gris clair
U	silex gris clair à blanc à grain grossier
V	"silex" blanc à taches rouges ("saucisson")
W	silex gris clair panaché, parfois piqueté de roux
X	silex brun-grisâtre piqueté de blanc et brun
Y	silex beige piqueté de rouge
Z	silex noir piqueté de blanc
AA	silex gris et brun-jaunâtre rayé de marron
AB	silex beige tacheté de blanc, brun, roux
AC	silex beige et brun marbré à cortex brun
AD	silex panaché, beige piqueté de noir, blanc, rouge
AE	silex brun-rose ou gris panaché, marbré de blanc, rose, brun, à zones foncées
AF	calcaire brun pâle
AG	andésite verdâtre porphyrique à phénocristaux de feldspath, amphibole, et mica
AH	silex gris clair à charophytes blancs d'environ 1 mm
AI	silex gris à brun avec de nombreuses petites taches blanches ("croûte de cantal")
AJ	silex blanc à taches rouges; altération gris (surface de galet)
AK	silex gris opaque à taches blanches (charophytes)

Type	Description
AL	silex patiné blanc avec petits charophytes blancs
AM	silex blanc à grain très fin, avec possibilité de clivage (plane) naturelle; galet
AN	silex brun ocre opaque
AO	silex translucide de couleur pâle
AP	silex tacheté brun foncé et gris foncé avec gastéropodes blancs d'à peu près 1 à 2 mm
AQ	quartzite noir à grain fin
AR	silex café-au-lait (patiné?) à gastéropodes silicifiés cristallins
AS	silex gris à noir tacheté piqueté de brun et blanc, à spicules d'éponge
AT	silex beige à gris marbré de taches noires, brunes et rouges, avec d'abondants spicules d'éponge

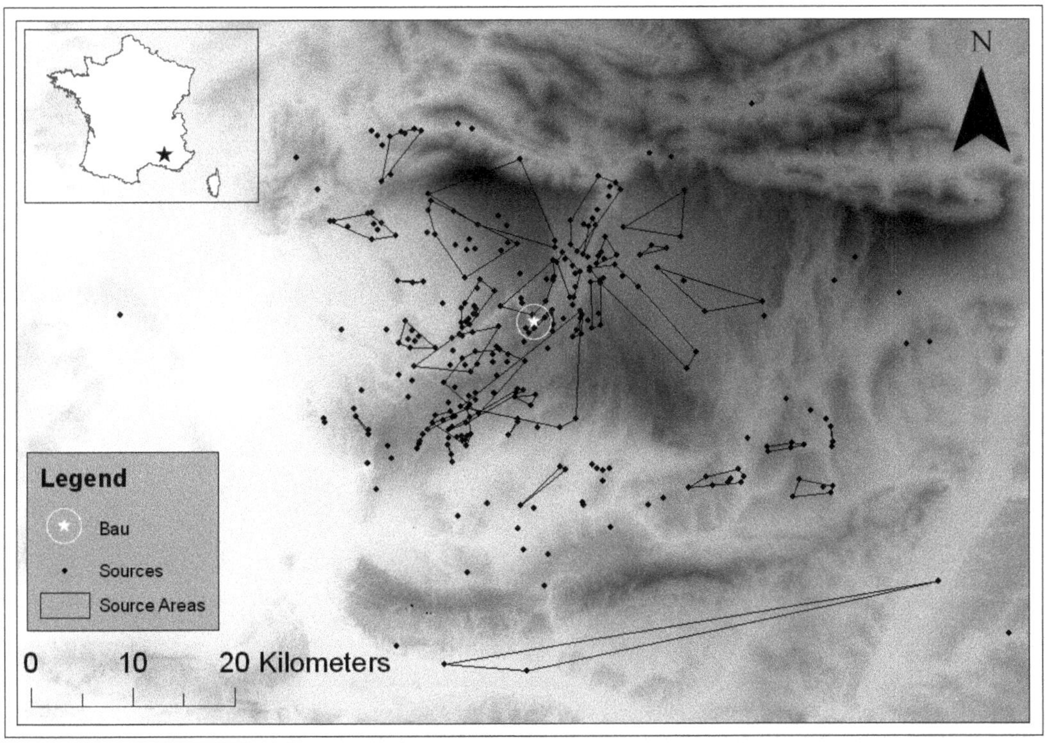

Figure 1.13. Map of the region showing potential lithic raw material sources. ***Carte de la région avec sources possibles de matières premières lithiques.***

flint types which are indistinguishable from each other, the sources were combined into source areas, containing from one to fifteen individual sources. Table 1.2 lists the used source areas, giving each a name which indicates roughly where it is. Table 1.2 also lists a distance value and a direction from the Bau to each source area. The distances are straight-line (Euclidean) distances to a representative source within the source area. In the resource use modelling (described further below) that Connie Browne and I have been doing for the last few years (Browne & Wilson 2011, Wilson & Browne 2014, Wilson et al. 2018), each source area has to be represented by a unique set of values, and we determined (by testing a whole lot of different models) that for the Bau, the best results come from using the source with the highest quality raw material available. That means that in many cases the representative source is not the closest source to the Bau, and the distances given in Table 1.2 are not necessarily minimum values.

Overall, many of the pieces were either patinated or burned, often to such an extent that the original rock type was no longer visible. I therefore could not say where those types originated from. The proportion of such pieces varies from layer to layer within the site, but often exceeds 50%. Of the remainder, almost all could be attributed to a source area. I examined nearly 40,000 pieces and could identify the flint type but not determine the source of only 0.4% of them. These are described below as "unknown", whereas the overly patinated and burnt pieces are "unidentifiable".

I must make it clear right away that the numbers of pieces per layer on the typological lists will never match the

Table 1.2. Used raw material source areas. ***Aires d'approvisionnement en matière première utilisées***

Source area	Direction	Distance (approximate)	Petrographic types
Unidentifiable	n/a	n/a	A, B, C, D, E, K, L
Unknown	n/a	n/a	F, M, T, Z AB, AD
Faraud etc.	SW	10.4	G, N, Y, AC, AS, AT
Méthamis etc.	W	8.0	H
Les Sautarels etc.	SW	10.7	I, O, AN
Sault	NE	9.1	J, R
Murs	SW	12.1	P, Q
Local	S	1.1	S, AF, AJ
St. Jean de Sault	NE	4.1	U, W, AM
Roussillon	S	18.2	V
Murs-Bezaure Olig.	SW	13.0	X
Ravin de la Treille	NW	4.9	AA, AI
St. Trinit	NE	14.7	AE
Durance	S	34.3	AG
Tertiary Calavon	S	14.6	AH
Tertiary Murs VdeV	SW	13.5	AK, AL
Mormoiron	W	11.4	AO
R de Guérin	SE	31.7	AP, AR
N.Aurel	NE	14.2	AQ

numbers on the raw materials lists. In some cases I did not see all of the pieces so the raw materials list has fewer pieces. In other cases, some of the pieces somehow did not get stored at the depository at Vaison-la-Romaine, so the raw materials list is longer. And in some cases, whether because of errors in marking the pieces at the dig, or because pieces have been separated from their identification (if they were with others in a labelled bag, for instance), I just was unable to match them all up. Most of the pieces do match, and the numbers are so large that they are useful, but no list will ever end up with exactly the same number of pieces as another list, even though they are supposed to be lists of the same things.

The raw material and typological information is presented layer by layer within this book, and I often refer to the results of the resource selection models mentioned above that Connie Browne and I have been working with for several years (Browne & Wilson 2011, Wilson & Browne 2014, Wilson et al. 2018). These are models that allow us to see what influenced prehistoric choices of which raw materials to use. Each potential source area is described in terms of factors such as the quality of the raw material, the size of pieces available there, the size of the source area, the difficulty of the terrain between the source and the Bau, the Calories (kCal) expended to travel from there to the Bau (which gives a better result than the distance in km), and so on (see Browne & Wilson 2011 for a full explanation). The models then calculate which combination of factors best matches the actual representation of raw materials within an assemblage, and weights them, so that we can see, for instance, that quality is more "important" (more highly weighted) in one layer than in another. The results also show whether a factor was positively or negatively important: for instance, a positive Beta value for quality means that sources with better quality material were used more, whereas a negative Beta value for terrain difficulty means that sources that are harder to get to were used less.

1.7. The scientific team

The synthesis provided in this book will include not only my own work, of course, but also the results obtained by many other researchers. Studies of fauna, sediments and pollen contribute to understanding the environment at the various times of occupation. Studies of fauna also contribute essential data to our understanding of how people behaved: what types of animals did they eat, and how did they obtain them? This needs to be combined with the stone tool data (typology and raw materials), and the characteristics of the occupation layers (use of fire, for instance), to give a fuller picture of hominin lives at the time. Faunal identification also gives us information on biochronostratigraphy, to combine with radiometric dating results in order to determine the age of the deposits. I am thus indebted to ALL of the contributors to the scientific study of the Bau de l'Aubesier, without whom this book would be impossible.

The multidisciplinary, international team that Serge put together included the following:

Serge Lebel: Director of excavations; with the collaboration in 1987-1988 of Robert Brandi, Direction des Antiquités, Aix-en-Provence
Alain Argant: Fauna, carnivores
Jacqueline Argant: Palynology
Bonnie Blackwell: ESR dating
Marie-Agnès Courty: Micromorphology
Martine Faure: Fauna
Philippe Fernandez: Fauna, ungulates
Bassam Ghaleb: U/Th dating
Claude Guérin: Fauna
Alfred Jaouich: X-ray diffractometry
Serge Lebel: Typology of the lithic assemblages
Jean-Claude Miskovsky: Sedimentology
Dominique Paquin: Geochemical sedimentology
Gilbert Prichonnet: Sedimentology
Henry Schwarcz: ESR dating
Hélène Valladas: TL dating
Lucy Wilson: Geology, archaeopetrography of the lithic assemblages
Manon Wilson: Micromorphology

In addition, Serge never released the full results of his typological study, so in order to obtain a complete record of the typology, I went through the collections, assisted by University of New Brunswick students Lauren Cudmore (2013), Lindsay Fleming (2014) and Leta Miciak (2013 & 2014). More detailed typo-technological analysis of part of the assemblage was undertaken by Leonardo Carmignani, as part of his Erasmus international doctoral programme (Carmignani 2017, Carmignani et al. 2017). My UNB student Peter Mears has been studying raw material use from the point of view of weights of materials, rather than numbers of pieces of them, for his Honours BSc degree (2020), and starting in 2020 for a Master of Science degree co-supervised by Connie Browne, which is now (summer 2021) rolling over into a PhD.

Serge obtained financing and logistical help from the Canadian Social Sciences and Humanities Research Council (SSHRC), the Ministère de la Culture et de la Communication de France, the Conseil Général du Vaucluse, the Service régional d'Archéologie, and UQAM. Students from Quebec were helped by UQAM and by the Office Franco-Québecois de la Jeunesse. In-kind help and goodwill were greatly appreciated, and in particular came from the Mairies of Monieux and Sault, the Ughetto family (Georges and Marie-Josée and their children, from Monieux, and Guy and Valérie and their children, from Sault), and Jean-Pierre Rey of Avignon. I am sure that Serge would want to mention other people, but I do not know whom. For my own research, I obtained funding from SSHRC (1991-1994), and later from the University of New Brunswick (UNB) and Dr. George R. (Rip) Rapp. I must especially thank Philippe Fernandez, for his help with this book: not only for reading and correcting each chapter, but for checking the French! I have also been greatly helped by Georges Ughetto (and the rest of the Ughetto families), by JP Rey and Monique Thiriot, by Muriel Bertrand, by Robert Russ of St.-Jean de Sault, and of course by my own family, who have been hearing all about this project for more than 30 years.

1.8. Sommaire du chapitre

Ce livre a pour objet l'étude de l'abri-sous-roche du Bau de l'Aubesier (le Bau) à travers sa situation et son contexte géographique, son évolution, et surtout son utilisation par des hominidés pendant une période de près de 300 000 ans. Le but du livre n'est pas de fournir un compte-rendu strict des fouilles qui ont eu lieu dans ce site, mais puisque ces dernières nous permettent de mieux comprendre les activités préhistoriques j'en parle souvent, et surtout des fouilles commencées en 1987, auxquelles j'ai participées. Le lecteur trouvera parfois dans le texte des commentaires personnels en marge du sujet principal; j'appelle ces anecdotes “From The Chest”.

Le livre est rédigé la plupart du temps en anglais mais il comprend beaucoup de français. En effet, j'ai souhaité conserver par exemple les extraits de rapports, et donner les titres des figures et tableaux ainsi que les sommaires de chaque chapitre, pour rendre l'ouvrage plus intéressant aux lecteurs francophones.

Mon but est de synthétiser les données obtenues à partir des études du site. Au lieu de découper l'ouvrage en chapitres thématiques (par exemple, un chapitre sur la faune, le lithique, etc...), j'essaie d'expliquer en commençant par le niveau le plus bas, donc le plus vieux, tout ce que nous savons sur chaque couche ou niveau à l'intérieur du site. Il s'agit pour moi d'essayer de comprendre les activités des préhistoriques qui vivaient dans ce lieu vraiment très privilégié et beau.

Le Bau est un grand abri-sous-roche d'environ 25 à 30 m de largeur, mesurant pas moins de 26 m de la paroi du fond jusqu'à l'extérieur des dépôts. Ce site contient au moins 13 m de dépôts de sédiments, souvent fortement cimentés en une brèche très dure qui a servi à protéger les niveaux archéologiques mais à rendre leur fouille très difficile. L'abri est orienté au nord, et se situe dans le côté sud des gorges de la Nesque. La rivière a actuellement un regime intermittent, et le fond des gorges de la Nesque est souvent à sec, mais aurait pu parfois contenir plus d'eau dans le passé, en particulier lors de périodes de climat plus humide.

Je ne répéterai pas ici l'histoire géologique de la région, étant donné qu'il existe déjà beaucoup de sources en français sur ce sujet. Il suffit de dire, j'espère, que les processus de l'histoire géologique ont créé un paysage calcaire karstique escarpé où l'eau était disponible seulement ponctuellement, par endroits. Dans ce paysage on trouve deux autres ressources très utiles aux hominidés: des gîtes de rognons de silex pour fabriquer leurs outils ainsi que des abris-sous-roche et grottes, pour les abriter.

Les fouilles archéologiques au Bau ont commencées en 1901-1904 avec les travaux de Franki Moulin, qui a découvert une industrie lithique très riche, des restes fauniques, et une dent de Néandertalien. D'autres études ont suivi dans les années 1950 et 1960, puis des fouilles scientifiques pluridisciplinaires ont commencé en 1987 sous la direction de Serge Lebel (qui était alors affilié à l'Université du Québec à Montréal). Serge était un personnage visionnaire qui obtenait les permis et trouvait les moyens necessaires à la fouille d'un site de brèche si difficile d'accès. Tous les deux canadiens, nous avions été étudiants ensemble en France et il m'a offert de participer à ses fouilles, ce que j'ai chaudement accepté.

Nous avons divisé le site par mètre carré (voir Figure 1.11) et commencé le travail dans la partie du gisement où toutes les autres fouilles avaient été menées, ce que nous appellons la Tranchée Moulin. Après quelques temps, nous avons également ouvert une deuxième tranchée dans les dépôts supérieurs, la tranchée L (dans la bande L de notre carroyage). Plus tard encore nous avons ouvert des fouilles dans la partie inférieure des dépôts, dans la pente du remplissage, là où auparavant personne n'avait soupçonné l'existence de couches archéologiques. Le site était donc plus riche, plus grand, et plus compliqué que prévu. Nous avons utilisé des lettres pour marquer les niveaux archéologiques dans la Tranchée Moulin ainsi que devant cette tranchée, dans la pente du remplissage. Nous avons utilisé des chiffres pour nommer les couches dans la tranchée L, en attendant de comprendre comment et si les couches des différentes zones étaient liées les unes aux autres. Serge a dirigé une douzaine de missions et fouilles pendant la période de 1987 à 2006, mais a par la suite coupé tous contacts avec ses collègues et arreté ses travaux sur ce site (voir From The Chest: Serge).

J'ai participé à toutes les fouilles, sauf la mission courte de 2006. Dans le même temps j'ai entrepris une étude des matières premières lithiques du matériel (voir Tableau 1.1) ainsi que leurs sources possibles dans la région: j'en ai échantillonné plus de 350, pendant plus de 20 ans de prospections. Bon nombre d'autres chercheurs ont contribué à l'étude du Bau (voir "The scientific team"), et sans eux ce livre n'aurait pas été possible. Plusieurs étudiants à moi et à Serge y ont également contribué par leurs efforts sur le terrain ou dans leurs travaux universitaires, et je les en remercie.

Les campagnes de fouilles dirigées par Serge Lebel ont été soutenues financièrement et par des aides logistiques par le Conseil de Recherches en Sciences Humaines (CRSH) du Canada, le Ministère de la Culture et de la Communication de France, le Conseil Général du Vaucluse, le Service régional d'Archéologie, l'Université du Québec à Montréal, l'Office Franco-Québecois de la Jeunesse, et les Mairies de Monieux et de Sault. L'aide et l'amitié offertes par les familles Ughetto (Georges et Marie-Josée et leurs enfants, de Monieux ainsi que Guy et Valérie et leurs enfants, de Sault) de même que celles de Jean-Pierre Rey ont été précieuses pour nous. J'imagine que Serge aimerait remercier d'autres gens, mais je ne sais pas qui; ils se reconnaitront.

Quant à moi, j'ai obtenu de l'aide financière du CRSH (1991-1994) et plus tard de l'University of New Brunswick ainsi que de Dr. George R. (Rip) Rapp. Je tiens surtout à remercier Philippe Fernandez, qui a non seulement étudié la faune du site, mais m'a beaucoup aidé dans la redaction de ce livre, tant en lecteur averti qu'en "correcteur" de mon français! (Je suis responsable des erreurs qui restent.) Mes travaux typologiques en 2013 et 2014 ont été facilités par Robert Russ et sa femme Martine et j'apprécie l'amitié qui nous lie désormais. L'aide et l'amitié indéfectibles depuis les années 1980 de JP Rey et Monique Thiriot, de Georges Ughetto et sa famille, ainsi que de Muriel Bertrand et sa famille, me seront toujours très chères. Et je remercie bien sûr ma propre famille, qui ont vécu "Bau, Bau, Bau!" avec moi depuis plus de trente ans.

2

The first sedimentary deposits to layer K2

2.1. Introduction

This chapter describes as much as possible of the initial configuration of the rock shelter when sediments started to accumulate in it, along with reasons why sediments actually *did* accumulate in the Bau – they did not have to! It discusses what those lowest sediments can tell us about what was going on at the site, and when that might have been. It describes and discusses the first signs of hominin use of the site that we know of: the layer K2. Finally, it concludes with a description of the finds from a test pit in the squares G23 and G24, which reached levels equivalent to the levels K2 and J4/K1 (chapter 3), but whose stratigraphy was never definitively linked to those layers. Their small assemblage size, and their uncertain link to the rest of the site stratigraphy, explain why they are not shown on the site map (see Figure 1.11).

2.2. The initial rock shelter

Almost nothing is known about the initial configuration of the rock shelter. We have not found its bedrock base anywhere within the site, although we have explored parts along the western and southern (back) walls, and we dug a deep test pit in the square metres C24 and C25. These are in the most forward, therefore lowest, part of the site (Figure 2.1), which should have the thinnest sediment cover. Even there, however, although we dug down close to three metres below the current surface of the sediments, we did not find bedrock. This is particularly surprising because in that area the western wall of the rock shelter curves towards the centre, making it look as if the base must be near the surface. Instead, the wall must take another steep downward curve.

What we can say is that the site must have undergone a continuous raising of its floor level, as sediments accumulated, and at the same time of its ceiling level, since the greatest part of the sediment was falling down from there. The side and back walls must also have been affected by erosion and frost cracking, but to a lesser extent. As loose sediments accumulated, they were periodically affected by percolating water carrying dissolved calcium carbonate. This reprecipitated both within and on the surface of the sediments, creating breccia and stalagmitic levels. We know that in many places such layers were cemented right on to the rock walls, which demonstrates that those were the walls at that time. However, we do not actually know the configuration of the walls in the very lowest levels, because all we have is the test pit.

As the ceiling was gradually raised by this process of bits cracking off, the floor must have risen more quickly. The loose plaquettes and rocks of all sizes from the ceiling were much less well packed as sediments than they were when they were part of solid rock: they therefore took up more room on the floor than they did on the ceiling. Occasionally very large rocks fell. We have examples that measure more than a metre in length by several tens of cm thick within the main deposits, although nothing as big as the huge boulders on the surface. These would not have taken up more space on the ground than in the overhang, but there are not many of them. Almost all of the deposit consists of fragments less than 15 cm long, and the great majority are less than about 5 cm long. The initial porosity of these deposits was high, although cementation has greatly reduced it, sometimes to zero. There also was some input of sediments from outside of the rock shelter, primarily as wind-blown sand and silt grains. Eventually, too, there is a certain component of infilling due to hominins bringing rocks, animals, and other materials in.

Some of the rock fall resulted in the retreat of the overhang, making the rock shelter less deep from front to back. It seems likely that the rock shelter has always extended as far back as it does now; it would have been

Figure 2.1. Photo of the lowest part of the site, including the location of the test pit in C25 (under the planks), and the curve of the western wall. Photo taken in 2015. ***Photographie de la partie la plus basse du site, avec localisation du sondage dans C25 (sous les planches) et la configuration de la paroi occidentale de l'abri. Photographie prise en 2015.***

part of an intricate channel system formed by groundwater circulation, only later exposed to the open air. We explored partway along the back wall at the level of layers 2 through 5 (down to about 3 m below datum), and although we found the wall, we also found that there were many irregularities in its shape in the centre of the site (bands L through P). It is certainly possible that it extends farther back in the unexplored area to the east of band P, and its configuration in deeper levels towards what we currently consider to be the back of the cave is entirely unknown. On the other hand, in 2006 Serge Lebel's team dug a test pit in the squares B17 and C17 (Lebel 2006), in which they did encounter the wall of the site, which must curve inwards towards the east in that area. We already knew that in band 18 square A18 existed in level H1 (down to nearly 7 m below zØ), but the rock wall moved eastwards below that level so that A18 disappeared and the sediments were found only starting in B18. In the B-C 17 test pit, the rock wall was encountered at z = -872 cm in the southwestern corner, at -911 cm in the southeastern corner, and at -956 cm in the northeastern corner. This shows how irregular the wall really is, dropping 45 cm in less than two meters horizontally.

What did the place look like before people came? Although the floor and ceiling were lower than they are now, it was probably always a very large rock shelter, wide and tall. The overhang may originally have extended out to cover more of the bands 24 and 25. Now the edge of the overhang is at about band 25 right at the western wall, but curves rapidly inwards to band 14 or 15 in the centre and east of the site. As mentioned above, the more extensive overhang must have cut down even more on sunlight entering this north-facing rock shelter, and the back of the site must have been dim even on sunny days.

The cliffs along the gorges of the Nesque contain many shallow rock shelters, with apparently very few of them containing any sediments. One exception is the Neolithic site of Castelleras (Courtin 1974, Sauzade 1983). There are also reports of other sites, such as "la grotte Jarle" which is supposedly similar to the Bau but has not been found again since it was described in the 1920s. In the late 1980s I found one tiny rock shelter with breccia in it, but the sediments contained no visible archaeological material. Beyond these potential exceptions, the fact is that most of the rock shelters in the gorges are empty of sediments, and the simple reason for that seems to be, in most cases, because their floors are tilted towards the gorge. Any sediments that fall into them get carried out again by water or gravity. At the Bau, clearly the configuration of the floor of the site must have allowed sediments to build up, but we do not know when that process started. It is possible that sediments were deposited then washed out several times before any permanent accumulation, because the rock shelter must have been opened up when the gorges opened, and that may have been as much as five million years before the earliest dates we have for these deposits. I would be curious to know whether there might be some sort of large rock fall under the front of the site, which served as a barrier to subsequent outward movement of sediments, but for now I have no information about that.

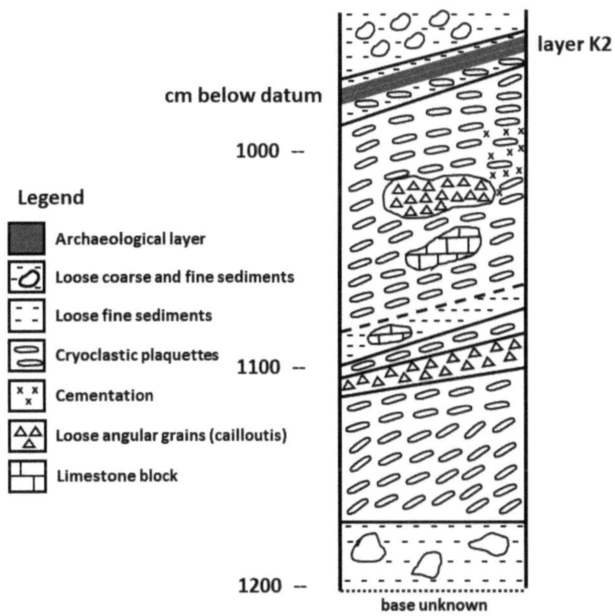

Figure 2.2. Stratigraphic section for the C25 test pit.
Colonne stratigraphique du sondage en C25.

The earliest sedimentation that we know of is from the bottom of the C25 test pit. Figure 2.2 shows the stratigraphy of the test pit, redrawn from a sketch in my personal notes from 1995. According to my notes, "(1) La couche de base est composée d'un mélange de sédiments fins sablo-argileux de couleur pâle, et de gros blocs. Epaisseur visible – 30 cm (sa base n'a pas encore été trouvée). Sa limite supérieure est horizontale. Les éléments constituants sont horizontaux." It is important that these sediments are horizontal, because it suggests that they are undisturbed, and that they probably accumulated by a combination of wind blowing in the fine sediments and frost cracking blocks off of the roof, i.e. in a cold, possibly very cold climate (lack of vegetation allowing the wind to blow fine grains around). If there was an obstruction at the front of the site, it may have stopped sediments from slipping out and allowed them to build up horizontally, or it may be that they were building up on top of a horizontal base. They are not cemented, which suggests that the climate continued to be cold (no water percolating through them).

Assigning an age to the lowest sedimentary layer can only be based on its stratigraphic position and characteristics. The oldest dated layers higher in the Bau's stratigraphic sequence have been shown to be from Marine Isotope Stage (MIS) 7 (an interglacial). The sediments at the bottom of the test pit, on the other hand, represent a cold period. They are horizontal, which suggests that they are in stratigraphic position, and they are well below the dated layers, so they may date to MIS 8, which ended approximately 300,000 years ago, and they are probably even older than that age, because several other layers above the lowest sediments (and below the earliest dated layers) are also from a cold period.

On top of the lowest sediments, again according to my personal notes, come (from lower to higher):

"(2) Plaquettes mesurant de 2 à 15 cm, avec quelques plus gros blocs. Les orientations des plaquettes sont variables, mais il y a une tendance d'avoir [sic] un pendage de 10° vers le Nord. La surface supérieure de la couche a un pendage de 15° à 20° vers le Nord. La transition entre cette couche et celle au-dessus est assez nette. La couche est poreuse. Il y a de la cristallisation ou précipitation de calcite autour des plaquettes. Epaisseur: 66 cm.

(3) Cailloutis légèrement aéré à petits éléments (0.5 à 4 cm). Un peu (10%) de matrice sableuse. Orientation des cailloux est variable; avec une légère tendance possible à un pendage vers le Nord de ~5°. La limite supérieure est assez nette et a un pendage de 20° à 30° vers le Nord. Il y a un peu de cimentation. Epaisseur – 18 cm.

(4) Cailloutis ou plaquettes semblables à la couche décrite ci-haut numéro 2. Le pendage de la limite supérieure de la couche est de 20° vers le Nord. La transition entre cette couche et celle au-dessus est nette. Epaisseur – 6 cm.

(5) Sédiments fins sablo-argileux, de couleur pâle, légèrement indurés. Quelques gros blocs mesurant jusqu'à 50 cm; leur pendage est de 20° Nord. La limite supérieure est transitionnelle, un peu diffuse, un peu irrégulière. Epaisseur – 15 cm.

(6) Plaquettes de 1 à 15 cm, normalement autour de 5 cm. Parfois gros blocs (>60 cm), parfois zones à petits cailloux. Ces zones peuvent être légèrement cimentées. Orientation variable; tendance à un léger pendage vers le Nord. Il y a de grandes zones lessivées, sans matrice, et d'autres zones indurées ou cimentées. L'épaisseur est variable, autour de 100 cm. La surface supérieure a un pendage variable vers le Nord ou Nord-Est, de 10° à 40°, mais normalement autour de 20°. La transition avec la couche supérieure est assez nette.

(7) Sédiments fins, sablo-argileux, un peu granuleux, contenant 10 à 15% de cailloux mesurant jusqu'à 4 cm. La couche est compacte, avec peu de porosité. Elle est noircie sur un épaisseur de 4 cm, à partir de 13 cm au-

dessus de la limite inférieure. L'orientation des éléments est variable, mais ils sont souvent horizontaux. Le niveau noir a un pendage de 10° vers le Nord. Epaisseur – 18 cm. La transition avec la couche sus-jacente est diffuse. La limite supérieure de la couche a un pendage de 10° vers le Nord."

These sediments, from what I called here layer numbers 1 through to 7, have an approximate thickness of at least 253 cm, and at least in our test pit contained no archaeological material at all. That includes the thin dark horizon in layer 7, which was rather disappointing at the time of excavation. Their variations in grain size presumably relate to climatic variations, as do the variations in their degree of cementation (or, usually, absence thereof). All in all, though, they are due to frost action during a prolonged cold to very cold climatic period, with occasional warmer periods allowing water to percolate through the deposits and precipitate calcite. The other striking feature is that starting with the second layer they are all inclined towards the North, that is, out of the site, even though the first layer was horizontal. The sediments probably accumulated as a talus slope, building up both upwards and outwards, and sliding over each other until they found stable positions. Water percolating through helped to consolidate them (precipitating calcite cement or crusts) but also washed away interstitial sediments from some places (poches lessivées), which may have caused slumping in those areas.

The reason for the colour of the darker horizon is unknown, but may be related to percolation of burned material from above. Manon Wilson, a UQAM student, did a micromorphological study of the site as part of her Master's degree (Wilson 1997). She took a micromorphology sample from this horizon (Figure 2.3), but did not provide any description or analysis of it in her Master's thesis. The fine material above the darker horizon corresponds to the archaeological layer K2, situated approximately 960 to 970 cm below zØ.

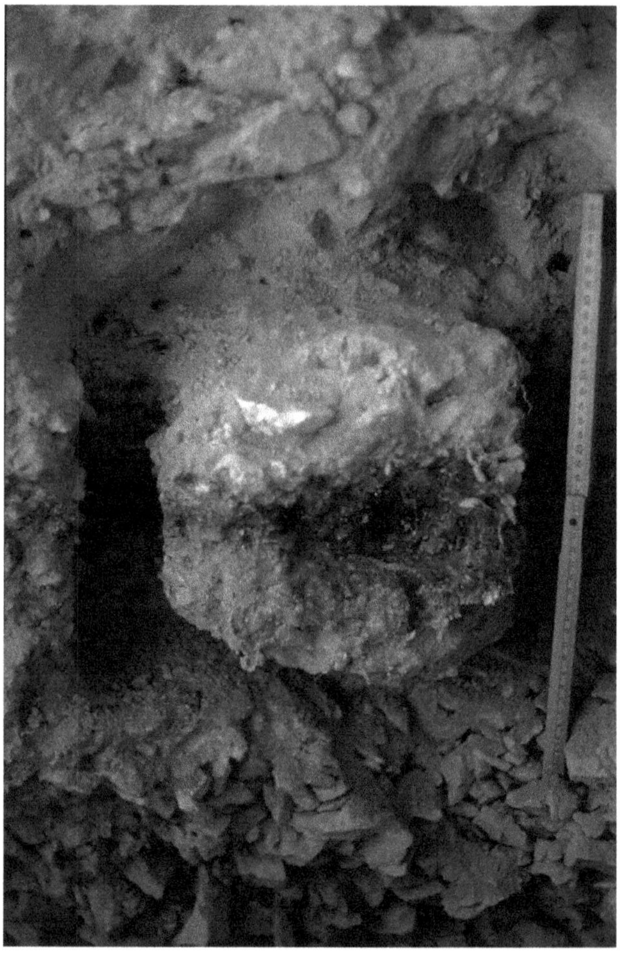

Figure 2.3. Prepared micromorphology sample from near the top of the deep C24/C25 test pit, prior to being wrapped in plaster and extracted, showing the thin black horizon. ***Echantillon préparé pour étude micromorphologique, avant d'être encaissé en plâtre et enlevé, montrant le niveau mince noir près de la surface du sondage profond en C24/C25.***

2.3. Layer K2

Layer K2 was first encountered as darker sediments within the sediments of B25, in July 1989. This colour change was accompanied by relatively abundant archaeological material, which is why it was disappointing that the darker level in the C25 test pit was completely sterile. The sediments of K2 are generally fine-grained, with some small rocks. They have not been directly dated, but have been interpreted as dating to a temperate climatic period, most likely during MIS 7 (Carmignani et al. 2017).

K2 was ultimately excavated in five squares: C24, B25 and C25, and B26 and C26 (see Figure 1.11). In adjacent areas farther into the site, even though some zones were excavated below the level at which K2 "should" be found, at best only occasional archaeological material was found; the layer proper seems to be absent. For instance, in my notes from 1990 I say that on the 14th of August Jean-Pierre Rey was digging in C21 and encountered several pieces of a stalagmitic floor deposit. "Il les a contourné le plus possible et dans un coin il a pu descendre jusqu'à ~960 cm, bien plus bas que le niveau théorique de K2, sans trouver de trace de ce dernier."

In band 26 the archaeological material disappeared on the northward side of the squares, due to the erosion of the slope at the front of the site, but it was concentrated and horizontal. In an area of 75 cm^2 in these two squares we excavated 794 lithics and 687 bones (Lebel 1990). Unfortunately, Fernandez (2001, 2006) in his thesis does not mention any faunal remains from K2, but instead has K1 as his lowest layer. Fernandez recently informed me (pers. comm.) that he has no record of any material at all (identifiable or otherwise) from K2, so it is possible that he was never shown these bones. In any case, I cannot say anything about the fauna.

My list of raw materials includes a total of 1290 pieces, and the typological list that we acquired in 2013-2014 by studying the material in storage at Vaison-la-Romaine includes 1166 pieces. Most of this discrepancy is due to variation in the numbers of pieces in the tamis/collectif category – for K2 that means material found in the sieves (tamis), consisting of small flakes and chips. (The "collectif" concept will be explained in a later chapter, because it was used only in the richer layers higher up the sequence.) Table 2.1 shows the numbers by typological category for the typological list and the raw materials list, cross-referenced by zone/couche/numéro identifiers as much as possible. "Liste Bordes" refers to pieces that can be assigned a number on the typological list established by François Bordes (Bordes 1961), such as racloirs and éclats levallois, or are otherwise classified into specific tool types such as choppers. Excluding the 11 pieces which appear on the raw materials list and for which we have no indication of typology, there is a total of 1279 pieces for which we know both the typology and the raw material type.

In terms of typology only, we have a large proportion of small flakes and chips, and a much smaller proportion of typologically identified pieces and cores. There are many larger flakes which were individually coordinated at the dig, with an average length of just over 30 mm, and an average weight of approximately 5 g. Blades make up a notable proportion of the assemblage (12% of a combined flakes+blades category, 3.2% overall) (Figure 2.4). These numbers almost certainly underestimate the proportion of blades, as some may have been grouped in with the flakes and/or with the tamis. Of those recorded, their average length is 50 mm, by 21 mm wide; their average weight is 11 g. Both flakes and blades show little or no cortex in most cases. Carmignani et al. (2017, p. 22) describe the blades as "triangular or rectangular, consistent with the pyramidal cores, and the rectangular blades can be linked to the semi-rotating unidirectional system. . ." Carmignani

Table 2.1. Lithic assemblage of layer K2, summarised by typological category. ***Assemblage lithique sommaire de la couche K2, par catégorie typologique***

Category	Typology list (n)	Typology list (%)	Raw materials list (n)	Raw materials list (%)
Liste Bordes	19	1.6	15	1.2
Debris	7	0.6	7	0.5
Eclats	267	22.9	245	19.0
Lames	37	3.2	34	2.6
Tamis/Collectif	818	70.2	963	74.7
Nucléus	18	1.5	15	1.2
non-identifié	n/a	n/a	11	0.8
Total	1166	100	1290	100

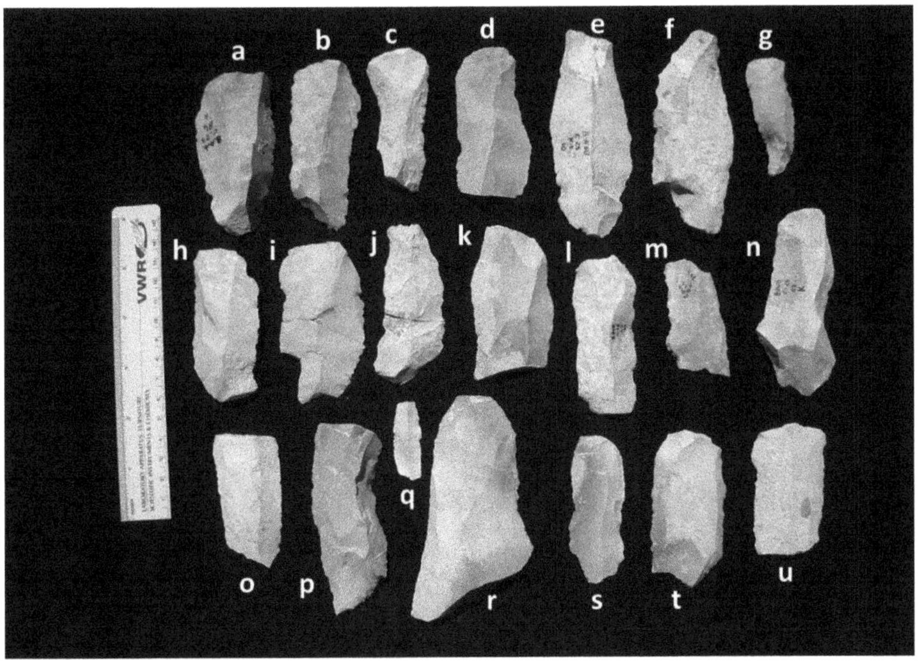

Figure 2.4. Blades and other elongate forms from the lowest levels of the Bau (K2 to J4) Specimens g, j, k, l, o, q, and r come from layer K2. ***Lames et autres formes allongées des couches inférieures du Bau (K2 à J4). Pièces g, j, k, l, o, q, et r proviennent de la couche K2.***

studied almost all of the known K2 items and gives a much more detailed technological and typological report on them (Carmignani 2017; Carmignani et al. 2017) than is given in my results, but for simplicity's sake I will use my results as much as possible throughout this book.

According to my list, the typologically identified pieces include 4 atypical Levallois flakes, 1 Levallois point, 10 various side-scrapers (of which 7 are part of multiple tools, and 4 have blades as blanks), 1 naturally-backed knife, 1 notch, and 2 other multiple tools which do not include a side-scraper (one denticulate on two edges, the other with two removals on the ventral face and various other retouch). As with the flakes and blades, most of the tools retain little or no cortex, except for one straight side-scraper with more than 50% cortex covering its dorsal face.

The 18 cores have an average length of 60 mm, an average weight of 82 g, and have traces of 12.7 removals remaining, on average (but with a great deal of variation). One of them can also be classified as a macro-tool (L. Carmignani, pers. comm.). According to my summary study, 6 of them are Levallois, 2 are non-Levallois blade cores, 5 are prismatic, 4 discoidal, and the other is globular. Carmignani (2017; Carmignani et al. 2017) classifies them somewhat differently and describes them in much more technical detail, but also finds that they include prismatic and discoidal forms, as well as both Levallois and non-Levallois types.

Carmignani et al. (2017, pp. 27-29) report that "At the Bau de l'Aubesier, reduction systems were performed on both the surface and the volume of cores, in order to produce both flakes and blades. . . At the Bau de l'Aubesier, the tools are on flakes and blades and are retouched by a marginal retouch that only slightly modifies the pieces, except for some truncated pieces and some partial bifacial tools (rostrum). This reduced importance of the tool kit at the Bau de l'Aubesier can be explained by the use of core technologies based on predetermined systems, such as Levallois, which produce a blank form that does not need to be modified."

The presence of the Levallois core technology in layer K2 is not unusual for an assemblage of this age. However, the existence of volumetric blade production in MIS 7 is evidence that this technology is older than was previously known for southern France, where it had previously been dated to the end of MIS 6 and the beginning of MIS 5 (Blaser et al. 2012; Carmignani et al. 2017; Gagnepain & Gaillard 2005; Gagnepain et al. 2004; Moncel 2005).

Together, all of this shows that the prehistoric people imported raw material and knapped it at the Bau (plenty of knapping debris is present, and a good number of cores), using several techniques (Levallois or not, blade production or flakes). The results concerning the raw materials used, and their origins, enrich our comprehension of this matter. Table 2.2 presents the number of pieces in each typological category which comes from each source area, as well as those which are not identified or identifiable. Table 2.3 shows the percentage values, rounded to one decimal point (totals between 99.9 and 100.1 are shown simply as 100).

Several results stand out. First of all, 10 of the 15 identified pieces on the Bordes list are made in flint from Murs, and one is in the flint from around Les Sautarels, which is close to Murs. Murs is located about 15 km to the south-west of the Bau, and represents the southwestern limit of the territory used for flint in K2. It has very good quality flint and seems to have been greatly appreciated for making tools. It is also present, often dominant, in all other typological categories, including cores, flakes and blades, so it was imported and knapped at the Bau.

The rest of the area used for raw materials in K2 (Figure 2.5) extends WSW towards Méthamis, following the gorges de la Nesque, across the gorges to the NW a few km (Ravin de la Treille), and to the NE to Sault and even St. Trinit, about 12 km away (as the crow flies). Most of the material comes from sources along a SW-NE axis through the Bau, and we will see that this pattern is repeated over and over in the Bau assemblages.

Overall, more than fifty percent of the pieces are too patinated to be identified as anything other than flint (very few are burnt). This number is due to the influence of the large number of small patinated flakes and chips found in the sieves (590 of the 1290 pieces). Patination affects small pieces more because they have a greater surface area relative to their volumes; this is why the larger pieces in the other typological categories are less patinated. It is also more difficult to recognise the flint type of a small, partly patinated piece than that of a larger, partly patinated piece, because so little is left to see. This proportion of patination is not exceptional for the Bau, but may also be partly explained by the fact that K2 is located at the front of the site, where it has been unprotected by any overhang for some long time, and hence has been more exposed to dripping water than other industries farther inside the site.

Almost no material identifiable as local was found, which seems very odd. There is good flint freely available as colluvial deposits on the hillsides above the Bau (Saume Morte), in large nodules often between 20 and 30 cm long. Why was it not used? Or was it used, but it can no longer be recognised? One potential explanation for it not being used is that the inhabitants of the Bau chose not to use it, for some reason. Another possibility is that the flint-bearing levels in the limestones on the hillsides were not exposed at the time. The current deposits are at least partly due to agricultural action in recent times. Maybe the flint was still enclosed in limestone during the Pleistocene? Or, was it used but then became preferentially patinated, compared to other rock types, so that can no longer be detected? I have tried many different approaches to try to detect whether patination has affected any rock types more than others, such as comparing the distribution of

Table 2.2. Number of pieces by typology and source area in layer K2. ***Nombre de pièces par catégories typologiques et groupes de sources des matières premières dans la couche K2***

K2 n	Liste Bordes	Debris	Eclats	Lames	TamisColl	Nucléus	non-id.	Total
Unidentifiable	2	1	77	8	590	4	2	684
Unknown	0	0	4	3	41	0	1	49
Faraud etc.	1	0	44	4	78	0	1	128
Méthamis etc.	0	0	5	3	59	1	1	69
Les Sautarels etc.	1	0	1	1	1	0	0	4
Sault	0	3	32	3	64	2	1	105
Murs	10	3	75	12	115	8	5	228
Local	0	0	1	0	0	0	0	1
St. Jean de Sault	0	0	0	0	0	0	0	0
Roussillon	0	0	0	0	0	0	0	0
Murs-Bezaure Olig.	0	0	4	0	11	0	0	15
Ravin de la Treille	0	0	1	0	3	0	0	4
St. Trinit	1	0	1	0	1	0	0	3
Durance	0	0	0	0	0	0	0	0
Tertiary Calavon	0	0	0	0	0	0	0	0
Tertiary Murs VdeV	0	0	0	0	0	0	0	0
Mormoiron	0	0	0	0	0	0	0	0
R de Guérin	0	0	0	0	0	0	0	0
N. Aurel	0	0	0	0	0	0	0	0
Total	**15**	**7**	**245**	**34**	**963**	**15**	**11**	**1290**

rock types in areas with high patination versus areas with low patination (within the same layers if possible), and have never been able to find that patination makes any difference to proportions of identifiable types. Therefore I have no real explanation to suggest for this result. All I can say is that for some reason, there seems to be almost no local raw material in K2.

Another surprising thing about the K2 assemblage is that identifiable flint of unknown sources is relatively abundant: a total of 49 pieces, mainly in the sieves but also in the larger flakes and blades. This represents about 3.8% of the total assemblage. In comparison, the complete raw material assemblage that I studied for the Bau, nearly 40,000 pieces, contains only 0.4% unknowns. This means that at least one other source was used in K2, and material from it brought to the Bau for knapping – but I have no information about where it is.

As mentioned, very little of the material from K2 was burnt, and there were no traces of hearths or fire structures, although the sediments did take on a darker colour in some squares. The lithic analysis recorded at least 111 burnt lithics, which represents 8.6% of the assemblage. This number represents pieces classified into raw material types for which the definition includes that they were burnt. Most of these are therefore unidentifiable as to source, but there is one type (N) which is still recognisable as coming from Faraud, despite being burnt. In addition, it is possible that some pieces were burnt but their original nature was not obscured. They would be recorded under that rock type, and do not figure in the 111 burnt pieces mentioned here, which therefore is most likely an underestimate. Still, the results suggest that fire was used, but not necessarily within the area we excavated of K2.

For the resource selection function work (Wilson & Browne 2014) we eliminate unidentifiable and unknown pieces, and disregard typology, which left us with 556 pieces from 9 source areas. In this analysis, we found that the best model was the one using all of our usual ten variables (see chapter 1 for more on how this method works). Table 2.4 gives the results for that model for K2, and Table 2.5 gives the Wald Chi-Square values converted to percentages, to make it easier to judge the relative weight of each variable.

Table 2.3. Percentages of pieces by typology and source area in layer K2. ***Pourcentages des pièces par catégories typologiques et groupes de sources des matières premières dans la couche K2***

K2 %	Liste Bordes	Debris	Eclats	Lames	TamisColl	Nucléus	non-id.	Total
Unidentifiable	13.3	14.3	31.4	23.5	61.3	26.7	18.2	53.0
Unknown	0	0	1.6	8.8	4.3	0	9.1	3.8
Faraud etc.	6.7	0	18.0	11.8	8.1	0	9.1	9.9
Méthamis etc.	0	0	2.0	8.8	6.1	6.7	9.1	5.3
Les Sautarels etc.	6.7	0	0.4	2.9	0.1	0	0	0.3
Sault	0	42.9	13.1	8.8	6.6	13.3	9.1	8.1
Murs	66.7	42.9	30.6	35.3	11.9	53.3	45.4	17.7
Local	0	0	0.4	0	0	0	0	0.1
St. Jean de Sault	0	0	0	0	0	0	0	0
Roussillon	0	0	0	0	0	0	0	0
Murs-Bezaure Olig.	0	0	1.6	0	1.1	0	0	1.2
Ravin de la Treille	0	0	0.4	0	0.3	0	0	0.3
St. Trinit	6.7	0	0.4	0	0.1	0	0	0.2
Durance	0	0	0	0	0	0	0	0
Tertiary Calavon	0	0	0	0	0	0	0	0
Tertiary Murs VdeV	0	0	0	0	0	0	0	0
Mormoiron	0	0	0	0	0	0	0	0
R de Guérin	0	0	0	0	0	0	0	0
N. Aurel	0	0	0	0	0	0	0	0
Total	**100**	**100**	**100**	**100**	**100**	**100**	**100**	**100**

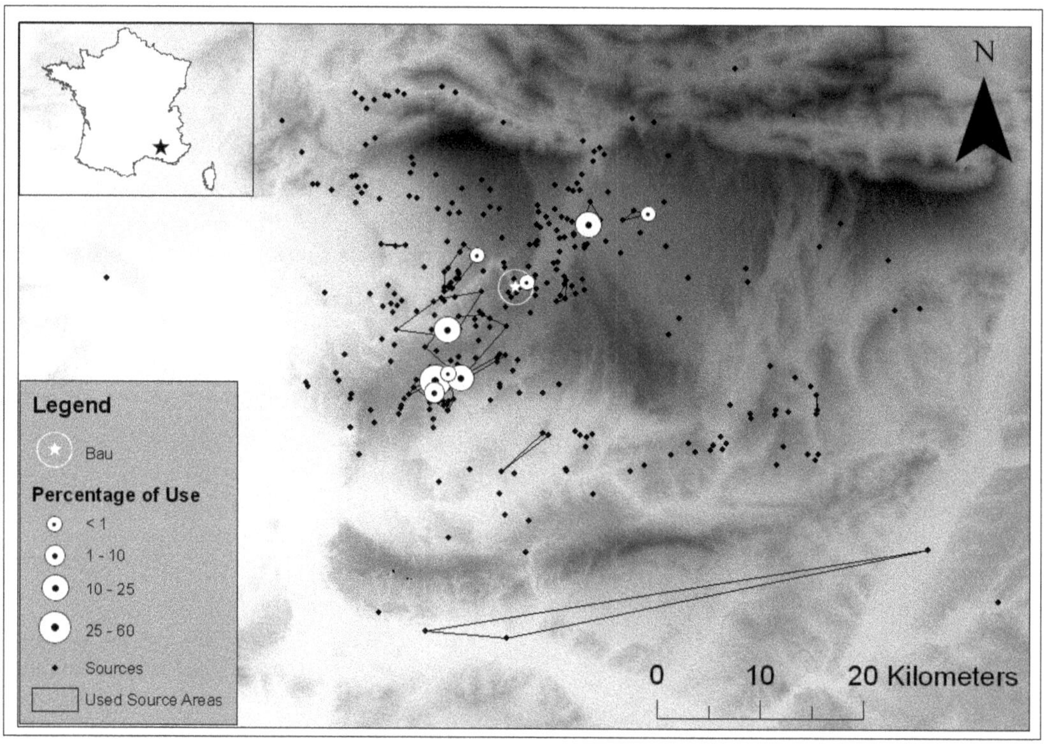

Figure 2.5. Map of the raw material source areas used for layer K2. ***Carte des aires de sources de matière première utilisées dans la couche K2.***

Table 2.4. Generalised linear model for raw material source choices in K2. P-values in bold font are statistically significant. See details of the model in chapter 1

Omnibus test: likelihood ratio chi-square = 496.015, df = 10, p = <0.001

Variable	Beta	S.E.	Wald Chi-Square	P-value	AIC
Quality	1.346	0.3932	11.723	**0.001**	122.902
Extent	2.136	0.6955	9.433	**0.002**	
Difficulty (Cal/km)	-0.064	0.0259	6.143	**0.013**	
Small rocks	-4.662	2.2776	4.191	**0.041**	
Medium rocks	-2.818	1.3003	4.698	**0.030**	
Large rocks	2.803	0.8151	11.828	**0.001**	
Very large rocks	-0.298	0.8529	0.122	0.727	
Calories	-0.002	0.0009	3.188	0.074	
AOSA	-1.5 e-7	7.5 e-8	4.002	**0.045**	
AOSISA	7.8 e-5	1.9 e-5	16.493	**<0.001**	
(Intercept)	-4.884	4.5019	1.177	0.278	

Table 2.5. Wald Chi-Square values as percentages, K2 assemblage

Variable	Wald Chi-Square	Wald Chi-Square as %
Quality	11.723	16.1
Extent	9.433	12.9
Difficulty (Cal/km)	6.143	8.4
Small rocks	4.191	5.7
Medium rocks	4.698	6.4
Large rocks	11.828	16.2
Very large rocks	0.122	0.2
Calories	3.188	4.4
AOSA	4.002	5.5
AOSISA	16.493	22.6
(Intercept)	1.177	1.6
Total	**72.998**	**100**

All variables except Very Large Rocks and Calories are significant, despite the small sample size. The most highly weighted variables are AOSISA, Large Rocks and Quality. Rock sizes are classified as Small rocks, 0-5 cm; Medium rocks, 6-15cm; Large rocks, 16-35cm; and Very large rocks, >35cm in size. AOSISA stands for Area Of Sources In the Source Area: it is the sum total of area covered by the individual sources that make up the source area. Having more source coverage in a source area made it more likely, for the K2 assemblage, that the source area would be used. This contrasts with AOSA: Area Of the Source Area, which is the total area of a minimum polygon covering all sources in the source area. Having a larger source area made it less likely that the source area would be used, presumably because too large of an area includes too much area with no flint in it. In terms of weighting, after AOSISA come the availability of large rocks and of better quality rocks, which also made it more likely that a source area would be used. In other words, the K2 people liked good big rocks that were fairly easy to find.

In our analysis of the resource selection models for the individual layers at the Bau (Wilson & Browne 2014), we combined the Wald Chi-Square percentages into two groups for each layer: the Terrain variables (Extent, Difficulty (Cal/km), Calories, AOSA and AOSISA), and the Raw Material variables (Quality, Small Rocks, Medium Rocks, Large Rocks, and Very Large Rocks). We ran the subset models for each layer, but in every case the overall model (with all ten variables) gave a better result than either of the subsets by itself. Thus we continue to use the overall model in every case. However, we can contrast the results for the two subsets as given within the overall model. As Table 2.5 shows, In the case of K2, the Terrain variables account for a total of 53.8% of the Wald Chi-Square values, while the Raw Material variables add up to 44.6%. As we will see, this is in agreement with the tendency in all of the lower layers of the Bau, where terrain variables matter more than the characteristics of the raw material itself, although both are always important.

In layer K2 we therefore have the first evidence of occupation of the Bau. Based on the sedimentary record, the layer accumulated during a warmer phase after a long cold climatic period. We have no direct dating of this layer, but it may represent MIS 7. The extent of the occupation is unknown, and some of it has probably been eroded away, due to its position at the front of the rock shelter. It may extend farther back than the area where we found it, but we were never able to find it farther south than band 24. It seems to have been an occupation that was concentrated near the front of the rock shelter. Given the likely greater extent of the overhang at that time, this makes some sense: the light would have been better there.

The hominins who created layer K2 did bring some faunal material into the site, but unfortunately we have no indication of what animals nor of how they were obtained or treated. However, their lithic industry was surprisingly advanced for that time period, with a strong presence of deliberate blade production, and both Levallois and non-Levallois techniques used for flakes and blades. They produced a variety of tools, especially scrapers. They obtained their raw materials mainly along a SW-NE axis, especially from the Murs area but also from other sources – but almost none identifiably of local origin. Their raw material choices seem to have prioritised ease of access, but were still aimed at obtaining good quality, large pieces. Their sourcing area extends to 15 km away from the site, but they brought at least some of their raw materials across that distance and knapped them at the Bau. They seem to have been a mobile group, who used the Bau for a fairly short stay.

2.4. From the Chest: Is K2 real?

I asked myself for a long time if K2 is a real layer. The situation at the Bau is that a lot of people dug at the top of the site and then sifted through each other's debris, over a period of decades. As they did so, they dumped their own debris all over the place. We constantly find pieces of knapped flint in the loose sediments on the surface in the unexcavated parts of the site. It therefore seemed possible to me that K2 is an accumulation of discarded material from the tranchée Moulin, that made its way to the bottom of the slope. The lack of faunal, floral and dating evidence does not help in this regard. However, two lines of evidence do reassure me. First of all, the layer presented as a concentration of material, well constrained within about ten cm thickness, and close to horizontal. It was also underneath other material (layers J4 and K1) in band 24. So, even though it is a small layer that disappears to the north and to the south, it does seem to make stratigraphic sense. The other reason I feel reassured is that the raw material analysis gives different results than for any of the upper layers, as we will see later. If it did result from discarded material from the tranchée Moulin, it would have had its "nice" pieces selected out, changing its typological character, but I cannot see how that could have changed its raw material proportions to this extent. Conclusion: yes, K2 is a real layer.

2.5. The test pit in G23/G24

G24 was dug in 1989 (Figures 2.6, 2.7, and 2.8), which is the only year for which I do not have Serge's excavation report; I only have my personal notes and the lithic data. In 1994 we dug G23 and found 87 lithics and 71 bones and teeth (Lebel 1994). The finds from G23 and G24 were not attributed to

Figure 2.6. Starting to excavate the test pit in zone G24 in 1989. ***Début de fouille du sondage dans la zone G24 en 1989.***

Figure 2.7. Test pit excavation in zone G24 in 1989, looking approximately to the north. The arrow points to the eastern wall of the pit. ***Sondage fouillé dans la zone G24 en 1989, vue vers le nord. La flèche indique la coupe Est du sondage.***

Figure 2.8. The G24 test pit, seen from the west (looking east), showing the stratigraphy in the eastern and southern walls of the pit. ***Le sondage en G24, vue de l'ouest, montrant la stratigraphie dans les coupes Est et Sud du sondage.***

a layer at the dig. The faunal remains are not described in Fernandez (2001, 2006), and Fernandez recently told me that he has no record of material from those squares, and therefore must never have seen it (pers. comm.): so once again I have no data on the fauna. In addition, since the assemblage is small, it was not used in the resource selection study of the lithics. This does not leave much to say about these finds, but they may as well be included here. I do not have information on the z values of the finds, so I cannot associate them with a layer or layers, but the following assemblage must have been excavated at depths similar to layers J through K2, or part of that extent. Table 2.6 summarises the lithic assemblage. As usual, the numbers of pieces on the typological list are not exactly the same as those on the raw materials list, but the pattern is similar.

The typologically identified pieces include 2 Levallois flakes, 2 Levallois blades, 1 atypical Levallois flake, 3 side-scrapers, 4 notches (one on a Levallois blade), and one flake with its bulb removed. The tools are fairly large, measuring 54.1 mm long on average, and weighing an average of 19.3 g. They have little or no cortex, with only one of them showing more than 10% of its dorsal surface covered in cortex. None of them show signs of burning. The flakes are a bit smaller, averaging more than 40 mm long but only 8.2 g in weight. They also show little or no cortex, and rarely have any signs of burning.

As we see in Table 2.6, blades are again present in notable proportions (nearly 4% of the complete assemblage). They range from 36 to 112 mm long (with an average of 58.2 mm), by 16 to 38 mm wide (average: 24.3 mm), and have an average weight of 19.8 g. One of them is cortical (>90% cortex on the dorsal surface), but the others show little or no cortex. None of them is burned.

The three cores are very different from each other. One is very small (18 g) and is a non-Levallois blade core with traces of 8 removals. Another is a small (51 g) prismatic core with at least 12 flake scars. The third is a large (280 g) Levallois core with 16 scars. They have little or no remaining cortex and are not burned.

Overall, the presence of abundant small flakes and chips (found in the screens), plus a strong presence of flakes and blades, some tools and three cores, suggests that some knapping was done on site. Tables 2.7 and 2.8 list the sources of the raw materials of the pieces, by typological category.

Compared with the K2 assemblage, G23/G24 stands out first of all by having a relatively small percentage of unidentifiable pieces: 24.3% overall, mainly (again) in the sieved material. Most of these are patinated; only 5.4% overall of the material is recorded as showing signs of burning (versus 53.0% unidentifiable, of which 8.6% are burned, in K2). G23/G24 also contrasts with K2 in having no pieces from unknown sources. On the other hand, it has a higher percentage of pieces from the local source: 3.8% is not very much, but it is better than 0.1%. The local pieces in G23/G24 include one notched Levallois blade, four flakes, three pieces in the sieves, and one core. The core is the smallest of the three, the non-Levallois blade core. Thus, local flint does seem to have been knapped on-site, and despite the small number of pieces we actually have remnants of at least three different trajectories: a Levallois blade, a non-Levallois blade core, and flakes. The visibility of local flint in this assemblage may mean that it was used (whereas it was not used for K2), or it may be due to the lower degree of patination of this material, so that it is still detectable (whereas it was obscured by patination in K2). In other words, its presence here still does not explain its absence in K2.

The most popular raw material source area for this assemblage is, again, the Murs area. Three source areas in that vicinity were used (Murs itself, plus Les Sautarels and Murs-Bezaure Oligocene), and together they account for 37.6% of the total assemblage. They are present in all of the typological categories of this assemblage, and material from Murs dominates the tools, blades and cores. Again, there is evidence that material from Murs was imported and knapped on the site, and there is evidence for several different technological trajectories (blades or flakes, Levallois or not. . .).

Table 2.6. Lithic assemblage of squares G23 and G24, summarised by typological category. ***Assemblage lithique sommaire du sondage en G23 et G24, par catégorie typologique***

Category	Typology list (n)	Typology list (%)	Raw materials list (n)	Raw materials list (%)
Liste Bordes	12	5.2	12	5.0
Debris	0	0	0	0
Eclats	67	29.3	79	33.1
Lames	9	3.9	9	3.8
Tamis	138	60.3	127	53.1
Nucléus	3	1.3	3	1.3
non-identifié	n/a	n/a	9	3.8
Total	**229**	**100**	**239**	**100**

Table 2.7. Number of pieces by typology and source area in squares G23 and G24. ***Nombre de pièces par catégories typologiques et groupes de sources des matières premières dans l'assemblage de G23 et G24***

G23/G24 n	Liste Bordes	Debris	Eclats	Lames	TamisColl	Nucléus	non-id	Total
Unidentifiable	2	0	13	2	41	0	0	58
Unknown	0	0	0	0	0	0	0	0
Faraud etc.	1	0	7	1	15	0	0	24
Méthamis etc.	1	0	5	0	10	0	2	18
Les Sautarels etc.	0	0	0	0	2	0	0	2
Sault	0	0	19	1	18	0	2	40
Murs	7	0	31	5	26	2	4	75
Local	1	0	4	0	3	1	0	9
St. Jean de Sault	0	0	0	0	0	0	0	0
Roussillon	0	0	0	0	0	0	0	0
Murs-Bezaure Olig.	0	0	0	0	12	0	1	13
Ravin de la Treille	0	0	0	0	0	0	0	0
St. Trinit	0	0	0	0	0	0	0	0
Durance	0	0	0	0	0	0	0	0
Tertiary Calavon	0	0	0	0	0	0	0	0
Tertiary Murs VdeV	0	0	0	0	0	0	0	0
Mormoiron	0	0	0	0	0	0	0	0
R de Guérin	0	0	0	0	0	0	0	0
N. Aurel	0	0	0	0	0	0	0	0
Total	**12**	**0**	**79**	**9**	**127**	**3**	**9**	**239**

The assemblage shows use of sources along the same SW-NE axis as in K2, but although there is considerable use of material from Sault (40 pieces overall), that seems to be the limit of travel in the NE direction. Sault is about 8 km from the Bau, in a straight line.

All in all, then, the G23/G24 assemblage shows both similarities and differences when compared to K2. It is a small assemblage, with some traces of burning of the lithics but no traces of fire structures (as far as I remember or have noted). The next chapter gives the results for the rest of the K and J levels. Will G23/G24 fit in better with them, or is it an ephemeral occupation of its own?

2.6. From the Chest: Measuring distance "as the crow flies"

I have spent many years investigating better ways of measuring the transport of raw materials across landscapes. There are many options: Euclidean distance (i.e. measured as a straight line); least-cost paths; measurements that stand in for distance, such as Calories expended; and so on. I started working on this question because I got so annoyed with papers that consider distance only as the length of a straight line drawn on a map: "as the crow flies". I walked thousands of km collecting flint samples over a time period of a couple of decades in the Vaucluse, and I know that hominins are not crows! Topography matters. I am therefore conscious of the irony of giving distances in this book using those straight-line measurements. However, for a quick indication of relative distance in these discussions, they are what is easiest to give. The reader can rest assured, however, that in the resource selection work we used a much better evaluation of distance, and did take topography into account.

2.7. Sommaire du chapitre

Dans ce chapitre nous commençons avec l'abri à son état initial, sans sédiment (d'autant que l'on puisse déterminer cette configuration). On explique comment et pourquoi les sédiments se sont déposés dans la partie inférieure du site, puis nous décrivons les premiers signes d'occupations par les hominidés: la couche K2. Enfin, on décrit le sondage dans les carrés G23 et G24, qui atteignent les niveaux équivalents aux couches K2 et J4/K1 (chapitre 3), mais avec lesquelles la stratigraphie n'a jamais été définitivement liée. Les zones G23 et G24 ne figurent pas

Table 2.8. Percentages of pieces by typology and source area in squares G23 and G24. ***Pourcentages des pièces par catégories typologiques et groupes de sources des matières premières dans l'assemblage de G23 et G24***

G23/G24 %	Liste Bordes	Debris	Eclats	Lames	TamisColl	Nucléus	non-id	Total
Unidentifiable	16.7	0	16.5	22.2	32.3	0	0	24.3
Unknown	0	0	0	0	0	0	0	0
Faraud etc.	8.3	0	8.9	11.1	11.8	0	0	10.0
Méthamis etc.	8.3	0	6.3	0	7.9	0	22.2	7.5
Les Sautarels etc.	0	0	0	0	1.6	0	0	0.8
Sault	0	0	24.1	11.1	14,2	0	22.2	16.7
Murs	58.3	0	39.2	55.6	20.5	66.7	44.4	31.4
Local	8.3	0	5.1	0	2.4	33.3	0	3.8
St. Jean de Sault	0	0	0	0	0	0	0	0
Roussillon	0	0	0	0	0	0	0	0
Murs-Bezaure Olig.	0	0	0	0	9.4	0	11.1	5.4
Ravin de la Treille	0	0	0	0	0	0	0	0
St. Trinit	0	0	0	0	0	0	0	0
Durance	0	0	0	0	0	0	0	0
Tertiary Calavon	0	0	0	0	0	0	0	0
Tertiary Murs VdeV	0	0	0	0	0	0	0	0
Mormoiron	0	0	0	0	0	0	0	0
R de Guérin	0	0	0	0	0	0	0	0
N. Aurel	0	0	0	0	0	0	0	0
Total	**100**	**100**	**100**	**100**	**100**	**100**	**100**	**100**

sur la carte du site (Figure 1.11) en raison de leur faible quantité de vestiges et leur stratigraphie incertaine.

Les contours initiaux de l'abri sont inconnus: nous n'avons jamais trouvé sa base rocheuse, même lors d'un sondage profond de 3 mètres dans les carrés C24 et C25. A cet endroit, la paroi rocheuse semble suggérer que la base doit être près de la surface, mais ce n'est pas le cas. La paroi plonge. De toute façon, il est clair que le niveau supérieur des sédiments (le plancher sur lequel on marche) a dû monter rapidement, au fur et à mesure que le plafond s'est délité, avec une accumulation de plaquettes et de rochers qui tombaient au sol. En même temps, il est probable que l'abri a toujours été grand et large.

Les plus vieux sédiments du site que nous connaissons se trouvent dans le fond du sondage de C25. D'après leur position stratigraphique et leur nature sédimentologique, ils pourraient dater du stade isotopique 8 (MIS 8, phase glaciaire), il y a plus de 300 000 ans – s'ils ne sont pas plus vieux encore. Au-dessus de ces sediments du fond de sondage, nous avons identifié plusieurs couches différentes, toutes sans vestiges archéologiques, totalisant au moins 253 cm d'épaisseur, avant d'arriver à la couche archéologique K2, située à environ 960 à 970 cm sous notre zØ (complètement en haut du remplissage de l'abri).

La couche K2 comprend des sédiments à grains fins et quelques roches. Elle n'a pas été datée, mais interpretée comme provenant d'une période tempérée, probablement du MIS 7. La couche a été fouillée sur cinq mètres carrés en avant du site; ailleurs, même quand on a atteint ce même niveau, on ne l'a pas trouvée. Quelques centaines de restes fauniques ont été mis au jour, mais malheureusement P. Fernandez ne les a jamais vus et je ne peux rien dire de leur nature. Par contre, j'ai le détail typologique et pétrographique de quelques centaines de pièces lithiques (Tableau 2.1). On voit ici, et on verra dans chaque chapitre, que la liste typologique ne correspond jamais exactement à la liste pétrographique, pour plusieurs raisons, mais surtout parce que les deux listes n'ont pas été établies en même temps. La liste typologique correspond à mon étude du matériel entreposé à Vaison-la-Romaine, pendant les étés de 2013 et de 2014. La liste pétrographique a été établie au fur et à mesure des campagnes de fouille de 1987 à 1999.

En K2 nous avons beaucoup d'éclats et débitage, et beacoup moins d'outils retouchés et nucléus. La proportion de lames est intéressante, surtout pour une couche si vieille (MIS 7), étant donné que la production systématique de lames dans le sud de la France avait auparavant été datée de la fin de MIS 6 au début de MIS 5. Les données pétrographiques (Tableaux 2.2 et 2.3) comprennent du matériel de plusieurs origines géographiques: Murs, Méthamis, Sault, etc. La plupart de ce matériel doit provenir de sources localisées le long d'un axe orienté du sud-ouest au nord-est, et nous verrons la même chose dans les autres couches du Bau. Ainsi, on peut dire que les hominidés de la couche K2 avaient une industrie avancée pour l'époque et utilisaient un vaste territoire. Leurs choix de matière première suggèrent qu'ils recherchaient de grands morceaux de roche, de bonne qualité pour la taille, et surtout qu'ils préféraient des gisements facile d'accès, minimisant ainsi leurs dépenses énergétiques. Leur grande mobilité sur leur territoire, comme indiquée par la provenance des matières lithiques, correspond bien aux indications de séjours de courte durée au Bau.

Le sondage en G23/G24 a été fouillé en deux saisons: G24 en 1989, et G23 en 1994. Malheureusement, une fois encore, P. Fernandez n'a pas vu les restes fauniques découverts lors de ce sondage et je ne peux rien dire de plus sur cet aspect. En revanche, j'ai les données lithiques (Tableaux 2.6, 2.7 et 2.8). On voit qu'il y a encore des lames en quantité notable, et assez de débitage pour suggérer des activités de taille sur le site. Il y a un peu plus de matériel d'origine locale que dans K2, mais en général les sources se trouvent encore sur l'axe SW-NE, de Murs à Sault. Ainsi, pour résumer, le matériel du sondage ressemble à celui de K2 sans être identique et on ne peut pas dire si les deux zones de fouille devaient être réunies ou non.

Dans le "From The Chest" je parle de l'idée de distance "à vol d'oiseau". Dans la grande majorité des publications archéologiques, lorsqu'on mentionne l'origine des matières premières, on donne ces distances en ligne droite: à vol d'oiseau. Pendant des années, des décennies maintenant, j'ai marché des milliers de kilomètres dans le Vaucluse lors de mes prospections, et j'ai constaté une chose importante: je ne suis pas un oiseau! La topographie est importante! Je monte et je descends les collines: je ne vole pas par-dessus. Les hominidés ne volaient pas non plus. J'ai donc cherché la meilleure manière de calculer la distance des sources: en ligne droite, ou en suivant le terrain le plus facile pour minimiser la dépense d'énergie ("least cost"), les dépenses caloriques nécessaires, etc. Cependant dans ce livre, pour exprimer simplement les distances relatives des gisements, je dois bien admettre que la distance la plus facile à concevoir est le kilomètre à vol d'oiseau – mais je suis consciente de l'ironie.

3

Layers from J4-K1 to J

3.1. Introduction

In the lower part of the site we arbitrarily divided the layers by metre depth. The limit between layer J and layer K is 905 cm below zØ, and the surface of layer J was fixed at -805 cm. The excavations showed, however, that there was a scattering of material throughout the bulk of the J metre, then a concentration towards the bottom, which continued without perceptible gaps down into the K layer, to a depth of about 920 cm. Finds in the upper parts of the layer were attributed to archaeological layers J, J1, J2 and J3. The concentration of material was called J4, and its continuation in the upper part of K was called K1. J4-K1 is considered to be a single archaeological layer. However, to be consistent with the presentation of results in previous publications (e. g. Fernandez 2006, Wilson & Browne 2014), this chapter combines information from J4-K1 with the smaller quantities of finds from the rest of layer J.

With J4-K1 we enter into the major archaeological levels of the Bau. This layer is rich in fauna and lithics, with thousands of pieces of each, and we also found one hominin tooth, Aubesier 4 (Lebel et al. 2001, Trinkaus et al. 2000a). Based on the faunal species present, and by extrapolation from the radiometrically dated layers higher in the sequence, K1 through J correspond to the later Middle Pleistocene (Lebel et al. 2001), MIS 7 through 6 (from Riss II into III in the Alpine sequence), roughly 240,000 to 200,000 years old.

3.2. The sediments, and evolution of the site from K2 to J

The sediments included in the section from the top of K2 to the top of layer J are uncemented and were therefore fairly easy to dig. They are also extensive, and not as constrained by the rock wall and the eroded front of the site as the lower level was. Our excavation at this level eventually included 29 squares, in bands A (23-25), B (17, 21-25), C (17, 21-25), D (21-24), E (21-24), F (21-24) and G (21-22) (see Figures 1.11 and 3.1). They show a great deal of lateral and vertical variation in granulometry, with some concentrations of larger rocks, and also a central portion in which the layers seem to have been affected by water flow and slumping. In that area the sediments contain no fine particles, and the bases of the layers dip towards the centre of the area excavated.

In my stratigraphic notes, written in 1995, I have the following descriptions of the layers, starting with the first layer above K2 (described in chapter 2), and moving up:

"(8) Sédiments constitués de 50% fraction fine sablo-argileuse avec granules, 50% cailloux jusqu'à 10 cm, normalement autour de 5 cm. Quelques gros blocs ou plaquettes jusqu'à 40 cm. Compact, avec peu de porosité. Orientation des cailloux variable, souvent horizontale. Surface supérieure a été partiellement fouillée et partiellement érodée; l'épaisseur restant est de 51 cm. Cette couche semble disparaître quelque part entre les bandes 21 et 23, probablement près de 21.

La couche (8) contient latéralement de grandes poches lessivées (au moins 50 x 40 cm); sus-jacents à ces poches il existe des niveaux plus fins, compacts, indurés. L'épaisseur de ce niveau est de 22 cm. Latéralement la couche (8) se mèle avec une couche à gros blocs.

(9) Niveau à gros blocs (20 à 30 cm), parfois lessivé, parfois avec une matrice à cailloux et matériel archéologique. Dans les bandes 21 à 23 cette couche est horizontale à sa base mais sa surface supérieure a une pendage de 15° vers le Nord. Dans la bande 21 sa base monte abruptement par 30 cm. Sa base reste plus ou moins horizontale latéralement après cette remontée, sauf près de la paroi où elle remonte et disparaît. L'épaisseur varie de 55 à 60 cm.

Figure 3.1. An overall view of the excavation area for layers J-K1, looking towards the exterior of the site. ***L'aire de fouille des couches J-K1, vue vers l'extérieur du site.***

Dans la coupe E/F 21-24, cette couche est surmontée des couches suivantes:

(10) Couche à cailloux de 2 à 4 cm (60%) et sédiments fins sablo-argileux avec granules. La couche contient quelques poches lessivées. Sa surface a un pendage de 20° vers le Nord. La transition à la couche supérieure est un peu diffuse. Epaisseur – 16 cm.

(11) Couche indurée à cailloux de 2 à 4 cm (30%) et sédiments fins sablo-argileux avec granules. La couche contient quelques poches lessivées. Sa surface a un pendage de 30° vers le Nord. L'orientation des éléments est variable. La couche contient quelques cailloux et blocs jusqu'à 10 cm. Epaisseur – 35 cm.

En regardant ces couches, on remarque un pendage général vers le Nord et l'Est; on remarque également, dans la coupe en G23, que ces mêmes couches ont des pendages à la fois vers le Nord et vers l'Ouest. On voit alors très bien que les couches forment une cuvette, où beaucoup d'eau est passée (en sortant du site vers le Nord dans la bande E). Il y a, en plus, une graduation de granulométrie, du fin à l'extérieur de la cuvette, à un cailloutis lessivé, et à un niveau à gros blocs sans sédiment interstitiel. . . . Ces couches représentent une période d'accumulation suivie d'une période d'érosion accompagnée de beaucoup d'eau . . ."

I interpret the succession as follows: the ensemble starts "avec une couche à sédiments fins et quelques cailloux: période chaude? Ensuite il y a eu accumulation encore de cailloux et de grandes roches, représentant probablement une période froide à très froide. Ces couches ont subi l'action de beaucoup d'eau, qui les a remaniées, lessivées, et a formé la cuvette. Cet action date probablement d'une période chaude et peut-être pluvieuse (ou à une période de fonte . . .)." The section shown in Figure 3.2 shows how the level of the larger rocks gets lower towards the centre-left (east) of this area.

In his report on activities in the short excavation of 2006, Lebel (2006) says that there is a hiatus in discoveries of archaeological material for 50 to 80 cm below layer I4, that is, in the top section of J (J to J3). This is not entirely true, but the material is sparse and scattered throughout those layers. In 2006, Lebel's team excavated a test pit including parts of B17, C17 and B18: ". . . dans deux secteurs adjacents B-17 (70 x 100 cm) et C-17 (30 x 100 cm) pour couvrir une surface de 1m^2. . . . pour éviter . . . la brèche très dure dans la bande 18, la dalle d'effondrement très épaisse en D-16 et pour localiser le sondage à au moins 1 m de la paroi rocheuse occidentale. . . . Nous avons également effectué un décapage sous la couche I-4 en B-18 . . . près de la paroi calcaire. . . . Le sondage a atteint une profondeur de -96 cm dans le coin sud-ouest (z= -8,72 m), -1,35 m dans le coin sud-est (z= -9,11 m) et -1,80 m

Figure 3.2. The section wall to the south of excavations in J-K1, showing that the level of large rocks gets lower towards the east (between the two diggers on the left of the photo). ***La coupe sud de la fouille dans les couches J-K1, montrant le pendage de la base du niveau à gros blocs vers l'Est du site (entre les deux fouilleurs dans la partie gauche de la photographie).***

(z= -9,56 m) vers le nord (coin nord-est). . . . Ensuite il faut admettre à la suite de ce sondage de la disparition des vestiges archéologiques et des couches anthropiques de J-1 à K-2 (mise à part quelques rares pièces lithiques). . . . Au cours du creusement et à -8,20 m de profondeur, une grande dalle calcaire tabulaire avec un pendage accusé vers le nord occupait le côté nord-ouest du sondage. La paroi rocheuse remonte rapidement et a été atteinte sur toute la surface du sondage. Elle occupe la moitié ouest et plonge rapidement en direction nord-est."

This shows how complex these deposits are. It is a shame that despite several sedimentologists agreeing to study aspects of the site over the years, no serious, complete study was ever done. Still, it looks like the succession from K1 through J may record at least three climatic phases: a relatively warm one (fine sediments) followed by a cold period (large rocks), followed by another warm phase, when abundant water flowed through the site. None of the deposits are cemented, despite that water. This may be due to the location of the overhang. Outside of the roof, the sediments were affected by rain (or meltwater), but inside the overhang the water dripping on them had come through the limestone roof, and was carrying dissolved calcium carbonate with it. That reprecipitated to form the cement in the breccia, which starts at band 20 and extends back from there throughout the rest of the site.

The archaeological presence in these levels is concentrated at the front of the site (bands 21 to 25), between about 880 and 920 cm below zØ. Not only is it sparse above that level, but it was almost entirely absent at those levels in the test pit in band 17. This may again mean that the hominins chose to stay in the better-lit front portion of the site, but it is also possible, as Lebel (2006) speculates, that the configuration of the rock shelter walls inhibited occupation further back.

3.3. Pollen and faunal remains

Jacqueline Argant took a series of samples for pollen during the summer of 1999 (Argant 2000). According to her report three samples were taken at a depth of -833 cm in square D19: number 25 (layer J1, in sterile sediments), number 26 (layer J4, fine rocks, bones, including rhinoceros), and number 27 (layer K1, sterile sediments). I can not explain the discrepancy between the z value and the layer attributions. Only sample 26 was analysed, and Argant states (2000) that the number of pollen and of taxa found in it is too small to be analysed statistically, which limits the strength of any interpretation. However, she did find 9 *Pinus* pollen grains, 1 *Juniperus*, 1 Cichorioideae pollen grain (a subfamily of the family Asteraceae of flowering plants, including lettuce, dandelions, and many other species), 1 Apiaceae (also known as Umbelliferae, a family of mostly aromatic flowering plants such as celery, carrots and parsley), 1 undetermined pollen grain, and 1 monolete spore. Pine is often a common pollen type, because pine trees are widespread and their pollen is abundant and blown easily on the wind. This result means that pine, juniper and some flowering plants may have grown in the area of the site, but otherwise gives us very little environmental information.

The faunal analysis was undertaken first by Martine Faure and Claude Guérin, and then by their student, Philippe Fernandez. In his summary of the fauna (Fernandez 2006), Philippe lists the ungulate species and their proportions in K1 through J as is shown in Table 3.1. Faure et al. (1996) add the non-ungulate *Ursus arctos* to the list. Alain Argant (A. Argant 2006) studied the carnivore material and does not list any from layers J through K, but does mention two pieces of *Ursus arctos* whose location numbers were illegible. Argant states that for the Bau overall, carnivore material is in small fragments, particularly including isolated teeth, with a few broken long bones and some phalanges. He does not specify which elements come from which layers, simply numbers of pieces from various species. However, in the case of the bear remains, both A. Argant (2006) and Faure et al. (1996) are certain that all of the bear remains throughout the Bau's sequence are from *Ursus arctos*, the brown bear, and not *Ursus spelaeus*, the cave bear.

The layer contains many fragments of large mammal bones (Figure 3.3). The majority of identifiable material comes from *Bos primigenius*, the aurochs. This large bovine preferred a prairie-park environment, open but with occasional clusters of trees. It is followed closely in abundance by *Equus mosbachensis*, a large horse species, which signals a fairly open and cold, rigorous, environment. The presence of the rhinoceros *Dicerorhinus hemitoechus* and the Irish elk *Megaloceros giganteus* reinforces this interpretation of a rigorous climate and an open or lightly forested landscape (Faure et al. 1996, Fernandez 2006). The other species are more ubiquitous, tolerating a wide variety of environments and habitats. All together, Fernandez assigns this layer to the MNQ (Mammal Neogene Quaternary) 24 biochronological zone, and dates it to MIS 6, with perhaps MIS 7 for the lowest levels (Fernandez 2006, p. 97).

Table 3.1. Faunal remains in layers K1 through J combined, after Faure et al. (1996), Fernandez (2001, 2006) and A. Argant (2006). ***Faune des couches K1 jusqu'à J combinées, d'après Faure et al. (1996), Fernandez (2001, 2006) et A. Argant (2006)***

Species	Common name	% of assemblage
Bos primigenius	aurochs	53.2
Equus mosbachensis	horse	35.3
Megaloceros giganteus	Irish elk	2.3
Cervus elaphus	red deer	2.3
Hemitragus cedrensis	tahr	2.3
Dicerorhinus hemitoechus	rhinoceros	1.7
Dama dama	fallow deer	1.1
Capreolus capreolus	roe deer	1.1
Rupicapra rupicapra	chamois	0.6
Ursus arctos	brown bear	n/a

Both Fernandez (2001, 2006; Fernandez & Legendre 2003; Fernandez et al. 1998) and Faure et al. (1996) attribute the abundance of aurochs and horse in this assemblage to selective hunting by the hominins who created it. Fernandez & Legendre (2003) analysed the horse teeth from the Bau and determined that in layer J, the teeth primarily come from adult animals, with juveniles also present, but almost no animals over about 15 years of age. They show that this must result from selective hunting, rather than scavenging or encounter hunting, and that two sorts of groups of horses must have been targeted. Horses divide themselves into bachelor groups (young males, sometimes with older males) and family groups (one stallion, some females and their young). In some cases, herds come together into larger groups for part of the year. The large proportion of adults represented in layer J, combined with the still significant proportion of juveniles, suggests that either both type of sub-group must have been hunted, or that perhaps the hominins were taking advantage of the combined herds. Fernandez & Legendre also studied the wear on the teeth, and suggest that this indicates that the horses died during the summer to early autumn (Fernandez & Legendre 2003).

Fernandez (2001) states that most of the aurochs found at the Bau are also of prime-aged adults. Aurochs were opportunistic, able to live in a variety of environments (Crégut-Bonnoure & Guérin 1996). They apparently lived most of the year divided into groups of females and young, and solitary males. They came together in the mating season, August and September, when the males could have violent confrontations (van Vuure 2002). It is tempting to think that that would be a good time of year to avoid the aurochs, and go hunt horses instead!

Faure et al. (1996) and Fernandez et al. (1998) discuss the significance of such a high proportion of aurochs and of horses within several layers at the Bau, including layer J. To have two main animals in a hunted assemblage, and two animals that differ so much in their behaviour, is surprising. As Faure et al. (1996) say, aurochs and horses have two very different defensive behaviours, and require different hunting strategies. In most European Middle Palaeolithic sites, hominins specialised in hunting one large mammal, or occasionally two, but in that case the two would be ones who behaved in much the same way. At the Bau, the hominins displayed flexible behaviour, whether seasonally or by taking advantage of the diversity of habitats available around the site: rocky cliffs, open plateaux, etc., in order to hunt these two species.

Figure 3.3. Faunal remains in J-K1. The striped arrow (cm scale) points north. ***Restes fauniques en J-K1. La flèche rayée (échelle centimétrique) indique le nord.***

The other animals present were used in small numbers, and may represent opportunistic hunting of animals encountered rather than selectively targeted. It is also possible that some of them may have been scavenged. The evidence is clear, however, that the majority of the meat obtained by these hominins came from deliberate, selective hunting of herds of gregarious animals. This probably could not have been effectively done by individuals acting alone, but rather required the concerted (and repeated) action of a well organised group.

3.4. Human remains

From my 1995 notes: "12 juillet. . . . en regardant un sac de dents de l'année dernière, Philippe [Fernandez] a sorti une dent humaine: de la couche K1. Elle est très grande, épaisse (l'émail), et usée. C'est une incisive." This tooth, from the square E23, layer K1, number 169, excavated in 1994, has since been named Aubesier 4 in the hominin remains sequence (Lebel et al. 2001, Trinkaus et al. 2000a). Aubesier 4 (Figure 3.4) "consists of most of the crown and root of a moderately worn left I^2" (Lebel et al. 2001, p. 11098), that is to say an upper second incisor from the left side. Based on dental wear, Lebel et al. (2001) consider that Aubesier 4 was from a prime-aged adult. It shows a "combination of strong shovel shape with lingual tubercles and strong labial convexity" which "is a characteristic of Neandertal incisors" (Trinkaus et al. 2000a, p. 215), as well as of their Middle Pleistocene predecessors (Lebel et al. 2001). "While each of these characters can be found to some degree in modern human populations, their combined form may be unique to the Neandertals" (Trinkaus et al. 2001, p. 215). The labiolingual diameter is large for a later archaic *Homo* (Lebel et al. 2001): equivalent to the later Middle Pleistocene sample studied by Trinkaus et al. (2000a), modestly larger than the Neandertal sample, and much larger than Upper Palaeolithic and modern samples. It can be considered a pre-Neandertal specimen, from the lineage which was eventually going to develop into the Neandertals (Lebel et al. 2001).

"The Aubesier 4 I^2 is notable for the high degree of beveling of its occlusal wear . . . This beveling angle is elevated

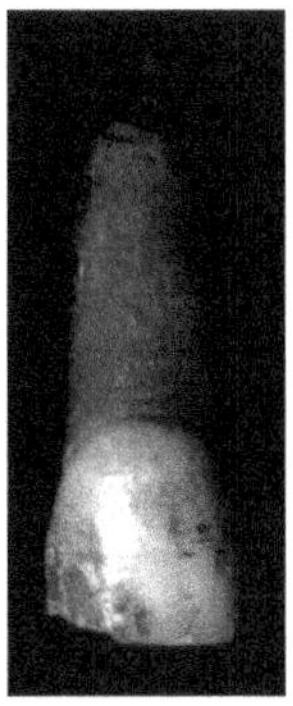

Figure 3.4. Labial view of the Aubesier 4 incisor. Photo E. Trinkaus. ***Vue labiale de l'incisive Aubesier 4. Cliché E. Trinkaus.***

compared with recent humans with similar degrees of occlusal attrition, and it is high even relative to values for Krapina and last glacial Neandertal anterior maxillary teeth. . . It implies a pattern of attrition, and hence use, of the anterior dentition similar to that of recent humans and Neandertals but at least as pronounced as that which leads to the labially open incisor occlusion in older Neandertals. The reasonably well-preserved labial surface of the crown does not exhibit distinctive transverse scratches, such as would be made by a lithic implement" (Lebel et al. 2001, p 11100). Trinkaus et al. (2000a) mention that fine scratches are present but show no predominant orientation. The combination of bevelling and scratches is taken to imply "considerable use of the anterior dentition for nonmasticatory purposes" (Lebel et al. 2001, p.11100). This person depended on his or her teeth to perform jobs for which later humans tend to use tools.

3.5. The lithic industry and raw materials

The raw materials list includes 2084 pieces from K1 through J, while the typological list includes 2546. As Table 3.2 shows, this discrepancy is mainly due to the greater number of sieve (tamis) pieces on the typological list, although there are other differences between the lists. In both cases, however, there is a reasonable proportion of finished tools, a lot of flakes and sieved material, a noteworthy percentage of blades (at least 3.5% of the full assemblage), and a good number of cores. Both blades and cores are more abundant (as numbers of pieces and as percent) in these layers than in K2. Figure 2.4 (chapter 2) showed some elongate pieces; the pieces not identified as coming from K2 are from K1 through J.

The typologically identified pieces (e.g., Figure 3.5) include 11 Levallois flakes, 5 atypical Levallois flakes (including blades), 3 Levallois points, 2 pseudo-Levallois points, 21 various sidescrapers, 1 atypical endscraper, 7 naturally-backed knives, 2 notches, 5 denticulates, 8 pieces with retouch on their ventral faces (usually removal of the bulb), 1 bifacially retouched flake, 1 chopping tool (the only one found in the entire Bau sequence; Figure 3.6), 34 multiple tools which include at least one sidescraper, and 13 multiple tools which do not include a sidescraper (and are very diverse). Most of the pieces have little or no cortex present, but there are 21 with between 10 and 50% cortex left, and 2 with more than 50%.

The cores range in size from 32 to 109 mm long, with an average of 65 mm, and in weight from 5 to 403 g, with an average of 116 g. They show an average of 13.5 removals. Both Levallois and non-Levallois techniques were used, and there are both blade and flake styles of core present. According to Carmignani et al. (2017, p. 26), "The pyramidal system disappears in unit J, replaced by the development of a semi-rotating system and in particular by the convergent method, which is absent in unit K."

This abundance of cores and knapping debris implies importation of raw material for tool production. Table 3.3 gives the numbers of pieces from each source area, by typological category, and Table 3.4 gives those results as percentages. The used source areas are shown in Figure 3.7. Of the 2084 pieces on the raw materials list, 855 (41.0%) are too patinated or burnt to be identified to their source area, and 11 (0.5%) are from an unknown source. Both of those numbers are much lower than their equivalents in K2, but higher than in the assemblage from the G23/G24 test pit. As was the case with those other assemblages, most of the unidentifiable pieces are patinated rather than burned: only 75 pieces (3.6%) are noted as burned in the catalogue, although this is surely an under-estimate. This value is the lowest of any assemblage considered this far.

The most abundantly used source area is again Murs, and there is a minor amount of material from its surrounding source areas (Les Sautarels, etc.). Raw material from that area is dominant in the cores. As with the previous assemblages, material was obtained primarily along a SW-NE axis, from Murs to St. Trinit, and in this case extending farther to the NE, to the quartzite source north of Aurel, about 15 km from the Bau. That piece is also the first non-flint piece in any assemblage mentioned so far. It

Table 3.2. Lithic assemblage of layers K1 through J, summarised by typological category. ***Assemblage lithique sommaire des couches K1 jusqu'à J, par catégorie typologique***

Category	Typology list (n)	Typology list (%)	Raw materials list (n)	Raw materials list (%)
Liste Bordes	114	4.5	98	4.7
Debris	11	0.4	10	0.5
Eclats	545	21.4	658	31.6
Lames	89	3.5	90	4.3
Tamis/Collectif	1714	67.3	1088	52.2
Nucléus	73	2.9	59	2.8
non-identifié	n/a	n/a	81	3.9
Total	**2546**	**100**	**2084**	**100**

Figure 3.5. Piece D21 J4 467, a concave sidescraper made on a Levallois point, in raw material type R, also from the area of Sault. ***Pièce D21 J4 467, racloir concave sur pointe levallois, dans le type de matière première R, provenant de Sault.***

Figure 3.6. The chopping tool, F23 K1 85, made of raw material type Q, from the Murs area, approximately 12 km south-west of the Bau. ***Le chopping tool, F23 K1 85, confectionné du type de matière première Q, de la région de Murs, à 12 km environ au sud-ouest du Bau.***

Table 3.3. Number of pieces by typology and source area in layers K1 through J. ***Nombre de pièces par catégories typologiques et groupes de sources des matières premières dans les couches de K1 jusqu'à J***

J-K1 n	Liste Bordes	Debris	Eclats	Lames	TamisColl	Nucléus	non-id	Total
Unidentifiable	15	1	199	29	581	7	23	855
Unknown	1	0	4	0	5	0	1	11
Faraud etc.	11	1	57	10	95	2	7	183
Méthamis etc.	5	1	33	7	26	4	4	80
Les Sautarels etc.	0	0	4	0	0	1	0	5
Sault	14	0	89	16	101	2	5	227
Murs	38	7	215	24	225	41	31	581
Local	2	0	16	0	21	1	2	42
St. Jean de Sault	0	0	3	0	0	0	1	4
Roussillon	0	0	0	0	0	0	0	0
Murs-Bezaure Olig.	4	0	30	4	29	0	4	71
Ravin de la Treille	0	0	0	0	0	0	1	1
St. Trinit	7	0	7	0	4	1	2	21
Durance	0	0	0	0	0	0	0	0
Tertiary Calavon	0	0	0	0	0	0	0	0
Tertiary Murs VdeV	1	0	0	0	1	0	0	2
Mormoiron	0	0	0	0	0	0	0	0
R de Guérin	0	0	0	0	0	0	0	0
N.Aurel	0	0	1	0	0	0	0	1
Total	**98**	**10**	**658**	**90**	**1088**	**59**	**81**	**2084**

is the central fragment of a broken flake, 25 x 22 x 9 mm in size, weighing 6 g. It shows no cortex and is not burned.

The NE source areas used also include St. Jean de Sault for the first time (4 pieces). Whereas the SW material, especially from Murs, is important in all categories and especially so in the cores, suggesting that a considerable amount was brought to the Bau to be knapped, material from the NE tends to be more prevalent in the tools, flakes and blades, so although it is present in the cores, it seems to be more likely to have been preferentially knapped outside of the site, and imported as usable products.

The source areas used can be grouped together into geographic areas, as is shown in Table 3.5. The distances given are only approximate, measured as straight lines on the map. It is interesting to contrast the use not only of source areas, but of these larger geographic areas. Table 3.6 gives the percentages of identified pieces (excluding unidentifiable and unknown source pieces) from each geographic region, in each of the three assemblages considered so far: K2 and G23/G24 (see chapter 2), and K1-J.

Local material is not much used, especially (as discussed in chapter 2) in K2, and material from the northeastern source areas accounts for around 20% of each assemblage. Murs (SW area) is increasingly important as we move up through these assemblages, from 44.3 % in K2 to 54.1 % in K1-J, counterbalanced by is a decrease in the proportion of material from the near SW and from the WSW (Méthamis and area). As we have seen above, the SW material is present throughout the whole *chaîne opératoire* and must have been knapped at the site, so it is interesting that over time there is this increased emphasis on material from around Murs, and decreased use of other sources to the south and west of the site. On the other hand, the northeastern material, which was imported more commonly in a finished form or as blanks, maintains approximately the same proportion within each assemblage. Use of that part of the territory seems to have maintained the same pattern through time, with hominins arriving at the Bau, from that area, with their tool kit already made. Since that tool kit was left behind at the Bau, it may well be that they were replacing it with flakes and blades made from the Murs cores. This could be an example of provisioning of place (Kuhn 1995, 2004),

Table 3.4. Percentages of pieces by typology and source area in layers K1 through J. ***Pourcentages des pièces par catégories typologiques et groupes de sources des matières premières dans les couches de K1 jusqu'à J***

J-K1 %	Liste Bordes	Debris	Eclats	Lames	TamisColl	Nucléus	non-id	Total
Unidentifiable	15.3	10.0	30.2	32.2	53.4	11.9	28.4	41.0
Unknown	1.0	0.0	0.6	0.0	0.5	0.0	1.2	0.5
Faraud etc.	11.2	10.0	8.7	11.1	8.7	3.4	8.6	8.8
Méthamis etc.	5.1	10.0	5.0	7.8	2.4	6.8	4.9	3.8
Les Sautarels etc.	0.0	0.0	0.6	0.0	0.0	1.7	0.0	0.2
Sault	14.3	0.0	13.5	17.8	9.3	3.4	6.2	10.9
Murs	38.8	70.0	32.7	26.7	20.7	69.5	38.3	27.9
Local	2.0	0.0	2.4	0.0	1.9	1.7	2.5	2.0
St. Jean de Sault	0.0	0.0	0.5	0.0	0.0	0.0	1.2	0.2
Roussillon	0.0	0.0	0.0	0.0	0.0	0.0	0.0	0.0
Murs-Bezaure Olig.	4.1	0.0	4.6	4.4	2.7	0.0	4.9	3.4
Ravin de la Treille	0.0	0.0	0.0	0.0	0.0	0.0	1.2	0.0
St. Trinit	7.1	0.0	1.1	0.0	0.4	1.7	2.5	1.0
Durance	0.0	0.0	0.0	0.0	0.0	0.0	0.0	0.0
Tertiary Calavon	0.0	0.0	0.0	0.0	0.0	0.0	0.0	0.0
Tertiary Murs VdeV	1.0	0.0	0.0	0.0	0.1	0.0	0.0	0.1
Mormoiron	0.0	0.0	0.0	0.0	0.0	0.0	0.0	0.0
R de Guérin	0.0	0.0	0.0	0.0	0.0	0.0	0.0	0.0
N.Aurel	0.0	0.0	0.2	0.0	0.0	0.0	0.0	0.0
Total	**100**	**100**	**100**	**100**	**100**	**100**	**100**	**100**

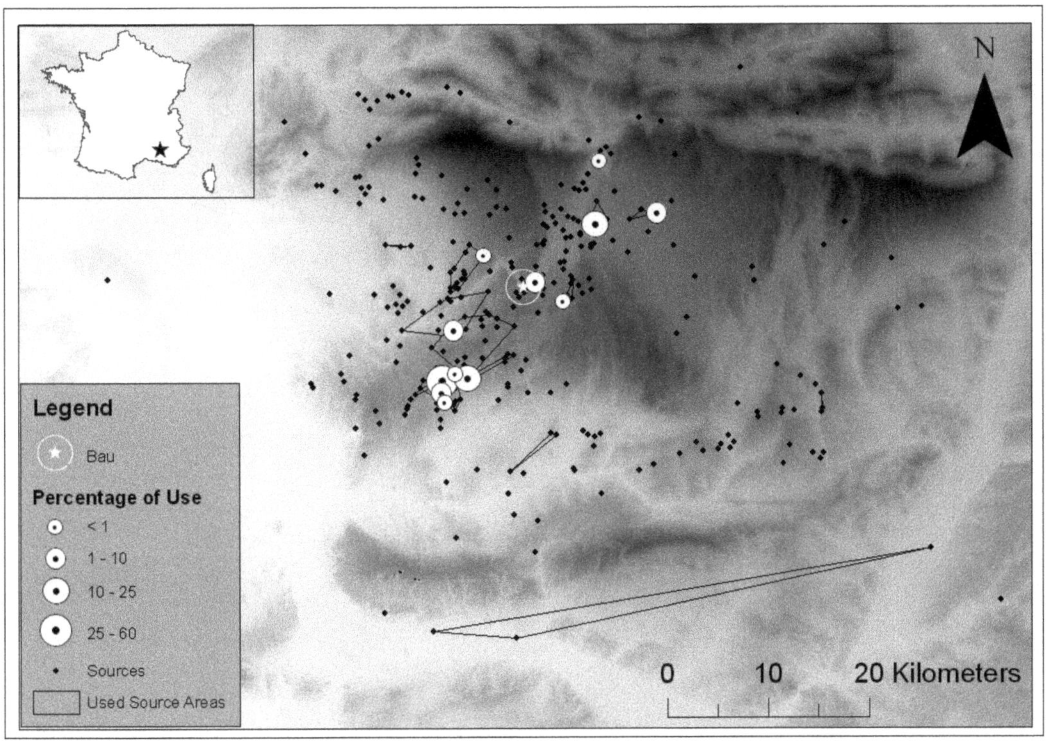

Figure 3.7. Map of the raw material source areas used for layers K1 through J. ***Carte des aires de sources de matières premières utilisées pour les couches K1 jusqu'à J.***

Table 3.5. Raw material source areas grouped by geographic region. ***Aires de sources de matière première regroupées par région géographique***

Geographic area	Approximate distance from the Bau	Source areas included
Local	1 km	Local
North-East	5 – 15 km	Sault, St. Trinit, N. Aurel
East	5 km	St. Jean de Sault
Far South-East	30 – 35 km	Rocher de Guérin
South	15 – 20 km	Tertiary Calavon
Far South	35 km	Durance
South-South-West	18 km	Roussillon
South-West	10 – 15 km	Murs, Les Sautarels etc., Murs-Bezaure Olig., Tertiary Murs V de V
Near South-West	3 – 10 km	Faraud etc.
West-South-West	7 – 15 km	Méthamis etc.
West	12 km	Mormoiron
North-West	5 km	Ravin de la Treille

Table 3.6. Percentages of identified pieces (excluding unidentifiable and unknown source pieces) from each geographic region. ***Pourcentages des pièces identifiées (excluant les pièces indéterminables ou de sources inconnues) de chaque région géographique***

Geographic area	K2 n.	K2 %	G23/G24 n.	G23/G24 %	K1-J n.	K1-J %
Local	1	0.2	9	5.0	42	3.4
NE	108	19.4	40	22.1	249	20.4
E	0	0	0	0	4	0.3
SW	247	44.3	90	49.7	659	54.1
Near SW	128	23.0	24	13.3	183	15.0
WSW	69	12.4	18	9.9	80	6.6
NW	4	0.7	0	0	1	0.1
other	0	0	0	0	0	0
Total	**557**	**100**	**181**	**100**	**1218**	**100**

where hominins bring a supply of raw material to a place where they know they will want to use it. This contrasts with provisioning of the individual (Kuhn 1995, 2004), where the individual hominin would carry what is needed in his or her own toolkit.

The resource selection function study (Tables 3.7 and 3.8) found that for K1-J there were several variables of approximately equal importance: Quality, Calories and Extent, followed by Difficulty (Cal/km) and Large rocks. The Beta values (Table 3.7) were unsurprising. In particular, Quality and Extent are positive (so better quality and larger extent of source correspond with greater likelihood of use), while Calories and Difficulty are negative (harder to get to means less likely to be used). The only positive Beta value in the rock size categories is for Large rocks (16-35 cm in size), suggesting that the hominins preferred and targeted this size range.

Although no single variable dominates the results for K1-J, this assemblage repeats the pattern seen in K2, where the total weight of the terrain variables (Extent, Difficulty, Calories, AOSA and AOSISA) outweighs that of the raw material variables (Quality, and the rock sizes). In this case, the difference is slightly greater than it was for K2: 56.8% for the terrain variables versus 42.4% for the raw material variables. Once again, ease of access to the raw materials is more important than the materials themselves, but raw material characteristics do still matter – especially quality and large size.

In the K1 to J level, therefore, we have evidence of a cold climate, probably MIS 6, with an open to lightly forested landscape. As is generally the case in such sites, the layer

Table 3.7. Generalised linear model for raw material source choices in K1-J. P-values in bold font are statistically significant. (Sample: 1069 pieces from 12 source areas)

Overall:
Omnibus test: likelihood ratio chi-square = 582.745, df = 10, p = <0.001

Variable	Beta	S.E.	Wald Chi-Square	P-value	AIC
Quality	1.572	0.3821	16.924	**<0.001**	170.484
Extent	2.512	0.6382	15.490	**<0.001**	
Difficulty (Cal/km)	-0.083	0.0224	13.852	**<0.001**	
Small rocks	-7.945	2.5164	9.968	**0.002**	
Medium rocks	-3.240	1.5041	4.641	**0.031**	
Large rocks	2.540	0.7650	11.023	**0.001**	
Very large rocks	-0.326	0.9289	0.123	0.725	
Calories	-0.003	0.0009	16.680	**<0.001**	
AOSA	-1.3 e-7	7.4 e-8	3.029	0.082	
AOSISA	4.6 e-5	1.6 e-5	8.227	**0.004**	
(Intercept)	3.506	3.8643	0.823	0.364	

Table 3.8. Wald Chi-Square values as percentages, K1-J assemblage

Variable	Wald Chi-Square	Wald Chi-Square as %
Quality	16.924	16.8
Extent	15.490	15.3
Difficulty (Cal/km)	13.852	13.7
Small rocks	9.968	9.9
Medium rocks	4.641	4.6
Large rocks	11.023	11.0
Very large rocks	0.123	0.1
Calories	16.680	16.6
AOSA	3.029	3.0
AOSISA	8.227	8.2
(Intercept)	0.823	0.8
Total	**100.78**	**100**

represents a palimpsest of material that accumulated over a lengthy period and is due to the actions of many different groups of people. Pre-Neandertal hominins used the Bau and brought in supplies of flint from the Murs area, along with toolkit elements from the northeastern sources. They selectively hunted aurochs and horses, and also probably more opportunistically hunted a variety of other animals, some of which were also large (rhinoceros, Irish elk). Some meat may have been obtained through scavenging, too, but it is clear that they were very effective hunters. The very large number of bones found in the excavation translates into many animals, and presumably large quantities of meat: they may have been provisioning the site with food as well as with lithic raw material. There is little evidence of fire, but some of the lithics were burned, so fires may have been located outside of our excavation area. There is evidence that the site was used in late summer, and it could also have been used at other times of the year.

3.6. Sommaire du chapitre

Dans la partie inférieure du site nous avons nommé les couches arbitrairement par mètre de profondeur. La limite entre les couches J et K était fixée à 905 cm sous zØ, et la limite supérieure de la couche J se trouvait à -805 cm. Cependant, en fouillant nous avons trouvé qu'il y avait du matériel dispersé le long de l'épaisseur de J, puis une concentration de matériel vers la limite inférieure. Cette même concentration se poursuivait sans différence perceptible jusqu'à l'intérieur de la couche K, à un profondeur d'environ 920 cm. Le matériel plus dispersé de la partie supérieure de J était attribué aux couches J, J1, J2 et J3. Le matériel plus concentré était attribué à la couche J4, et sa continuation en K était appelé K1. On considère maintenant que J4-K1 représente une seule couche archéologique, l'une des plus importantes du Bau. Ce chapitre comprend aussi le matériel moins abondant de J à J3 (comme cela a été fait dans des publications antérieures).

J4-K1 a livré des milliers de restes fauniques et lithiques, et une dent humaine, Aubesier 4. Selon la biochronologie, et en prenant en compte les datations radiométriques des couches supérieures, les couches de K1 à J doivent correspondre à la partie supérieure du Pléistocène moyen, MIS 7 à 6, ce qui leur donne un âge entre 240 000 et 200 000 ans.

Les sédiments ne sont pas bréchifiés, et étaient faciles à fouiller. Ils montraient beaucoup de variation granulométrique latéralement et verticalement, avec quelques concentrations de plus grandes roches. Dans la partie centrale de la zone fouillée, les sédiments semblaient avoir été affectés par de l'eau, qui a lessivé les grains les plus fins, et provoqué un effondrement des couches: les couches des deux côtés ont un pendage vers le centre. Ces effets sont liés sans doute aux variations climatiques ainsi qu'à la position de la limite du plafond rocheux. En effet, à partir de la bande 20 et en allant vers l'intérieur de l'abri, les couches ont été affectées non pas par de la pluie, mais par de l'eau qui avait traversé le plafond rocheux, et qui était chargée en carbonate. La reprécipitation de carbonate de calcium à l'intérieur de ces sédiments a créé le ciment et formé la brèche que l'on trouve si abondamment dans le site.

Si l'étude des pollens de cette couche n'a donné que très peu d'informations, l'étude de la faune nous donne une très bonne idée de l'environnement (froid et rigoureux, prairie-parc, donc ouvert mais avec quelques arbres épars) ainsi que des stratégies de chasse des hominidés. Il s'agit effectivement de la chasse, dans la plupart des cas, et non de charognage. La faune est dominé par l'aurochs, suivi par le cheval *Equus mosbachensis*. Les deux espèces ont des comportements défensifs très différents, et ont été chassées de façon sélective, ce qui nécessite deux stratégies de chasse différentes. De plus, les chevaux ont été chassés pendant l'été et le début d'automne.

La dent humaine, Aubesier 4, est une grande deuxième incisive supérieure légèrement usée, avec des traits semblables aux dents de Néandertaliens. Elle est considérée pré-Neandertalienne ou archaïque, provenant de la lignée qui allait devenir les Néandertaliens dits classiques. Les types et positions des traces d'usures sur cette incisive suggèrent que la personne utilisait ses dents pour accomplir certains travaux non-liés à la mastication, pour lequels les humains plus récents utiliseraient des outils. Malgré cela, nous avons trouvé beaucoup d'outils lithiques dans cette couche! L'abondance des nucléus et de débitage (Tableau 3.2) suggèrent que certains outils ont été fabriqués dans le site. C'est surtout le cas concernant les matières premières provenant de la région de Murs qui apparaissent dans toutes les parties de la chaîne opératoire, et dominent les nucléus. En revanche, les matériaux provenant de gisements au NE du Bau constituent plus les catégories d'outils finis, et ont probablement été taillés plus souvent à l'extérieur du site. L'assemblage comprend également des matériaux de quelques autres gisements dans la région (Tableaux 3.3 et 3.4) et il est probable que ces hominidés aient fréquenté un grand territoire qu'ils connaissaient très bien.

Dans le niveau J à K1 nous avons un climat froid, probablement du MIS 6, avec un paysage ouvert ou légèrement boisé. Les couches représentent des accumulations de restes laissés pendant plusieurs occupations différentes sur une longue période de temps; ces restes sont mélangés, ce qui est presque toujours le cas dans les sites préhistoriques. Si ce que nous avons trouvé ne donne pas une vision détaillée de la vie d'un seul groupe préhistorique, nous voyons cependant des tendances repetées par beaucoup de groupes à travers le temps, qui nous donnent des signaux forts. Ils pratiquaient une chasse sélective, orientée sur l'aurochs et le cheval, et chassaient de façon plus opportuniste peut-être quelques autres grands animaux (rhinocéros, mégacéros. . .). Ils amenaient du silex de Murs pour le tailler au Bau, et peut-être utilisaient-ils ces outils fabriqués pour remplacer ceux confectionnés dans du silex provenant des gisements du NE amenés déjà finis dans leurs trousses à outils, et qu'ils allaient laisser derrière eux au Bau. Il existe quelques traces de feu, mais nous n'avons pas trouvé de foyers dans les zones fouillées. Les hominidés fréquentaient le Bau pendant les étés, mais il est également possible qu'ils aient été présents pendant d'autres périodes de l'année. Il est possible qu'ils amenaient du silex et de la viande de grands animaux au Bau en surplus, sachant qu'ils allaient peut-être revenir.

4

Layer I

4.1. Introduction

This chapter includes all results from the metre-thick layer from 705 to 805 cm below zØ, which we called layer I. It was excavated over about 34 m^2 in bands A (15, 19, 20 and 22, all partial), B (15-22), C (15-22), D (17-23), E (19-23) and F (19-21) (see Figure 1.11), which gave us plenty of room to put people to work (Figure 4.1). Layer I includes sub-layers I, I1, I2, I3 and I4, with the bulk of the material coming from I2. For instance, the typological list includes 3250 lithic pieces, 1972 of which (60.7%) come from I2. Layer I was much richer in bones than in lithics, however. In the report for excavation seasons 1993 to 1995, Lebel (1995) reports that in 12 m^2 (bands A-F, squares 19-20) in layers I2 and I3 we found 1175 lithics and 10,177 bones and teeth. I do not have a final total number of bones and teeth found, but it is not unlikely that there were more than 30,000. Although many of these were undeterminable fragments, there was still a large amount of determinable material, as we will see below, requiring careful work of coordination and extraction (Figure 4.2). This layer also produced a pathological half-mandible of a hominin, and a separate hominin tooth, which will also be described in this chapter. The layer has not been radiometrically dated, but on the basis of the fauna and of its stratigraphic position, it likely corresponds to the later Middle Pleistocene (Lebel et al. 2001), and to MIS 6 and MNQ 24 according to Fernandez (2001, 2006), which puts it in the Riss III of the Alpine glacial sequence. Since level I is stratigraphically below layer H1, it must be older than the TL dates obtained for that layer (minimum 169 ± 17 ka, maximum 191 ± 15 ka) (Lebel et al. 2001).

4.2. The sediments, and evolution of the site in I

The metre-thick I level consisted of pale-coloured calcareous sediments of variable granulometry and archaeological content. It was divided as we excavated into sub-layers based on either archaeological or sedimentological criteria, depending on the case. Finds were attributed simply to layer I in the case of surface discoveries at the right level, but in most cases as we dug from H into I, the layer was named I2 right away. Layer I1 was located only near the rock shelter's western wall, in A19 and B19-21, and was sedimentologically differentiated from I2 on the basis of its fine grain size (silt to sand-sized calcareous grains, without many larger clasts). It was of variable thickness (up to perhaps 15 cm), unconsolidated and easy to dig. I2 was a mixture of small and large clasts (some up to several tens of cm in size) and a fine matrix, with very abundant bones, teeth and some lithics. Its uppermost level included several large slabs of limestone which had fallen from the rock shelter ceiling. The bones were often fragmented, but still many large, identifiable pieces survived, as will be discussed below. Its thickness varied but reached 60 cm in band E. I2 was sometimes cemented into breccia, especially in band D. In my notes on stratigraphy from 1995 I remark that the base of I2 is horizontal, where it meets I3, and that the transition with the overlying layer is not distinct.

Layer I3 is described in my stratigraphic notes from 1995 as "Couche à 60% de cailloux de 2 à 4 cm et sédiments fins sablo-argileux avec granules." In one area this layer also included a large block which had fallen from the cave roof, which was in turn covered by layer I2.

Layer I4 was defined on the basis of a change in sediment, with its base being the arbitrary limit of 805 cm below zØ. It was part of a sedimentary layer that extended into level J and was above the "cuvette" (chapter 3), described as:

"Couche indurée à cailloux de 2 à 4 cm (30%) et sédiments fins sablo-argileux avec granules. La couche contient quelques poches lessivées. Sa surface a une pendage de 30° vers le Nord. L'orientation des éléments est variable.

Figure 4.1. Excavations in progress, layers H (bottom of photo) and I, summer of 1993. ***Fouilles en cours, couches H (au premier plan) et I, été 1993.***

Figure 4.2. Coordinating objects (mainly faunal material) in layer I. ***Prise de coordonnées des restes (surtout fauniques) de la couche I.***

La couche contient quelques cailloux et blocs jusqu'à 10 cm. Epaisseur – 35 cm."

Lebel (2006), while describing the test pit in B17/C17, remarks that: "Le début des fouilles pour le sondage a été amorcé entre les élévations (Z) de -7,69 et -7,80 m sous la couche I-3. . . . La couche archéologique I-4 s'est manifestée rapidement sous la couche I-3 par la présence de témoins lithiques (surtout des produits de débitage) et osseux nombreux."

Layers I through I4 thus start above the large rocks of the "cuvette", which had been subjected to copious water flow, and include a relatively coarse, indurated to cemented layer (I4); a finer, sometimes cemented, layer (I3); and a coarser layer with a fine matrix, again occasionally cemented, especially in band D (I2). Near the rock shelter wall were the fine, loose sediments of I1. There were occasional large slabs of rocks which had fallen from the roof, and the transition from I into H (discussed in more detail in chapter 5) was marked by the presence of several of these slabs (Figures 4.3 and 4.4). In general, the boundaries of the layers were horizontal, as were the orientations of the clasts within the layers. The abundance of cryoclastic clasts of various sizes, including the large slabs in the upper layers, suggests that the climate was cold throughout the period of accumulation of these layers. The cementation that exists dates to a later, warmer, period when water dripped onto and flowed through the deposits. Band D seems to have been located under the most consistent area of dripping water during that later period, because it is always the most strongly cemented.

4.3. Pollen and faunal remains

Argant (2000) describes taking samples for palynological analysis along a stratigraphic column in C15. Five of the samples came from layer I: numbers 20 (722 cm below zØ), 21 (-746 cm), 22 (-762 cm), 23 (-780 cm), and 24 (-802 cm). Only numbers 22 and 23 (both attributed to

Figure 4.3. Layer I2 sediments, bones and lithics, under a large slab of limestone at the top of the picture. ***Sédiments, os et lithiques de la couche I2, sous une grande dalle calcaire en haut de l'image.***

Figure 4.4. Large slabs of rock marking the surface of layer I2 in zones B-C 15-16, with the rock shelter wall visible on the left. North is towards the top of the photo. ***Grandes dalles de roche démarquant la surface de la couche I2 dans les zones B-C 15-16, avec la paroi rocheuse visible à gauche. Le Nord est vers le haut de la photo.***

layer I4) were analysed, and number 23 was sterile. In her preliminary analysis of number 22, Argant (2000) says that the pollen were scarce and poorly preserved, but she did find a predominance of *Pinus*, along with some *Cedrus*, some unidentifiable conifer pollen grains, and from one to three grains each of *Abies*, *Alnus*, *Corylus*, *Fagus* and *Quercus*. Small numbers (from one to five each) of pollen from a variety of non-arboreal taxa were also found (Poaceae, Cichorioideae, Apiaceae, Rosaceae, Ericaceae, Ranunculaceae and Cyperaceae).

In her 2006 report, more samples had been analysed (J. Argant 2006) and she interprets the samples from layer I as reflecting a steppe environment: cold, dry, with few trees and a lot of open space. The trees that are present are "héliophiles et xérophiles", meaning that they like sunlight and dry conditions:

"A 746 cm et 690 cm on observe la disparition du cèdre, tandis que les Cichorioideae apparaissent en plus grand nombre. Ceci laisse supposer une sécheresse accrue et, globalement des conditions plus rigoureuses favorisant une végétation steppique. A la base de la séquence de la coupe C15, les taxons héliophiles et xérophiles dominent. . ., permettant d'imaginer un paysage steppique où les arbres sont très certainement rares. La représentation élevée du pin pourrait être due, dans ce cas, en partie à des apports lointains. Rien ne s'oppose pourtant à ce que des pins soient également présents dans les lieux abrités des gorges de la Nesque, malgré les conditions probablement froides et sèches. . ."

This agrees well with the faunal results (Fernandez 2001, 2006). Layer I (including I through I4), produced very abundant bone and teeth remains (Figure 4.5), and even though many of them were fragmentary, there were still many bones and teeth which could be identified to taxon (Fernandez 2001, 2006), and in some cases age at death (Fernandez & Legendre 2003). For example, Fernandez (2001) describes a total of at least 30 aurochs teeth and 113 identifiable aurochs bones (Figure 4.6), and at least 96 identifiable horse bones from a total of no less than 23 individual horses (Figure 4.7). This richness makes the Bau one of the premier palaeontological sites for the later Middle Pleistocene in southern France, in addition to its archaeological importance.

Figure 4.5. A variety of faunal remains in layer I. ***Restes fauniques dans la couche I.***

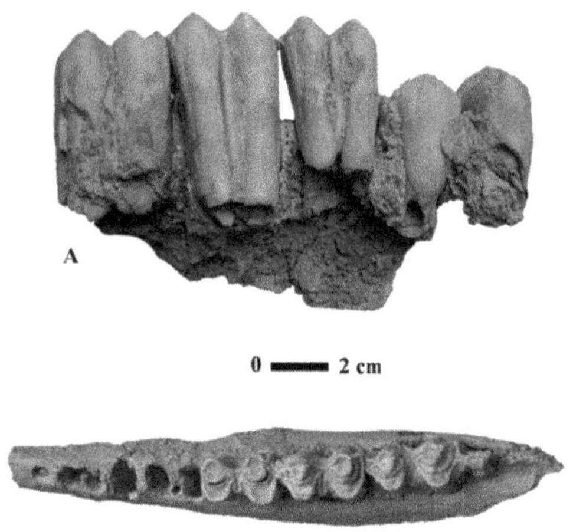

Figure 4.6. ***Bos primigenius primigenius*****; A. Top right dental row (C20 I4 925) with from left to right M3/, M2/, M1/, P4/ and P3/, lingual face; B. Lower left dental row (D20 I4 458) with from left to right M/1, M/2 and M/3, occlusal face. Photos P. Fernandez.** ***Bos primigenius primigenius; A. Rangée dentaire supérieure droite (C20 I4 925) avec de gauche à droite M3/, M2/, M1/, P4/ et P3/, face linguale; B. Rangée dentaire inférieure gauche (D20 I4 458) avec de gauche à droite M/1, M/2 et M/3, face occlusale. Clichés P. Fernandez.***

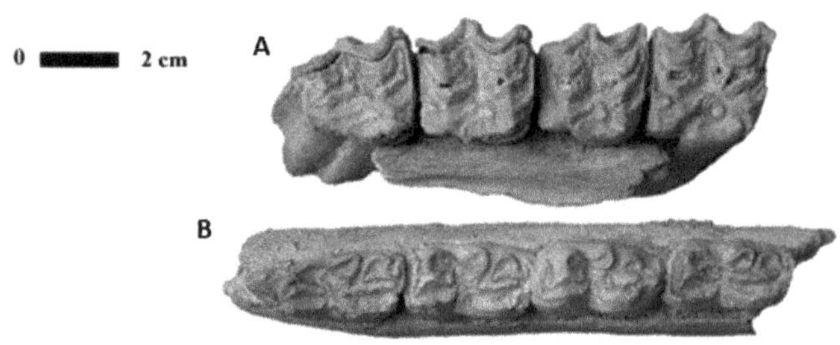

Figure 4.7. ***Equus mosbachensis*****; A. Top right dental row (C20 I4 896) with from left to right M3/, M2/, M1/ and P4/; B. Lower left dental row (C20 I2 429) with from left to right P/2, P/3, P/4 and M/1. (A and B occlusal face). Photos P. Fernandez.** ***Equus mosbachensis; A. Rangée dentaire supérieure droite (C20 I4 896) avec de gauche à droite M3/, M2/, M1/ et P4/; B. Rangée dentaire inférieure gauche (C20 I2 429) avec de gauche à droite P/2, P/3, P/4 et M/1. (A et B face occlusale). Clichés P. Fernandez.***

The species present (Table 4.1) are similar to those of layer J, with the aurochs most abundant, followed by horses, although both are present in smaller proportions than they were in layer J (43.2% for aurochs, versus 53.2% in J; 31.0% for *Equus mosbachensis*, versus 35.3% in J). Those decreases are compensated for by a marked increase in the abundance of red deer (10.3% versus 2.3% in J), and by the presence of wild boar, which is absent from all other layers at the Bau except H1 and 4. In fact, of 7 individual boars identified in the Bau's assemblage, 5 come from layer I. There are also other small similarities and differences between the faunal assemblages of the two layers I and J, such as Irish elk being again present in I (Figure 4.8). All in all the same interpretation is justified: these remains are present due to selective hunting by hominins, targeting two distinctly different large mammals, with the possibility of occasional encounter hunting or scavenging for the other animals.

The evidence for selective hunting is once again bolstered by the detailed analysis of age of death for the horse teeth (Fernandez & Legendre 2003; Fernandez et al., 2006). Although layer I includes animals from all age classes, and unlike J includes animals over 18 years old, the bulk of the meat came from prime-aged adults, 6 to 12 years old, hunted during the period from July to September.

The brown bear remains were described by A. Argant (2006) as consisting of 10 pieces: 2 from I2, 5 from I3 and 3 from I4. Subsequently, E. Crégut-Bonnoure (Crégut-Bonnoure et al., 2011) was able to examine all of the material, and determined that layer I included remnants of three individual bears, one each from layers I2, I3 and I4.

Table 4.1. Faunal remains in layers I through I4, after Fernandez (2001, 2006) for ungulates, and A. Argant (2006) and Crégut-Bonnoure et al. (2011) for *Ursus arctos*. *Faune des couches I jusqu'à I4, d'après Fernandez (2001, 2006) pour les ongulés, et A. Argant (2006) et Crégut-Bonnoure et al. (2011) pour* Ursus arctos

Species	Common name	% of assemblage
Bos primigenius	aurochs	43.2
Equus mosbachensis	horse	31.0
Cervus elaphus	red deer	10.3
Hemitragus cedrensis	tahr	5.4
Megaloceros giganteus	Irish elk	4.0
Sus scrofa	boar	2.3
Dama dama	fallow deer	1.9
Dicerorhinus hemitoechus	rhinoceros	0.8
Capreolus capreolus	roe deer	0.6
Rupicapra rupicapra	chamois	0.6
Ursus arctos	brown bear	n/a

The remains from I2 consist of an upper canine tooth and an upper first molar, both from the right side of the mouth. Layer I3 provided premaxillary, maxillary and mandibular bone fragments and several teeth, one of which (the upper left canine), although matching this individual, was actually found in I2. Layer I4 had fragments of the left and right mandibles along with a few teeth or fragments of teeth, and the upper half of the left ulna.

This combination of animals reflects a rigorous climate, and an age falling within MIS 6. We saw in chapter 3 that *Equus mosbachensis* is associated with a cold climate, and that *Bos primigenius* can also tolerate such conditions. The rhinoceros and the Irish elk, even if not abundant, do confirm that there must have been open areas (Fernandez, 2006). Wild boar, although typically associated nowadays with forested areas, can tolerate a wide range of temperatures, and are most constrained by the fact that they find it difficult to cope with a thick snow cover (Fernandez 2006), so their presence in I corroborates the suggestion that this was a dry period, as well as being cold. On the other hand, there must have been some patches of forest, because the deer species would require it. Fernandez (2006, p. 96) says, for instance, that the scarcity of *Capreolus* in I, relative to layers higher up, supports the idea that its climate was less clement, because although roe deer live nearly anywhere, they always have at least some forest in their territory. All together, the pollen and faunal remains present a coherent picture of a cold, dry, steppe environment with patches of forest in protected areas, such as within the gorge itself. Well-organised selective hunting by the hominins allowed them to exploit the variety of animals that lived in this environment, with emphasis on aurochs and horses.

4.4. Human remains

Two human remains were found in level I: a tooth from I2 (number D18 I2 1210, known as Aubesier 10) and a partial mandible from I3 (D16 I3 561, Aubesier 11), both found

Figure 4.8. *Megaloceros giganteus*. Pathological case of anodonty involving a missing P3/. Left maxillary fragment (C20 I3 800) with P2/, P4/ and M1/. Vestibular face above, inferior face below. Photos P. Fernandez. *Megaloceros giganteus. Cas pathologique d'anodontie dentaire de la P3/. Fragment maxillaire gauche (C20 I3 800) portant P2/, P4/ et M1/. En haut face vestibulaire, en bas face inférieure. Clichés P. Fernandez.*

in July 2000 (Lebel & Trinkaus 2001, 2002a; Lebel et al. 2001). I will deal with the tooth first.

The Aubesier 10 tooth (Figure 4.9) was found at z=-745 cm, and is a slightly worn right upper first or second molar (M^1 or M^2) of an adult. There are no distinguishing characteristics which would differentiate between M^1 and M^2 for this isolated tooth (Lebel & Trinkaus 2002a). Its size is not unusual for a late Middle Pleistocene maxillary molar, but it has a pronounced degree of taurodontism: its roots are fused more than usual, and this is due to expansion of the pulp cavity, not just bridging between the roots (Lebel & Trinkaus 2001, 2002a). Taurodontism is known to exist in Neandertal and pre-Neandertal hominins, although it is not always present; it is also known to be present in other hominin samples, including some modern humans. However, Aubesier 10 does display it to an unusual but not exceptional degree (Lebel & Trinkaus 2002a). The tooth also displays some faceting of its surfaces due to wear, but not to an excessive degree, and no pathological characteristics.

The most noteworthy aspect of Aubesier 10 is that it has a "toothpick" groove: a shallow groove worn into its side near the base of the enamel, formed by frequently using some small thin object to poke out material stuck between the teeth. Lebel & Trinkaus (2001, p. 17) describe it as "shallow, polished, has sharp margins, and exhibits fine buccolingual wear striae within it". Lebel & Trinkaus (2002a, p. 665) further suggest that "the complete feature was probably a parallel-sided abrasion area, which formed a shallow sulcus into the softer root but only produced fine striations on the harder enamel." Toothpick grooves are known from other Pleistocene human teeth, and indicate that this hominin engaged in some degree of dental hygiene (Lebel & Trinkaus 2002a, Lebel et al. 2001).

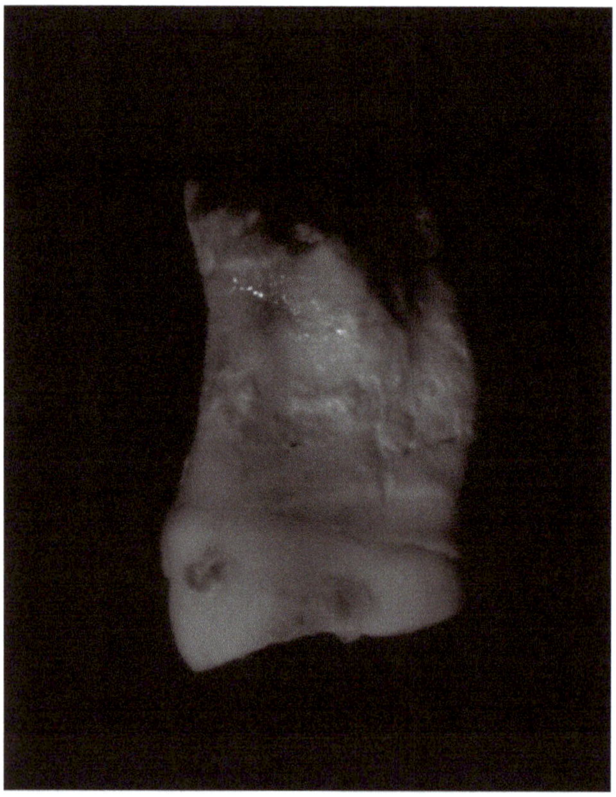

Figure 4.9. Aubesier 10 in mesial view. Photo E. Trinkaus. ***Aubesier 10 en vue mésiale. Cliché E. Trinkaus.***

The mandible, Aubesier 11 (Figure 4.10), was found in layer I2 at z=-720 cm. It is a bit more than half of a complete mandible, extending from the left canine tooth socket around to the back on the right side (Lebel & Trinkaus 2001). Its morphological features "align it closely with later Middle Pleistocene Neandertal-lineage specimens" (Lebel & Trinkaus 2001, p. 17). It is of a fairly small size, making it slightly more likely to have been a female than a male (Lebel & Trinkaus 2002a). Age at death is unknown. It has some postmortem damage, but what happened to it before the death of the hominin is what makes it exceptional.

Lebel & Trinkaus (2002a) fully describe the condition of the piece, and it makes for almost frightening reading. Much of the alveolar bone (which contains the tooth sockets) has been affected by resorption: that is, been re-absorbed by the body, so it is missing. The bone which lines the tooth sockets is missing in some cases, as are other parts of some tooth sockets – or the entire socket. In addition to that, the apex of the left second incisor (I_2) socket was enlarged by an abscess, which continued into the left I_1 socket; the entire right I_1 socket was occupied by an abscess cavity (so there is little or no trace of that socket left); there is not much left of the right I_2 socket; a fragment of the root of the right canine is still present, as is a fragment of the neighbouring premolar (P_3), but they are abraded – which must have been due to them being exposed and used before death; the socket of the next tooth, P_4, is only partially present; the distal root socket of the first molar (M_1) is open and enlarged, suggestive of an abscess cavity; the M_2 was lost long enough before death that the socket is completely resorbed and bone has grown across part of it; and the M_3 socket is tilted (because of the earlier loss of the M_2) and has a hole in it due to an abscess (Lebel & Trinkaus 2002, pp. 666-669).

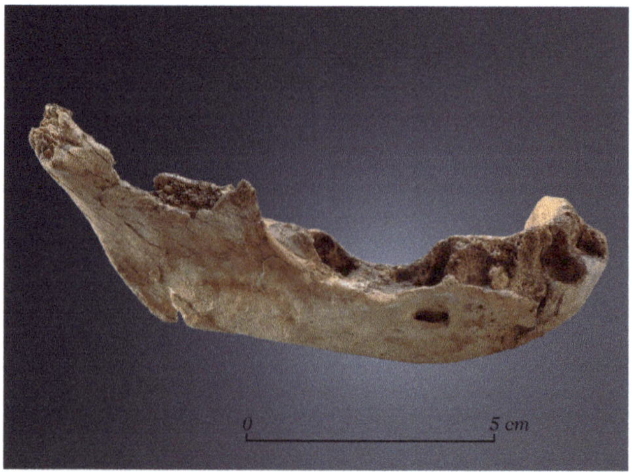

Figure 4.10. The half-mandible, Aubesier 11. Photo M. Olive, DRAC-PACA. ***L'hémi-mandibule Aubesier 11. Cliché M. Olive, DRAC-PACA.***

Why was this hominin's mouth in such bad shape? Lebel & Trinkaus (2002a, p. 669) suggest that the lesions must have been "the product of extensive periodontal and endodontal abnormalities" and furthermore that "Given the tendency of lesions to spread through the periodontal and gingival tissues once present . . ., it is likely that the preserved portion of the alveolar process is representative of the general state of the alveolar bone throughout the mandible and maxilla". They then suggest two possible causes (Lebel & Trinkaus 2002a). First, if this was an older individual whose teeth had been worn down enough to expose the pulps, they could have become infected. That would make this "one of the older known Middle Pleistocene humans" (p. 670). Alternatively, although dental caries are rare in Pleistocene hominins, they could have been involved here, and in that case there is no way of knowing the hominin's age at death (Lebel & Trinkaus 2002a, p. 670).

In any case, this individual had lost nearly all of his or her teeth over a lengthy period before death. Only the roots of the right canine and one premolar are known to have been present and exposed. Of the rest, most were probably absent and some certainly were. The first molar might still have been present, but if so it was secured by only one-quarter of its roots. The third molar might also have been present, but unstable, and tilted so that any upper tooth could not chew properly against it (Lebel & Trinkaus 2002, p. 670). This is extremely unusual among Palaeolithic hominins. According to Lebel & Trinkaus (2002a, p. 682) "Aubesier 11 presents the most pathologically altered pre-Late Pleistocene human mandible known."

This individual's capacity to chew would have been severely limited. Some chewing was done (the canine and P_3 roots are worn), but it would have been very ineffective. If other teeth were present, they would have been unstable, so could not have withstood much force. Eating hard or tough foods would have been impossible. And of course, this individual must have been in a lot of pain. Lebel et al. (2001) and Lebel & Trinkaus (2002a) suggest that to have stayed alive as long as this hominin did, he or she must have eaten only soft foods, or else must have cooked food. There is little evidence of fire in I2, but fire was extensively used in the layer above that, H1, as we will see in chapter 5, so it may well have been within the behavioural capacities of the hominins in I2.

In addition, Lebel et al. (2001) and Lebel & Trinkaus (2002a) suggest that this individual must have received assistance from his or her social group, such as preferential allocation of the softer foods, or extra preparation of foods before eating. There are other known cases of Middle Pleistocene and Neandertal hominins surviving for some length of time despite serious physical injuries or abnormalities (Lebel et al. 2001; Lebel & Trinkaus 2002a), so this is not impossible. Lebel & Trinkaus (2002a, p. 681) point out that "Given the density and variability of archeological remains in layer I, especially faunal remains, it is unlikely that Aubesier 11 was not part of a substantial, closely interacting social group. . . However modest the aid might have been, it may well have made the difference between rapid starvation and prolonged life". Lebel et al. (2001, p. 11102), speaking of these remains and of the older incisor Aubesier 4 (chapter 3), state that "these remains reinforce several paleobiological aspects of these human populations, including their frequent nondietary use of the anterior dentition and the prolonged survival of individuals with serious impairments. These human populations therefore had achieved a level of sociocultural elaboration sufficient to maintain debilitated individuals and to provide the motivation to do so. However, they also were using their biology to accomplish daily activities, reflecting behavioral patterns and a level of technological development significantly different from those of Upper Paleolithic and recent humans."

Beyond their unusual characteristics, the human remains are important because they are an addition to the relatively small European late Middle Pleistocene hominin sample. Lebel et al. (2001) and Lebel & Trinkaus (2002a) report that the three hominin remains found in the lower levels of the Bau show a combination of characteristics that places them within the lineage that was developing into the Neandertals, while at the same time demonstrating that these Middle Pleistocene hominins had retained a great deal of variability (Aubesier 10 displays an unusual degree of taurodontism, for instance, but Aubesier 11 was not taurodont at all) (Lebel & Trinkaus 2002a). ". . .the Aubesier 4, 10, and 11 later Middle Pleistocene human remains are all fully compatible with, and reinforce the existence of, the emergence of a Neandertal craniofacial morphological during the second half of the Middle Pleistocene in the northwestern Old World. The I2 crown, the M1–2 root, and the mandibular lateral corpus and ramus all exhibit features that are observed in their highest frequency in the late Middle and early Late Pleistocene archaic *Homo* of the northwestern Old World. . . At the same time,. . . it is unlikely that the Neandertal morphological complex evolved in isolation in Europe. It is more probable that these features increased in frequency through a combination of genetic drift and isolation by distance across the northwestern Old World, as one aspect of the general regional differentiation of hominid craniofacial morphology as the genus *Homo* became increasingly geographically dispersed during the Middle Pleistocene" (Lebel et al. 2001, p. 11100).

4.5. The lithic industry and raw materials

Table 4.2 shows the breakdown of the lithic assemblage from layer I, all sub-layers combined. There is a larger difference than usual between the numbers in the typology list and those in the raw materials list. This appears most in the tamis (sieved material), but there are differences in other typological categories, too. Most of the differences are probably due to the fact that a lot of work was done excavating layer I during the 2000 season, and I did not catalogue the raw materials for those pieces (unlike every previous year).

Table 4.2. Lithic assemblage of layers I through I4, summarised by typological category. ***Assemblage lithique sommaire des couches I à I4, par catégorie typologique***

Category	Typology list (n)	Typology list (%)	Raw materials list (n)	Raw materials list (%)
Liste Bordes	160	4.9	179	8.6
Debris	16	0.5	11	0.5
Eclats	571	17.6	422	20.3
Lames	62	1.9	51	2.5
Tamis/Collectif	2366	72.8	1221	58.8
Nucléus	74	2.3	56	2.7
Galets	1	<0.1	3	0.1
Rognon	0	0	1	<0.1
non-identifié	n/a	n/a	133	6.4
Total	**3250**	**100**	**2077**	**100**

The assemblage includes a good proportion of typologically identifiable pieces (Liste Bordes) and a large proportion of small chips and flakes (tamis), as well as larger flakes and cores. The proportion of blades is down, compared to the lower layers, but there still are some. Table 4.2 also adds two new categories: galets, which are rounded cobbles brought into the site from the river bed, and "rognon", which is an unworked flint nodule.

The typologically identifiable pieces include a high proportion of sidescrapers of various types: 113 (70.6 %) of the pieces are sidescrapers, either individually or as part of a multiple tool. There are 56 multiple tools that include a sidescraper, and 7 multiple tools that do not include a sidescraper. The list also includes Levallois flakes, Levallois points, endscrapers, naturally backed knives, notches, denticulates, and a variety of other minor types. A total of 24 pieces have had their bulbs partially or completely removed. This corresponds to a Typical Mousterian, rich in sidescrapers, assemblage. The abundance of multiple tools suggests some degree of curation of the assemblage, where pieces are kept, re-used, and as necessary converted into new tools. However, as a percentage of all typologically identifiable pieces, multiple tools are less common in I (39%) than they were in K1-J (42%) or K2 (50%). The abundance of sieved material, flakes and cores makes it clear that lithics were produced at the site, and all parts of the *chaîne opératoire* are present – at least for some raw materials.

The cores weigh an average of 62.4 g (range from 8 to 325 g) and show an average of close to 12 removals, but with considerable variation. They include both Levallois and non-Levallois types: discoid, prismatic, and more amorphous types, and also blade cores.

The flakes average 14 g in weight and 50 mm in length, and more than a quarter of them show more than 10% cortex on their dorsal faces, suggesting that they came from the early stages of reduction of the cores. The blades are on average a bit longer (58 mm), lighter (13 g), and tend to have less cortex than the flakes (only about 11% of the blades have 10% or more cortex).

The raw material results are very interesting (Tables 4.3 and 4.4). As was the case with K2 (chapter 2), there is a high proportion of unidentifiable pieces, especially in the sieved material, most of which are patinated. A total of 152 pieces are identified as burned (7.3%), lower than in K2 (8.6%) but higher than K1-J (3.6%). There is again, as in K2, almost no local material identifiable in the assemblage. In fact, 3 of the 11 local pieces are the galets, which are limestone from the river bed, and not flint at all.

In terms of the main source areas used (Figure 4.11), layer I does contrast with the other layers discussed previously. The sources in the Murs area are generally the most commonly used (except in the tamis material), but they do not dominate to the same extent as we have seen before. Their highest value is in the cores, at 46.4%, but their next-highest value is in the flakes, at 25.1% - still important, but leaving lots of room for use of other source areas. Overall, the Murs source area accounts for 13.1% of the pieces, and it is closely followed by the Faraud area, at 12.0%. The Méthamis area was used much more than we have seen before, at 9.1% overall, and represents 17.9% of the cores (10 pieces), as well as the one unworked nodule. This is a pattern that we have not seen before, with use diverging somewhat from the main SW-NE axis seen in other assemblages.

The north-eastern source areas, and in particular Sault, were used much less than in the lower layers, but again do seem to more commonly represent later parts of the *chaîne opératoire* (tools, flakes, blades, rather than knapping by-products and cores) (Figure 4.12). As Table 4.5 shows, recalculating the numbers after excluding unidentifiable and unknown source pieces, and grouping

Table 4.3. Number of pieces by typology and source area in layers I through I4. *Nombre de pièces par catégories typologiques et groupes de sources des matières premières dans les couches I à I4*

I-I4 n	Liste Bordes	Debris	Eclats	Lames	TamisColl	Nucléus	Galets	Rognon	non-id	Total
Unidentifiable	86	3	164	22	845	7	0	0	85	1212
Unknown	1	0	2	0	2	0	0	0	0	5
Faraud etc.	19	6	68	4	131	8	0	0	14	250
Méthamis etc.	24	0	56	7	83	10	0	1	7	188
Les Sautarels etc.	1	0	0	1	0	1	0	0	0	3
Sault	12	0	15	5	57	2	0	0	4	95
Murs	29	2	106	10	87	26	0	0	12	272
Local	2	0	0	0	4	0	3	0	2	11
St. Jean de Sault	0	0	0	0	0	1	0	0	0	1
Roussillon	0	0	0	1	0	0	0	0	0	1
Murs-Bezaure Olig.	2	0	4	0	9	1	0	0	9	25
Ravin de la Treille	0	0	1	0	0	0	0	0	0	1
St. Trinit	3	0	5	1	2	0	0	0	0	11
Durance	0	0	1	0	0	0	0	0	0	1
Tertiary Calavon	0	0	0	0	0	0	0	0	0	0
Tertiary Murs VdeV	0	0	0	0	1	0	0	0	0	1
Mormoiron	0	0	0	0	0	0	0	0	0	0
R de Guérin	0	0	0	0	0	0	0	0	0	0
N.Aurel	0	0	0	0	0	0	0	0	0	0
Total	**179**	**11**	**422**	**51**	**1221**	**56**	**3**	**1**	**133**	**2077**

source areas by geographic region (see Table 3.5 for a list of groups), whereas previously we had seen roughly 20% of each assemblage coming from the north-east, in I-I4 only 12.3% does so. The south-west (Murs etc.) is down to 35.0%, while the near SW (Faraud etc.) is nearly doubled over what it was in K1-J (29.1% versus 15.0%). The WSW, Méthamis area, at 21.9%, is nearly twice as abundant as it was in K2, and more than three times as common as it was in K1-J. Furthermore, it was supplying raw material to the Bau: the provisioning of place (Kuhn 1995, 2004) in this layer was based not only on Murs but also on Méthamis. Cores from Faraud (the near SW) are a bit less common than those from Méthamis and Murs, which may reflect ease of access. The Faraud area is close enough to the Bau that it would not take long to go there on purpose, and it would also be covered frequently when hominins were hunting or travelling further afield. Perhaps that meant that there was less impetus to bring cores from that area in to the Bau. On the other hand, none of that explains why material from there is more common in this layer than in the others.

It is not just the larger pattern that is noteworthy for layer I, however. This is the layer that includes the one and only piece from the Durance river, and is also the first layer to show any use of the source at Roussillon. The piece from the Durance (C19 I 375) (Figure 4.13) is a large flake, 82x54x27 mm, weighing more than 113 g (it was weighed after having had a chunk cut out of it for thin-sectioning). It is the only volcanic rock specimen in the site, an andesite, and must have come from the alluvial deposits of the Durance river, which carried it down from the Alps. It shows rounding on its dorsal side, which is evidence of river transport: it was flaked off of a cobble. It shows no signs of having been burnt. Since the Durance curves around the region (Figure 4.11), its alluvial deposits

Table 4.4. Percentages of pieces by typology and source area in layers I through I4. ***Pourcentages des pièces par catégories typologiques et groupes de sources des matières premières dans les couches I à I4***

I-I4 %	Liste Bordes	Debris	Eclats	Lames	TamisColl	Nucléus	Galets	Rognon	non-id	Total
Unidentifiable	48.0	27.3	38.9	43.1	69.2	12.5	0	0	63.9	58.4
Unknown	0.6	0	0.5	0	0.2	0	0	0	0	0.2
Faraud etc.	10.6	54.5	16.1	7.8	10.7	14.3	0	0	10.5	12.0
Méthamis etc.	13.4	0	13.3	13.7	6.8	17.9	0	100.0	5.3	9.1
Les Sautarels etc.	0.6	0	0	2.0	0	1.8	0	0	0	0.1
Sault	6.7	0	3.6	9.8	4.7	3.6	0	0	3.0	4.6
Murs	16.2	18.2	25.1	19.6	7.1	46.4	0	0	9.0	13.1
Local	1.1	0	0	0	0.3	0	100.0	0	1.5	0.5
St. Jean de Sault	0	0	0	0	0	1.8	0	0	0	<0.1
Roussillon	0	0	0	2.0	0	0	0	0	0	<0.1
Murs-Bezaure Olig.	1.1	0	0.9	0	0.7	1.8	0	0	6.8	1.2
Ravin de la Treille	0	0	0.2	0	0	0	0	0	0	<0.1
St. Trinit	1.7	0	1.2	2.0	0.2	0	0	0	0	0.5
Durance	0	0	0.2	0	0	0	0	0	0	<0.1
Tertiary Calavon	0	0	0	0	0	0	0	0	0	0
Tertiary Murs VdeV	0	0	0	0	0.1	0	0	0	0	<0.1
Mormoiron	0	0	0	0	0	0	0	0	0	0
R de Guérin	0	0	0	0	0	0	0	0	0	0
N.Aurel	0	0	0	0	0	0	0	0	0	0
Total	**100**	**100**	**100**	**100**	**100**	**100**	**100**	**100**	**100**	**100**

could have been exploited at any point along its length. However, the shortest way to the Bau from the Durance goes due north, through the gap between the Petit and the Grand Luberon mountains.

Interestingly, travel along that N-S line between the Durance and the Bau passes very close to Roussillon. The one piece from Roussillon is a blade (B21 I4 98) (Figure 4.14), 49x22x5 mm in size, and weighing 8 g. Roussillon is known nowadays for its ochre deposits, but the upper surface of the ochres preserves, in some places, an iron-spotted siliceous crust of distinctive appearance. We called it "la roche saucisson": the sausage rock. It is not a great raw material for stone tool making, but it was used and provides further evidence that the hominins of layer I had some access to the southern part of this region.

The resource selection function study (Wilson & Browne 2014) showed that Terrain variables were again more important than Raw Material variables in this layer, as in the lower ones, but that raw material characteristics did still matter. Table 4.6 shows the overall results, and Table 4.7 shows the Wald Chi-Square values converted to percentages. We see in Table 4.6 that the Beta values are negative for Difficulty (Cal/km), Calories and for Small and Medium rocks, meaning that source areas with more small or medium rocks would be less likely to be used, especially if they were hard to get to. The highest Wald Chi-Square values are for Quality, but that is only a bit higher than the value for Extent, and these are followed fairly closely by Large rocks, AOSISA and Difficulty. In other words, the source areas most likely to be used were those with better quality, bigger (but not too big) rocks, especially if the sources were big and covered a lot of the source area, and were not too hard to get to. This is a similar pattern to the results for the other layers, although with some switching around of the variables in importance (AOSISA up, Calories down, for instance). In this case,

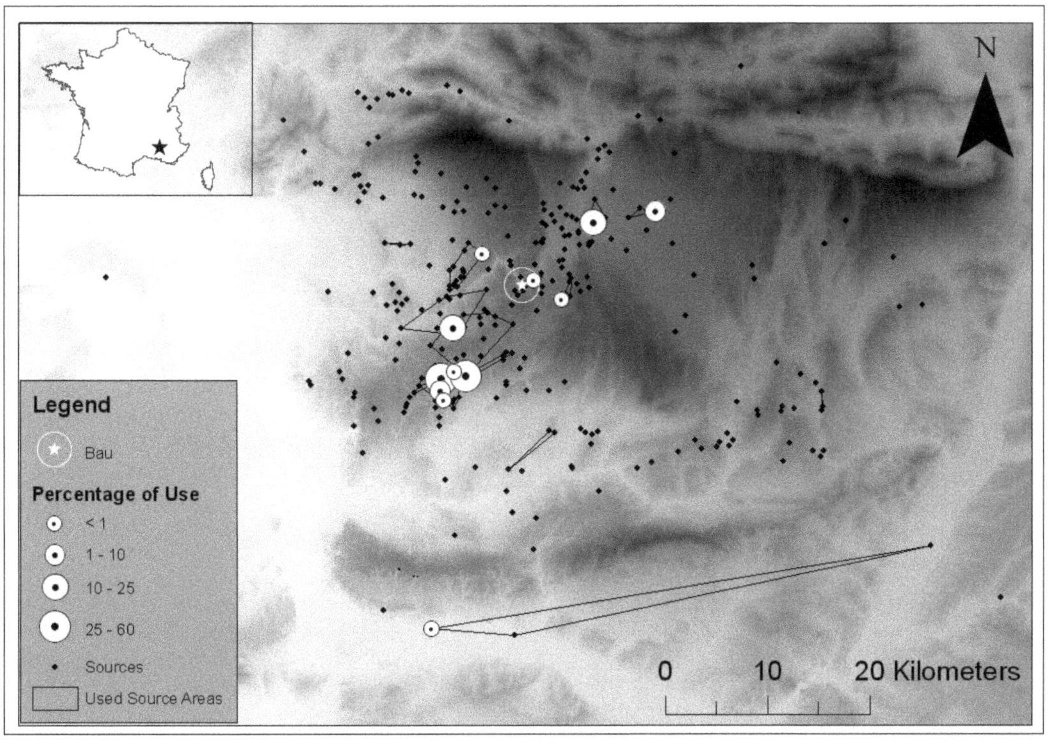

Figure 4.11. Map of the raw material source areas used for layers I through I4. ***Carte des aires de sources de matières premières utilisées pour les couches I à I4.***

Figure 4.12. Piece B20 I4 658, a blade or elongate flake, made in raw material type AE, from the St. Trinit area, approximately 14.7 km NE of the Bau. ***Lame ou éclat allongé B20 I4 658, confectionné dans la matière première type AE, provenant de la région de St. Trinit, 14.7 km environ au NE du Bau.***

the total weight of the Terrain variables is 54.6% (versus 56.8% in K1-J) and the Raw Material variables add up to 45.0% (versus 42.2% in K1-J).

In summary, layer I accumulated during a cold, dry climatic period, MIS 6. The pre-Neandertal hominins who occupied the Bau at this time were skilled selective hunters using a Typical Mousterian industry, rich in sidescrapers, with a few blades. They made their tools using both Levallois and non-Levallois technologies, using raw material procured primarily from the SW and WSW of their territory, and only to a small extent from the NE. They exploited sources that were relatively easy to get to, which provided large pieces of good quality, and they imported raw material to provision their campsite. They favoured hunting large mammals in the open plains around the site, but also caught a fair number of deer and a few wild boar, possibly from the more sheltered areas where there might have been patches of forest. The horses were hunted preferentially in the summer to early fall period, so these hominins may have used the Bau seasonally. At least one of the hominins engaged in dental hygiene often enough to leave a toothpick groove. Another one had such extensive tooth loss and infection that he or she must have been very ill, barely able to eat, and in a lot of pain for a considerable time before death. The social group took care of that individual, long enough for extensive bone resorption and even new growth across a tooth socket, showing that they supported group members, even if that person could no longer contribute productively. Caring for family has a long history.

4.6. Sommaire du chapitre

La couche I se situe à une profondeur située entre 705 à 805 cm sous zØ, et les fouilles ont eu lieu sur une surface d'environ 34 m^2. Cette couche était divisée en sous-couches I, I1, I2, I3 et I4, et la plupart du matériel provenait de la sous-couche I2. La couche I a livré dans sa totalité quelques milliers de vestiges lithiques, et quelques dizaines de milliers de restes fauniques. Cette

Table 4.5. Percentages of identified pieces (excluding unidentifiable and unknown source pieces) from each geographic region for layers K2, K1-J and I. ***Pourcentages des pièces identifiées (excluant pièces indéterminables ou de sources inconnues) de chaque région géographique pour les couches K2, K1-J et I***

Geographic area	K2 n.	K2 %	K1-J n.	K1-J %	I-I4 n.	I-I4 %
Local	1	0.2	42	3.4	11	1.3
NE	108	19.4	249	20.4	106	12.3
E	0	0	4	0.3	1	0.1
Far S	0	0	0	0	1	0.1
SSW	0	0	0	0	1	0.1
SW	247	44.3	659	54.1	301	35.0
Near SW	128	23.0	183	15.0	250	29.1
WSW	69	12.4	80	6.6	188	21.9
NW	4	0.7	1	0.1	1	0.1
other	0	0	0	0	0	0
Total	**557**	**100**	**1218**	**100**	**860**	**100**

Figure 4.13. The specimen of type AG, a flake of volcanic rock (andesite). ***Spécimen du type AG, un éclat de roche volcanique (andésite).***

Figure 4.14. Type specimen of raw material type V, nicknamed "*la roche saucisson*", a ferruginous siliceous crust from the Roussillon area, approximately 18 km south of the Bau. ***Spécimen du type V, "la roche saucisson", une croûte ferrugineuse siliceuse de la région de Roussillon, 18 km environ au sud du Bau.***

couche a également livré une hémi-mandibule humaine pathologique, ainsi qu'une dent humaine isolée. La couche est datée biochronologiquement du stade MIS 6.

Dans la couche I il y avait des sédiments de couleur pale et de granulométrie variable. La couche I1 était localisée près de la paroi rocheuse avec un sediment à grain fin (limons et sables). Dans la plupart des zones fouillées la couche H était directement posée sur I2, qui contenait d'abondantes roches ainsi qu'une matrice fine. En surface d'I2 on trouvait souvent de grandes dalles de roches tombées du plafond. L'épaisseur de la couche I2 pouvait atteindre 60 cm. En dessous, suivaient la couche I3 plus riche en cailloux, puis I4, qui était plus indurée, voire cimentée. La base d'I4 était notre limite arbitraire à -805 cm. D'une manière générale les couches en I comprenaient des cailloux et des roches en position plutôt horizontale. La nature cryoclastique des sédiments corrobore l'idée qu'ils se sont accumulés pendant une période de climat froid. Elle a été suivie par une période plus chaude et humide comme le suggère le ciment dans certaines zones, pendant laquelle de l'eau dégouttait du plafond et circulait dans les dépôts. C'est surtout le cas dans la bande D, qui est toujours la plus fortement cimentée.

Les pollens laissent supposer que le paysage était steppique et ouvert. La faune comprend beaucoup de

Table 4.6. Generalised linear model for raw material source choices in I-I4. P-values in bold font are statistically significant. (Sample: 850 pieces from 12 source areas)

Overall:
Omnibus test: likelihood ratio chi-square = 541.538, df = 10, p = <0.001

Variable	Beta	S.E.	Wald Chi-Square	P-value	AIC
Quality	1.031	0.2573	16.059	**<0.001**	163.967
Extent	1.965	0.5068	15.033	**<0.001**	
Difficulty (Cal/km)	-0.066	0.0204	10.593	**0.001**	
Small rocks	-3.816	1.8912	4.071	**0.044**	
Medium rocks	-1.520	0.8507	3.193	0.074	
Large rocks	1.905	0.5643	11.397	**0.001**	
Very large rocks	0.082	0.7798	0.011	0.917	
Calories	-0.001	0.0006	5.240	**0.022**	
AOSA	1.4 e-8	4.4 e-8	0.093	0.760	
AOSISA	4.0 e-5	1.2 e-5	11.150	**0.001**	
(Intercept)	-2.004	3.5126	0.325	0.568	

Table 4.7. Wald Chi-Square values as percentages, I-I4 assemblage

Variable	Wald Chi-Square	Wald Chi-Square as %
Quality	16.059	20.8
Extent	15.033	19.5
Difficulty (Cal/km)	10.593	13.7
Small rocks	4.071	5.3
Medium rocks	3.193	4.1
Large rocks	11.397	14.8
Very large rocks	0.011	<0.1
Calories	5.240	6.8
AOSA	0.093	0.1
AOSISA	11.150	14.5
(Intercept)	0.325	0.4
Total	**77.165**	**100**

restes d'aurochs et des restes identifiables d'au moins 23 chevaux, ce qui fait du Bau, entre autres aspects, l'un des sites paléontologiques les plus importants du sud de la France pour la période du Paléolithique moyen. La plupart des chevaux ont été tués pendant la période de juillet à septembre. La faune (Tableau 4.1) comprend également du cerf, et la plus grande proportion de sanglier au Bau: au moins 5 individus. Comme dans la couche J, les hominidés de la couche I ont pratiqué une chasse sélective orientée surtout vers deux espèces très différentes, avec probablement parfois une chasse plus opportuniste associée peut-être à du charognage.

Deux restes humains ont été trouvés dans la couche I en juillet 2000: une dent, en I2 (Aubesier 10), et la fameuse hémi-mandibule en I3 (Aubesier 11). La dent correspond à une première ou une deuxième molaire supérieure droite non pathologique. Elle appartenait à un adulte avec un taurodontisme plus marqué qu'à l'usuel, avec une élongation de la chambre pulpaire et une coalescence des racines. La dent est d'autant plus intéressante qu'elle a des traces d'utilisation (rainure) de cure-dent.

L'hémi-mandibule (Aubesier 11) est gracile, appartenant plus probablement à une femme, mais ni son sexe ni son age ne sont déterminables de façon certaine. Cette mandibule est probablement la plus pathologique connue du Pléistocène moyen. Pendant une période de quelques semaines ou mois avant sa mort, l'individu a subi la perte successive de ses dents, de l'arrière vers l'avant, ainsi que la perte ou resorption d'une partie de l'os des alvéoles dentaires. Cet individu aurait perdu presque toute capacité à la mastication, avec des tentatives de mastication ayant laissées des traces sur les racines exposées au niveau de la canine et de la prémolaire la plus antérieure. Une telle mastication a sans doute été infructueuse pour des aliments durs ou résistants et bien entendu douloureuse. La durée de vie de cette personne a probablement été prolongée avec l'aide de proches lui prodigant un accès préférentiel aux aliments les plus tendres ou peut être cuits, bien que l'on n'ait que très peu de traces de feu dans cette couche. Sa mort a sans doute mis fin à une souffrance difficilement imaginable aujourd'hui. En revanche, selon Lebel et al. (2001) pendant cette période les hominidés avaient la capacité et la motivation nécessaires pour prendre soin de leurs proches, malgré leur niveau technologique qui impliquait l'utilisation du corps, en particulier des dents, pour effectuer des travaux usuels, et en cela très différent du niveau technologique de populations plus modernes.

Parmi les pièces lithiques on constate moins de lames que dans les couches inférieures, et bon nombre de pièces retouchées (racloirs surtout) ainsi que des éclats, nucléus et débitage. Cela indique une chaîne opératoire complete, au moins pour quelques types de matière première, avec un assemblage correspondant à une industrie du Moustérien typique, riche en racloirs. Les matières premières proviennent préférentiellement de la région de Murs, mais cette origine ne domine pas autant que dans les couches inférieures. Elles sont suivies de près par des sources autour de Faraud et près de Méthamis, donc un peu à côté de l'axe normal SW-NE des couches inférieures.

La couche I nous donne également des indices de frequentation d'un territoire plus grand, avec la présence d'un grand éclat en roche volcanique (andesite) provenant de la Durance (la seule pièce du Bau provenant d'une source si éloignée), ainsi qu'une petite lame provenant de croûtes siliceuses près de Roussillon; cette dernière marquant pour la première fois l'utilisation de cette source dans le site. L'éclat volcanique provient d'un galet de rivière, arrondi sur un côté et qui a sans doute été ramassé dans les alluvions de la Durance qui l'aurait transporté des Alpes. On ne peut pas localiser avec certitude, l'endroit exact le long de la Durance où ce galet aurait été trouvé, mais le chemin le plus court entre la Durance et le Bau suit une ligne N-S entre le Petit et le Grand Lubéron – ce trajet passant d'ailleurs près de Roussillon.

Pour résumer, la couche I s'est accumulée lors d'une période climatique froide et sèche, et nous livre les vestiges de plusieurs occupations par des pré-Néandertaliens qui utilisaient une industrie moustérienne typique. Ils vivaient dans un vaste territoire, et nous ont laissé comme témoignage des pièces lithiques fabriquées dans des matériaux provenant surtout du sud-ouest du Bau, et moins du nord-est. Ils privilégiaient des roches provenant de gisements assez faciles d'accès, et qui pour la plupart offraient des matières premières de bonne qualité en rognons de grande taille. Ils pratiquaient une chasse sélective orientée vers les chevaux et l'aurochs dans un paysage ouvert, mais ont également consommé d'autres animaux, dont le cerf et le sanglier, qui devaient se trouver dans des endroits plus boisés. Les chevaux étaient tués préférentiellement en été et au début de l'automne, ce qui suggère que le Bau a pu être utilisé de façon saisonnière. L'un de ces hominidés a nettoyé ses dents assez souvent pour laisser des traces de cure-dent. Un autre individu avait la bouche en si mauvais état qu'il/elle a dû être très malade bien longtemps avant sa mort, comme en témoigne la reformation de l'os mandibulaire. Certains individus ont dû en prendre soin, même si cette personne ne pouvait plus contribuer aux besoins du groupe. La bienveillance familiale a donc une longue histoire.

5

Layer H

5.1. Introduction

The layers described in previous chapters consisted of mainly unburned, white calcareous sediments, varying in granulometry, matrix (absence or abundance), and degree of cementation (if any), but still essentially variations on a theme of "chunks of wall that fell down". In H we meet the first layer that is really something different. About 30 to 45 cm of the metre called H consists of highly burned, black, greasy sediments crammed with burned bones and flints. This black layer, H1, is thickest near the rock shelter wall, and disappears within about 4 or 5 metres of the wall, towards the east. We did not find its southern limit, since we did not excavate to that depth farther south than the band 14: H1 is well protected underneath nearly 2 m of strongly cemented and very dense breccia. (For instance, in 1994 a team of 7 people spent two full weeks working with jackhammers to open up 4 square metres down to the level of H1.) We do know, however, that the black sediments of H1 disappear farther north in the site around about band 18.

The metre of deposits that we arbitrarily named H extends from 605 to 705 cm below zØ and includes three sub-layer attributions. Anything found above the archaeologically rich layer would have been labelled as coming from H, but there was very little such material. The black layer was H1; its z values were generally between 655 and 695 cm below zØ, and in many cases this attribution was extended to less burned, less rich areas at approximately the same range of z values. The lowest part of the metre, in the excavated squares most distant from the burned area, was called H2, although in the richer area we became more inclined to call everything H1, down to the transition to layer I. My records also contain information on 2 pieces from a screen, labelled F21 H3 1. This is outside of the main excavation area, and those pieces were not thought to have any relationship to the layers H1 and H2, so H3 just indicates that they were found somewhere between 605 and 705 cm below zØ. The squares dug at the overall depth of H include A14-15 and A18-20; B15-20; C15-20; D14 and D17-20; E19-20; and F19-21 (see Figure 1.11). H1 was dated both radiometrically and on the basis of stratigraphy and fauna. The radiometric dates put it in the MIS 6, Riss III in the Alpine glacial chronology, with TL dates between 169 ± 17 and 191 ± 15 thousand years (ka) (Lebel et al. 2001). The faunal data put it either a bit younger, in the subsequent interglacial, MIS 5e (Fernandez, 2001) or in MIS 6 (Crégut-Bonnoure et al., 2011).

The archaeological material was not distributed evenly throughout this area. In addition to H1 being thicker towards the rock shelter wall, the archaeological deposits were more abundant at a small distance from the wall. As an example of this, Figure 5.1 shows the number of lithics (all layers combined) from each square in my lithic records (bearing in mind that I do not have data from the test pit in A14, and am probably lacking data from D14). These numbers are also influenced by the fact that the squares in band A are incomplete, because some of the metre is taken up by rock shelter wall; the absence of data in A16 and A17 is due to the fact that the rock wall entirely fills those squares at this level. The layout shows that the larger numbers of pieces seem to be in a line of squares moving eastwards (away from the rock shelter wall) as you go north. This may partly reflect the contours of the rock shelter wall, with the bulk of the deposits maintaining a distance of a metre or two from the wall. It is also clear that there is a rapid fall-off in the amount of material north of the band 18.

Within H1, the stratigraphy is complex. In summary (more details below) the very lowest part of H is effectively a continuation of the underlying layer I: calcareous sediments with little sign of burning. Beneath and above that is a series of large slabs of rock, likely fallen from the rock shelter ceiling. One set of slabs is in layer I2 (see

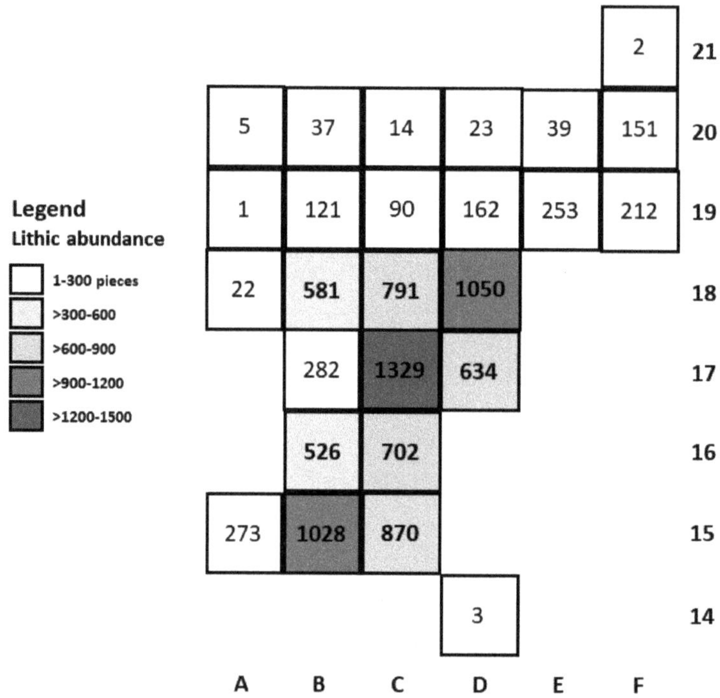

Figure 5.1. Distribution of lithic material in layer H: number of pieces in each square. Squares with no numbers were not excavated, or data are not available. See text for details. ***Distribution des pièces lithiques dans la couche H: nombre de pièces par carré. Les carrés sans indication correspondent à des données manquantes ou des secteurs non fouillés. Voir texte pour plus de détails.***

chapter 4). The base of the next set of slabs is at -695 cm in B15, and descends gradually to -715 cm (thus technically within layer I) in C18. There are at least two superimposed layers of slabs in H, within which (i.e., between the slabs both vertically and horizontally) is a layer rich in lithics and bones which we dug in two levels (*décapages*). On top of the slabs, the area was "paved" in rounded cobbles brought up from the stream bed, and that surface is embedded in and intermixed with burned sediments rich in archaeological material. Above *that*, we dug 3 or 4 *décapages* within a thickness of about 20 or more cm of highly burned, very black and greasy, sediments, again very rich in bones and lithics, and with in addition a large cracked rock which served as an anvil, upon which bones were broken open. Above that was a layer containing another slab, fragmented, plus small flat calcareous plaquettes, more or less cemented together or indurated, forming an apparent floor. Archaeological material was found on top of that floor, but it was not abundant and was not burned. The rest of the metre thickness above that consists of nearly completely sterile, highly cemented, calcareous breccia.

5.2. The sediments, and evolution of the site in H

The overview provided above demonstrates the complexity of layer H. This section will describe the layer's components and sedimentological evolution in more detail, starting at the bottom, with the level concerned in our 8th *décapage*. (We dug *down* through time, because we started at the top – here I explain *upwards* through time.) However, before starting that, I should explain that none of the black layer was visible when we started (Figure 5.2). Remember that the Bau site presented as an upper plateau and a lower slope. Layer H is within the lower slope area, and when we started excavating in 1987 and 1988, only light-coloured sediments were visible on that slope. As we later discovered, this is because the burned area peters out by band 18, and in the early years the slope sediments extended out to about band 20 at that level, as I noted in my personal notes from 1988 ("la ligne G/H tombe juste à l'intérieur du carré 20").

In 1988 I took detailed notes of the sediments visible on the slope, in stratigraphic order. Table 5.1 summarizes those notes for the section overlapping the level arbitrarily assigned to H, the upper level of which (-605 cm) fell within the stalagmitic layer (*plancher stalagmitique*). Most of the lowest level noted (*gros blocs, arrondis*) would have been in layer I, but the top of that could be in layer H, in places. We see from Table 5.1 that even in deposits outside of the bands where the black layer was to be found later, it was the lower parts of the layer where archaeological material was visible. It was, of course, the presence of visible material that encouraged us to dig there, but we had no idea of how rich the deposits were going to be once we got into them.

By 1990, excavation at that depth had proceeded back into the deposits to the band 19, and we were finding more and more traces of burning. Lebel (1990, pp. 13-14) reports "Localement, ces traces s'organisent en quelques concentrations variant de 10 à 20 centimètres de diamètre et discernables par leur couleur noir et l'association

Figure 5.2. The slope of the site in 1988, cleaned and ready for excavation to start. Layer H is approximately at the level of the jug attached to the wall (containing coloured water for use in a water-level z measurement system). *La pente du remplissage en 1988, nettoyée et prête pour la fouille. La couche H se situe approximativement au niveau de la bonbonne d'eau fixée à la paroi (contenant de l'eau colorée servant de niveau à eau pour mésurer le z).*

d'ossements calcinés à différents degrés. . . Elles sont en contact avec des pierres thermoclastiques noircies. . . . Ces lentilles noires sont visibles dans la coupe 18/19 et se poursuivent sous les dépôts." After a year without an excavation in 1991, the years from 1992 to 2000 saw several major efforts to uncover deposits down to layer H, and extend the excavation both laterally and towards the south (inside of the rock shelter), giving the results described below.

5.2.1. The lowest level of H:

As described above, this level resembles layer I. It consists of a few cm of unburned deposits, sometimes stained by the dark material percolating down from above. This level is on top of the slabs which sealed layer I2, and underneath the lowest set of slabs in the black layer, H1 proper. Water flow through the deposits was channeled by those slabs: it had to go around them, between them, and under them, so in some places it flowed faster and was more competent at removing any finer particles that might have been there.

Table 5.1. Summary of stratigraphic notes (1988) concerning the part of the slope of the deposits relevant to layer H, at the beginning of our excavations in this part of the site. ***Notes stratigraphiques sommaires (1988) concernant la partie de la pente du remplissage au niveau de la couche H, au début de nos travaux dans cette partie du site***

Thickness in cm	Description
15	plancher stalagmitique
8	vide sous plancher
15	blocs non-consolidés
7	petits cailloux, granules, non-indurés
28-30	cailloux non-indurés, petits à moyens, matériel archéologique
15-20 + zone fouillée = ~ 40	gros cailloux + blocs, un peu induré, matériel archéologique
~45	gros blocs, arrondis

This means that directly underneath the slabs we generally found coarse-grained sediments, bone fragments and some lithics, with little or no interstitial matrix, and no cement. The material was washed and lying there ready for us to pick it up (Figure 5.3). The archaeological material was considerably less abundant in an area about 75 cm wide, running N-S along the rock shelter wall, as Figure 5.3 shows. That "empty" area had not been excavated ahead of the picture: it was like that when we uncovered it. This partly explains the lower numbers of lithics near the wall, shown in Figure 5.1 (which includes all lithics from H, not just this décapage). Note also that the bottom (south) and right (east) sides of the picture are darker because of layer H1 proper, which is exposed in the section walls surrounding the excavated area, as well as because of some staining of the lowest sediments themselves.

5.2.2. The slabs:

Lebel (2000) reports that a series of six slabs, each 20 to 30 cm thick and in some cases more than a metre in diametre, covered roughly 10 square metres, especially in the bands 16 and 17. The surface they formed sloped slightly towards the north, and created a clear distinction between the black layer and the 8th décapage underneath the slabs. He speculates that it was not by chance that humans occupied this surface. The slabs certainly did create an apparent, if discontinuous, floor (Figures 5.4 and 5.5), and there was a great deal of archaeological material on and between them, which made it tricky to remove them (Figure 5.6). Note that I cringe every time I look at Figure 5.6. You are correct: no-one should be wearing sandals while doing such work, and we did know that. We just got carried away and keen to get it done, especially since we had been finding such treasure troves of remains underneath the slabs.

Figure 5.3. The 8th level (décapage) of layer H, beneath the slabs, grading into layer I2, viewed from above. North is towards the top of the photo. Zones B and C, 15 to 17, are visible, and a portion of D17 on the top right-hand side of the photo. ***Le 8è décapage de la couche H, sous les dalles, devenant la couche I2 vers le haut de la photo (vers le nord), vue d'en haut. Les zones B et C, 15 à 17, sont visibles, ainsi qu'une partie de la zone D17 en haut à droite de la photo.***

Figure 5.4. Slabs in place, 7th décapage of layer H1. Their removal gave the 8th décapage, shown in Figure 5.3. ***Les dalles en place, 7ème décapage de la couche H1. Leur enlèvement a donné le 8ème décapage illustré en Figure 5.3.***

Figure 5.5. Layer H1 face-on, looking south towards the interior of the rock shelter, with the slabs still in place. This view gives a good impression of the thickness and darkness of the overlying combustion layer, seen in the section wall. The rock shelter wall is to the right of the photo. ***La couche H1 de face, vers le sud (intérieur de l'abri), avec les dalles toujours en place. On voit bien dans la coupe l'ampleur et la couleur foncée de la couche de combustion. La paroi rocheuse se trouve à droite de la photo.***

Figure 5.6. Removal of a slab from layer H1, revealing many bone fragments underneath. ***Enlèvement d'une dalle de la couche H1, laissant apparaître beaucoup de fragments osseux en dessous.***

5.2.3. The cobbles:

Above the slabs we uncovered a large number of rounded cobbles, immersed in and stained with burned deposits, bones and lithics. There were dozens of these, but I do not have details of them available. I remember that we filled several large sacks with them, and Lebel (1996) reports that some were hammerstones, but gives no further information. Their rounded shape, however, indicates that they had been transported by water, so they presumably had been brought up from the Nesque. The frost-cracked rocks found in all other layers of the site have angular shapes. It is worth noting that although some rock shelters do have a stage in their histories where a stream flowed through them, there is no evidence of this at the Bau, and at the time of formation of layer H the rock shelter must have been several tens of metres above the river, so the only way such rounded cobbles can have arrived in the site is through human agency. Also, even if some showed signs of having been used as hammerstones, most did not. Some of them may well have been used to break open bones to extract the marrow, as Fernandez in Lebel (1996) suggests, but that does not explain why there are so many of them. So-called cobble pavements have been described at a number of other Palaeolithic sites (McNabb 2011), although there is certainly no consensus on their function or even (in some cases) their reality. What excavators see while digging does not always look convincing to others, later. However, the concentration of cobbles in this level of H1 was unusual for the Bau, and it deserves some explanation. For want of a better one, perhaps their function was as some kind of flooring.

5.2.4. The combustion layer:

When we started digging the slope deposits in 1988, we created an initial section wall at the southern side of band 21 (where it is in contact with band 20), and dug everything in front of that, trying to "straighten out" the deposits – convert a slope to a series of steps. As noted above, those deposits were unburned. Over the next couple of years we extended the excavated area back into the deposits, to the band 18, and started finding traces of burned material. In 1992 and 1993 we worked on bands 17 and 18, and discovered just how burned those deposits were: a thick, very black layer containing greasy, sticky, fine-grained interstitial sediments, frost- and fire-cracked rocks of various sizes, lithics, and bone fragments. The lithics did not seem to be very burned (although my data suggest that roughly 15% of them do show some signs of burning). The bones, on the other hand, were often burned, cracked and spalled, and ranged in colour from light to dark brown, to black, through to white again and even occasionally blue, which indicates temperatures of up to 800°C (Lebel 2006). We therefore pursued excavation of this layer, and from 1993 on spent a great deal of energy jackhammering down through the breccia in order to free up squares in bands 15 and 16, and in 2000 in band 14.

One of the early interesting finds in the combustion layer was a rock anvil. A large (48 x 42 x 15 cm, Buisson-Catil 1994) broken rock was found on the border between B17 and B18. Although it was composed of limestone, like the rock shelter, it stood out from the rest of the sediments because of its larger size, suggesting that it was not just part of the normal deposits. In addition it was broken into at least six large fragments, due to a mechanical shock on the centre of its upper surface: the fractures radiated out from the centre, and the fragments were still *in situ*. There was nothing in the overlying sediments that weighed enough to have caused the fractures just by falling on the rock, so it was clear that it had been broken due to some human action. The rock was located on the edge of the combustion area, surrounded by hundreds of burned bone fragments. Fernandez (2001, 2006), in his study of those bone fragments, concluded that the anvil had been used to break open the bones so that their marrow could be extracted, and that the bones had subsequently been used as fuel for the fire: "Après examen de centaines d'esquilles carbonisées à proximité du bloc, . . . le type de fractures et la combustion systématique en particulier de la surface corticale et interne des os a permis d'envisager, d'abord une récupération de la moelle et ensuite, une utilisation du matériel osseux comme combustible" (Fernandez 2006, p. 16).

The combustion area itself (Figures 5.7, 5.8 and 5.9) was flat, with no evident structure or border of rocks (for instance): it seemed to just be due to a series of fires being laid and maintained directly on the ground (Lebel 1996). Its only demarcation from the surrounding area was the black colour and the burned contents. The fires seem to have been repeatedly laid about 1.5 m from the rock shelter's wall (Lebel 1996), and burned deposits accumulated especially thickly near the rock shelter wall (up to 40 cm in some zones). We immediately realised that it was likely to be the result of many repeated events, rather than a single large one, and subsequent more detailed analysis corroborated this interpretation.

Analysis of these deposits included micromorphological and geochemical work by M. Wilson (1997), X-ray diffraction studies by Jaouich and Paquin (1996), and detailed micromorphological study of a series of samples from the test pit in A14 by M.-A. Courty (2006). M. Wilson (1997) found that the sediments were of medium porosity yet highly permeable. She reported that they had been affected by frost and by flowing water, but saw little sign of biological (i.e. microbial) action. She reported that there were abundant iron oxides (goethite and limonite), probably due to water moving through the sediments. Jaouich and Paquin (1996), on the other hand, did not find much evidence of goethite. They reported that their samples were composed of variable quantities of calcite, apatite (from bones), and quartz, with occasional presence of lime and portlandite (due to combustion of limestone),

Figure 5.7. Part of the combustion area in layer H1 in 1993. Zones B-C 17 and 18 are visible, with part of D18 in the upper centre of the photo. The slips of paper mark particular objects of interest, but I do not have the list of what those objects are.
Une partie de l'aire de combustion de la couche H1 en 1993. Les zones B-C 17 et 18 sont visibles, ainsi qu'une partie de la zone D18 au centre en haut de la photo. Les papiers marquent quelques objets intéressants, mais je n'ai pas la liste identifiant ces objets.

Figure 5.8. Layer H1 after some more excavation (compared to Figure 5.9), seen from above, showing the bands B and C from left to right, from band 15 at the bottom of the photo to band 17 shown partially at the top. Note the grey lens of breccia at the bottom of the photo; it was later removed by jackhammer. ***Une deuxième vue de la couche H1 après continuation de la fouille (après la Figure 5.9), vue d'en haut, montrant les bandes B et C de gauche à droite, de la bande 15 en bas jusqu'à la bande 17 en haut de la photo. La zone grise en bas était une lentille de brèche que nous avons par la suite enlevée en utilisant les marteaux-piqueurs.***

and some clays: primarily illite, with minor biotite and chlorite included with the clay fraction. The consistency in the clay signal in all of their samples suggested to them that the clays were derived from the surrounding limestone, rather than from any external source. Overall, then, the layer is composed of minerals derived from the limestone of the rock shelter, plus contributions from bone fragments and flint tools, and the whole combination has been thoroughly burned. For what little it is worth, I remember hearing at some point in the early to mid-1990s that someone had done an analysis of a sample from H1 and found haemoglobin and fatty acids in it. Unfortunately, I can find no mention of that in any of the reports. Clearly, however, a lot of blood, fat and meat must have been present, to match up with the bones. On the other hand, any pieces of charcoal or burned wood that we found were very small, scarcely identifiable. It may well be that the fat from the bones was the main fuel source for the fires.

Courty's work (2006) re-emphasises the complexity of the deposits. She took a series of samples in B15 and down through the test pit in A14, at various levels and also laterally from each other; three of these included some of the black layer, and sample H1-2, at z=-667 cm, seems to have been entirely within it. H1-2 is reported to be very finely stratified. It has cm-thick beds which are black and consist of clays and organics, alternating with orange-coloured beds rich in bones, and thicker beds with altered calcareous plaquettes, abundant flint and bones.

Overall Courty (2006) found at least 16 clearly distinguishable layers in a thickness of 42 cm. The layers are interstratified and imbricated, which leads to spatial discontinuities and makes the strata difficult to follow laterally. This is complicated by the occurrence of brecciated (cemented) layers and areas, which make it difficult to recognise compacted surfaces. However, the layers are definitely anthropogenic, as evidenced by the large quantity of phosphatic and carbonised material forming the matrix around abundant artifacts, and by the state of that material: pulverised, homogenised, and compacted (Courty 2006). Courty (2006) found no evidence of intense bioturbation, other than a few insect casings, nor any evidence of water flow which could have sorted and mixed the sediments. Signs of frost were found only in the uppermost, least anthropogenic, layer, and in two other strata within H1, but these signs were limited to a fine lamellar structure due to superficial freezing (Courty 2006). Therefore the breakage of the artifacts and other components of the strata must have been due to human action, such as walking on the deposits, not frost. Courty (2006) concludes that "La nature des débris organo-minéraux et leur micro-organisation sont caractéristiques de faciès anthropiques développés sur des surfaces naturelles d'occupation soumises à un piétinement intense", with the sediments enriched by the detritus of the various activities practiced there, such as using fire, cooking meat, eating, and burning or throwing away the remnants. Courty (2006) found that samples from A14 were better preserved than those taken from B15, which she suggests might be due to less trampling of the surfaces near the rock shelter wall. Courty (2006) also suggests that the good preservation of the individual layers, despite their minimal thickness, means that they must have originally been much thicker – perhaps as much as 10 times thicker than they are now.

Courty's (2006) samples are rich in evidence of combustion: she mentions burned bones, burned substrates, ash, and carbonised residues including RCF ("résidus carbonisés fibreux à structure végétale bien conservée"), which form in temperatures up to a few hundred degrees, and RCV ("résidus carbonisés vésiculaires ("char") formés de carbone vitreux homogène, sans structure histologique"), the formation of which requires pyrolysis at high temperatures (>1000°C) in a reducing atmosphere. RCV, the products of the highest temperature combustion, are found throughout the samples (Courty 2006).

Figure 5.9. Layer H1, zones B-C 15-17, at a later stage of excavation. Note the rock extending out of the section wall on the right; the top of it can be seen in Figure 5.8. ***La couche H1, zones B-C 15-17, plus tard pendant la fouille. Notez la roche qui s'étend de la coupe droite: sa surface supérieur est visible dans la Figure 5.8.***

Courty (2006) was also able to draw conclusions about the climatic conditions at the times of formation of the various strata within H1. At the base of H1, Courty (2006) finds evidence that the rock shelter wall must have had some herbaceous vegetation on it. The Mediterranean-type climate, warm and humid, would have had strong seasonal contrasts and a great deal of evaporation. The rock shelter walls were subject to substantial disaggregation. Moving upwards through the deposits, the climate became more humid and less markedly seasonal, with less evaporation. Moss and lichen then developed on the rock shelter walls, due to nearly permanent humidity and less precipitation. This led to reduced circulation of liquids within the deposits, but the solutions were very acidic, causing dissolution of the organic and phosphatic material (bones). This stage was followed by increased herbaceous vegetation and once again a climatic regime with strong seasonal contrasts: rain in the spring, a lot of evaporation in the summer, and a cold winter with significant snow cover. In the next stage, the winters became milder, but then at the top of H1 the climate once again showed strong seasonal contrasts, with rigorous winters, deep snow cover and strong frosts.

This complexity of strata and environmental variations suggests to Courty (2006) that the deposition of H1 took a relatively long time, during which the environmental conditions were changing rapidly. The repeated occurrence of burned microstrata followed by less markedly anthropogenic layers containing limestone plaquettes shows that human occupation, and the combustion activities associated with it, occurred on an occasional basis, rather than continuously, at about the same rhythm as the environmental changes: "L'analyse séquentielle de la couche H1 traduit une évolution rapide des conditions environnementales, à des pas de temps semblables à ceux des rythmes d'occupation" (Courty 2006). The post-depositional transformation of the deposits suggests that the occupations may have been separated by hundreds of years, but it is currently impossible to date the deposits precisely enough to corroborate that idea.

5.2.5. The uppermost deposits:

As mentioned above, the top of the metre attributed to layer H consisted of breccia and some loose sediments which contained very little archaeological material. My personal stratigraphic notes from 1995 mention a series of thin, unconsolidated, fine-grained layers pinched between coarser, often brecciated, layers; these were visible in the section wall between bands 18 and 19 at depths between 589 and 620 cm beneath zØ. I speculated at the time that they might represent one or more episodes of warming. At a depth of approximately 630 to 640 cm below zØ, archaeological material became more common, and this became our first H1 décapage. Lebel (1996) reports that we found large bone fragments, teeth and a few tools in these sediments, which were on top of a "dalle d'effondrement" at about -650 cm. This was a set of broken rock fragments which had fallen from the rock shelter ceiling or walls and subsequently been cemented back together into a solid but discontinuous floor covering at least 7 or 8 square metres. This provided an evident physical and temporal separation between the overlying deposits and the combustion layer underneath (Lebel 1996).

Courty (2006) took three samples that included the contact between the black layer and the overlying deposits, as well as two samples well above H1. The sediments consisted of cryoclastic debris in a brown or yellowish-brown matrix; the matrix consisted of a mixture of sands and silts in most cases, but the uppermost sample (H1supA, z=-615 cm) had a sandy matrix and yellowish-orange colour. There is thus evidence of a cold climate, causing frost action and perhaps also responsible for windblown silt. The more orange colour of the sandy matrix higher up may suggest a warmer climate, with more water percolating through the deposits, oxidising sediments and removing finer-grained particles. Water would of course also be responsible for bringing in dissolved carbonate material, which reprecipitated within the deposits to form the cement in the breccia.

5.3. Pollen

In 1999 J. Argant (2000, 2006) took a series of samples down a column in zone C15, with sample numbers from 9 to 19 corresponding to z values between -605 and -705 cm below zØ, therefore from layer H. As Table 5.2 shows, she identifies sample 9 (z = -610 cm) as coming from layer G, and sample 19 (z = -690 cm) as coming from layer I2, but does place the rest within layer H, and specifies that samples 14 through 17 (z values from -659 to -677 cm) came from the black layer with archaeological material. Only seven of these samples were treated to extract pollen. Of those, three (11, 17 and 18) had pollen contents too small to be analysed statistically, and the other four (12b, 13, 15 and 16) had sufficiently large numbers and diverse enough assemblages for statistical analysis. For those four, the number of pollen grains varied between 140 and 305, and the number of taxa between 23 and 25 (J. Argant 2000).

Some of the samples showed signs of having been subject to disturbance by wind or water, which may have removed some pollen, and which also brought in older pollen: J. Argant (2000) cites samples 16 and 17, for instance, as containing old-looking conifer pollen, which would not be contemporaneous with the deposits it was found in. However, she says this does not pose a problem in the interpretation of the pollen spectra, inasmuch as it can be distinguished from the "fresh" pollen, and disregarded. On the other hand, numerous fragments of pollinia (coherent masses of pollen grains which did not detach from each other) were also found (including in samples 12b, 13, 15, 16 and 17). These indicate short transport from nearby plants, and little post-depositional disturbance, which supports the validity of these spectra as indicators of the surrounding environment at the time.

Arboreal pollen is dominated by *Pinus* (pine), which is always present; it is generally accompanied by minor quantities of other conifers, such as *Abies* (fir). The presence of pine is not surprising, since pine trees produce prodigious amounts of pollen and it is very easily transported on the wind. It may have been blown in from some distance, but it is also possible that pine trees were growing nearby, possibly in the gorges de la Nesque (J. Argant 2000). Pollen from pine and other conifers always greatly outnumber the pollen from broadleaf trees, of which the most common is *Quercus* (oak). Other deciduous trees appear sporadically, but include *Fagus* (beech), *Alnus* (alder), *Corylus* (hazel), and *Tilia* (linden) (J. Argant 2000).

The proportions of arboreal (AP) and non-arboreal (NAP) pollen vary through the sequence of spectra. In the archaeologically richest part of the black layer, H1,

Table 5.2. Samples for palynological analysis taken from the section wall of zone C15 in 1999, excerpted from J. Argant (2000), Tableau 1. * = treated sample. *Echantillons pour analyse palynologique prélevés de la coupe de la zone C15 en 1999, extrait de J. Argant (2000), Tableau 1. * = échantillon traité*

N° échantillon	carré	z(cm)	couche	Caractéristiques
9	C15	610	G	brèche
10 11* 12 (+12a, 12b*, 12c) 13*	C15	630 643 648 654	H	couche noire, brèche
14 15* 16* 17*	C15	659 665 671 677	H1	couche noire archéologique
18*	C15	682	H1	stérile
19	C15	690	I2	

sample 16 (z = -671 cm) has an AP value of 58.82%, while sample 15 (z = -665 cm) has an AP value of 49.84% (J. Argant 2000). The NAP include *Poaceae* (grasses) and other herbaceous and flowering plants, some of which J. Argant (2000) classifies as steppic. The steppic proportion is relatively low in the rich part of H1. This time period seems to have had a partly wooded, partly open landscape. The samples lower in the sequence reflect a dry, cold, rigorous environment with a steppic vegetation (J. Argant 2006), but throughout layer H1 J. Argant (2006) sees an amelioration in climatic conditions. She suggests (J. Argant 2006) that this improvement may coincide with the change from MIS 6 to 5.

Sample 13 (z = -654 cm) has the highest AP value (70.54%) (J. Argant 2000). J. Argant (2000) describes this as coming from "couche noire, brèche", and from its z value it probably represents the top of layer H1 proper. It seems to correspond to the mildest, wettest climate in the sequence, which was able to support a woodland cover (albeit still with open patches) including conifers and broadleaf trees, especially *Pinus* (pine pollen make up 57.67% of the total) and *Quercus* (9.54%).

According to the pollen, the climate subsequently became much more rigorous, which accords well with the micromorphological results described above for the top of H1 (Courty 2006). Sample 12b (z = -648 cm) is, like sample 13, described by J. Argant (2000) as coming from "couche noire, brèche", but it provided the lowest AP value in this sequence: 29.29% (J. Argant 2000). The large increase in non-arboreal pollen includes a large proportion of steppic taxa, representing a cold climate. The trees present include *Pinus*, *Abies*, *Picea* (spruce), *Cedrus* (cedar), *Betula* (birch) and *Alnus*, which to this New Brunswicker sounds a lot like home! We do not have *Cedrus* in eastern Canada, but we have the rest of those genera, and long, cold, snowy winters. The importance of steppic taxa in sample 12b, however, suggests that the climate was dryer at that time around the Bau, compared to present-day New Brunswick.

5.4. From The Chest: How to "un-lump" an assemblage

Having gone through this detailed explanation of the internal stratigraphy, micromorphology and palynology of H1, and seen the variations that exist within the layer, it would be a shame to then consider all of the archaeological finds as if they were from a single undifferentiated assemblage: everything from anything called H, all in together. Ordinarily, the finds from a layer could be plotted by their X, Y and especially Z values, to see if natural concentrations appear – but for the reasons explained in chapter 1, I do not have those values.

It occurred to me as I was looking at the lithic data that, even though I do not have the site records, I do have the catalogue numbers (zone, couche, numéro) allotted to the pieces. As we dug down through each zone, we gave numbers in numerical order to the finds, so logically, the lowest numbers were found near the top of the layer, and the highest numbers were found at the bottom. I therefore found the highest catalogue number in H1 in each of the main zones for that layer, and divided them by ten. The numbers are given in Appendix 1 of this chapter, but, for example, the highest number in C17 H1 is 1380, so I divided that by ten and got ten tranches of 138 potential numbers each, which I labelled 10, 20, 30 and so on, up to 100.

Why divide the numbers, not the actual pieces? The way I did it, there can be more lithic pieces in some tranches than others, because numbers were also allotted to bones and teeth. I thought this was a more realistic division of the finds than arbitrarily deciding that one-tenth of the lithics had to be in each tranche, because there actually *were* some levels at which bones were more common.

Why 10 tranches? First of all, because 10 is an easy number to divide things by. Also, because we had 8 décapages, and if I had divided the numbers by 8 it might have been difficult to resist thinking that my tranches represent each décapage. They do not. I do not know where the décapages start and stop. Finally, 10 tranches in H1 gives me 10 data points, which I can analyse statistically, and which can show if there are trends. Any two points will form a line, and that line will have a trend. A third point will usually change that apparent trend. Having ten points will show whether the line is just zig-zagging, or is really going somewhere.

The tranches do not represent individual occupations (but nor do the décapages); each tranche may contain part of an occupation, more than one occupation, or parts of two different occupations. However, they are closer to one occupation than the entire assemblage is. To explain why this is useful, I use the analogy of a picture with pixels. Treating the entire assemblage as one lump gives a picture with one pixel. Using 10 pixels (or in fact, 11, because I added H2 as a separate tranche) may show some sort of image or structure. It will still be a blurry picture, but it will be a lot better than one pixel.

Having done that for the lithics, I went back and looked again at the faunal data in Fernandez (2001). He does not give a complete list of identified teeth and bones by zone, couche and numéro, but for some species he does list the individual pieces in certain categories, such as teeth which were used to determine age at death. I went through those and extracted the information about bones and teeth in the four most common species in the assemblage, and allotted them to my tranches. This is so far less than ideal that I am sure many people will want to dismiss it out of hand: how can I base anything on a count of pieces which might not even include all identified pieces, and certainly does not include all of the pieces that would have been present, and anyway, what good are pieces? Every animal consists of many potential pieces, but they did not all necessarily get preserved in an identifiable state. I should really use the Minimum Number of Individuals in each tranche, but calculating that is beyond my capabilities, with the information I have. So I just totted up how many pieces were present, excluding multiple teeth from one jaw (I counted the mandible as a single piece). I did then look at the tranches with the highest numbers of pieces for the two most common species (tahr and horse), and checked them in detail to see if they could have been one particularly abundantly preserved animal of each type, but in both cases the remains clearly had to have come from multiple animals. This gives me some confidence that I can use this information, tentatively, admitting how weak it is, as an indication of species abundance in each tranche – and combine it with lithic information for the same tranches.

5.5. Fauna

The faunal results are drawn primarily from Fernandez (2001, 2006). Table 5.3 gives a summary of the species found, including carnivores reported by A. Argant (2006). In addition, Fernandez (in Lebel 2006) reported on faunal material found in 2006. Although those remains are not included in the percentages given in Table 5.3, they would not materially change the results, so I have only included the new species identified: *Sus scrofa* (wild boar), and *Vulpes vulpes* (red fox). Red fox was indeed a new species for the Bau, but boar had already been reported for layer H1 in Fernandez' preliminary report on the H1 fauna in the site report for 1993-1994-1995 (Lebel 1996), only to disappear from the list in subsequent reports and publications. Its presence in the 2006 material is more like confirmation that it really was there, although very minor. Fernandez' preliminary report in Lebel (1996) differentiates between the material found at different levels: above the combustion area, the combustion area itself, and the cobble-bearing level below that (the lowest levels had not yet been excavated at that point), so some of those results are incorporated into the discussion below.

Table 5.3. Faunal remains in layers H, H1 and H2 combined, after Fernandez (2001, 2006), Fernandez in Lebel (2006), A. Argant (2006), and Crégut-Bonnoure et al. (2011). ***Faune des couches H, H1 et H2 combinées, d'après Fernandez (2001, 2006), Fernandez in Lebel (2006), A. Argant (2006), et Crégut-Bonnoure et al. (2011)***

Species	Common name	% of assemblage
Hemitragus cedrensis	tahr	34.6
Equus cf. *taubachensis*	horse	18.0
Bos primigenius	aurochs	15.7
Capreolus capreolus	roe deer	15.7
Cervus elaphus	red deer	12.0
Megaloceros giganteus	Irish elk	2.2
Dama dama	fallow deer	1.5
Rupicapra rupicapra	chamois	1.3
Rangifer tarandus	reindeer	0.1
Ursus arctos	brown bear	n/a
Sus scrofa	boar	n/a
Vulpes vulpes	red fox	n/a

The faunal results from layer H are interesting in a number of ways. They are chronological indicators: which species and which combinations of species are present can be characteristic of particular time periods. They are climatic indicators, since some species have, for instance, limited temperature tolerances, and they are environmental indicators because of the habitat characteristics the animals require. They are also indicators of human activity: both *what* humans were doing (hunting?, and if so – how?), and *when* they were doing it, through evidence of the season of the year at which the animals died.

In terms of chronology, Fernandez (2001) states that the association of *Hemitragus cedrensis* and *Equus* cf. *taubachensis* implies an Eemian age, that is, MNQ 25 or MIS 5e, the last interglacial period (Riss-W*ü*rm in the Alpine terminology), which lasted from about 130,000 to 115,000 years ago. This is younger than the TL dates cited above of 169 ± 17 to 191 ± 15 ka (Lebel et al. 2001), which would be within the last part of the Riss glaciation (Riss III), but seems to agree with the indications of climate from the micromorphological and palynological analyses. On the other hand, more recently Crégut-Bonnoure et al. (2010) associate H1 with MIS 6, in agreement with the TL dates.

The proportions of animals within the assemblage should not be taken as representative of their relative abundance within the surrounding area. As we will see below, they were actively and selectively hunted by the hominins. However, the environmental requirements of the animals present range from cold (reindeer) and open (Irish elk), to wetter and more wooded (deer), suggesting that we are looking either at an area which had a diverse assortment of habitats within a relatively small area, or at a palimpsest spanning several occupations over a length of time during which there were important climatic fluctuations. Both are probably true. The area around the Bau includes relatively low level (500 to 600 m asl) plains, high plateaus above 1000 m, the narrow and steep gorges of the Nesque, and the slopes of the Mont Ventoux, therefore many different species of animals can find suitable habitats within a few km of the site. We also know from the stratigraphic and the micromorphological evidence that the site must have been occupied and abandoned many times within the period of accumulation of layer H, with Courty (2006) speculating that there may have been centuries between each occupation: plenty of time for climate to change, as in fact it did. The faunal evidence by tranche, discussed below, corroborates some of these changes. There is also evidence within the faunal remains for periods of abandonment: although traces of carnivore activity are very rare at the Bau, Fernandez (2001) reports that there are two bones from H1 bearing such traces. One (B18 H1 174) is a femoral diaphysis of a horse, which had been broken, burned, and then chewed on (scored). It was found amongst the many bone fragments in the combustion area. The other (C16 H1 464) is a femoral articulation of a roe deer, which showed a number of tooth marks from a small carnivore. In both cases, Fernandez' (2001) interpretation is that the site and the remains had been abandoned by the hominins before being investigated by these carnivores.

Fernandez (2001, 2006; see also Fernandez & Legendre 2003, and Fernandez et al., 2006) devotes considerable energy to the analysis of age-at-death of the animals, based on their teeth. This is important in two ways. First, the proportions of animals in different age categories provide evidence of the way the hominins obtained them. Since most of these animals could not have come to the Bau by themselves (horses, for instance, are not generally found roaming up and down limestone cliffs), the hominins had to have acquired them somehow, but was that by scavenging or hunting? If it was by hunting, what kind of hunting? What were their strategies and goals? Although it is possible that the occasional carcass of any species was scavenged, Fernandez (2001) finds no evidence for systematic scavenging: the age distributions do not match the expectations of scavenging, where the hominins would have been finding and exploiting the carcasses of the weakest animals, i.e., the oldest and the youngest. They therefore were obtaining meat through hunting, but there were some differences of approach depending on the species, as we will see below.

Age at death can also be used to indicate the season of death. If, for instance, an animal was about six months old when it died, and we know that the species tends to give birth in early summer, we can assume that the animal likely died in early winter. That in turn tells us when the hominins were using the site, especially since the patterns of breakage of the bones show that they were fresh when broken. The meat itself may have been eaten right away, or smoked, dried or naturally frozen in winter for later use, but when that bone was broken, the hominin was at the site, and the animal had just been killed.

A species-by-species look at the fauna, as provided by Fernandez (2001), shows that the hominins were able to adapt their hunting strategies to best exploit the physical and social characteristics of their prey. As Table 5.3 shows, the most common animal in the assemblage is the tahr, an ungulate similar to ibex (Figures 5.10 and 5.11). Most of the remains derive from prime-aged and older adults, with only a few juveniles. Fernandez (2001) suggests that this is due to selective hunting which targeted the larger animals (perhaps because they provide more meat), and that this would have occurred during the fall and winter. The animals group together for the rut in the fall, making more adults available. In the winter, tahr, especially the larger males, are less mobile because they have trouble handling snow, and because they use up a lot of energy in the rut. They therefore stay within a restricted territory. This is especially true for the older males, making them not only easier prey, but easier to find. On the other hand, a male during or just after the rut would provide much less food, especially fat, than that male would have a couple of months earlier. The loss of fat may be significant. Humans can not use more than 40% protein in our diet: this is called the protein ceiling (Ben-Dor & Barkai 2020). The rest of the diet needs to consist of fat and carbohydrates. There is a loose relationship between the amount of fat in an animal and the size of the species: bigger species tend to have proportionally more fat (Ben-Dor & Barkai 2020). Tahr are a medium-sized animal, with the Bau's version similar in size to the extant Himalayan tahr species, *Hemitragus jemlahicus*, whose adults weigh about 35 to 90 kg (Fernandez 2001; Kennedy 2002). In the winter, a male tahr would not have much fat left on him, and hominins would not have access to plant foods to make up that other 60% of their diet, so the meat could not all be used unless a source of fat was found. There are two explanations for where that fat came from. First, bone marrow. The tahr

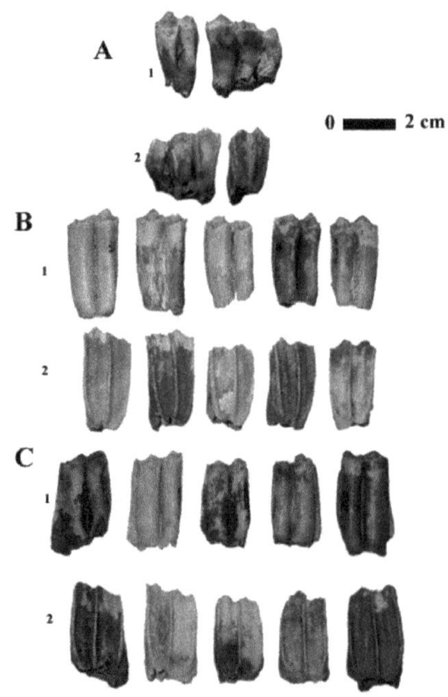

Figure 5.10. ***Hemitragus cedrensis, upper teeth.*** **A1, from left to right: M1/, P4/, P3/, and P2/ right (D17 H1 260), lingual face. A2, same series and same references, vestibular face. B1, upper M2, from left to right: D20 I2 189, right; D18 H1 571, right; C20 I2 412, right; B15 H1 19, right; and D18 H1 739, left; lingual face. B2, same series and same references, vestibular face. C1, upper M3, from left to right: B15 H1 336, right; B16 H1 28, right; C16 H1 516, right; C18 H1 997, left; and C16 H1 368, left; lingual face. C2, same series and same references, vestibular face. Photos P. Fernandez.** ***Hemitragus cedrensis, dents supérieures. A1, de gauche à droite: M1/, P4/, P3/, et P2/ droites (D17 H1 260), face linguale. A2, même série et mêmes références, face vestibulaire. B1, M2 supérieures, de gauche à droite: D20 I2 189, droite; D18 H1 571, droite; C20 I2 412, droite; B15 H1 19, droite; et D18 H1 739, gauche; face linguale. B2, même série et mêmes références, face vestibulaire. C1, M3 supérieures, de gauche à droite: B15 H1 336, droite; B16 H1 28, droite; C16 H1 516, droite; C18 H1 997, gauche; et C16 H1 368, gauche; face linguale. C2, même série et mêmes références, face vestibulaire. Clichés P. Fernandez.***

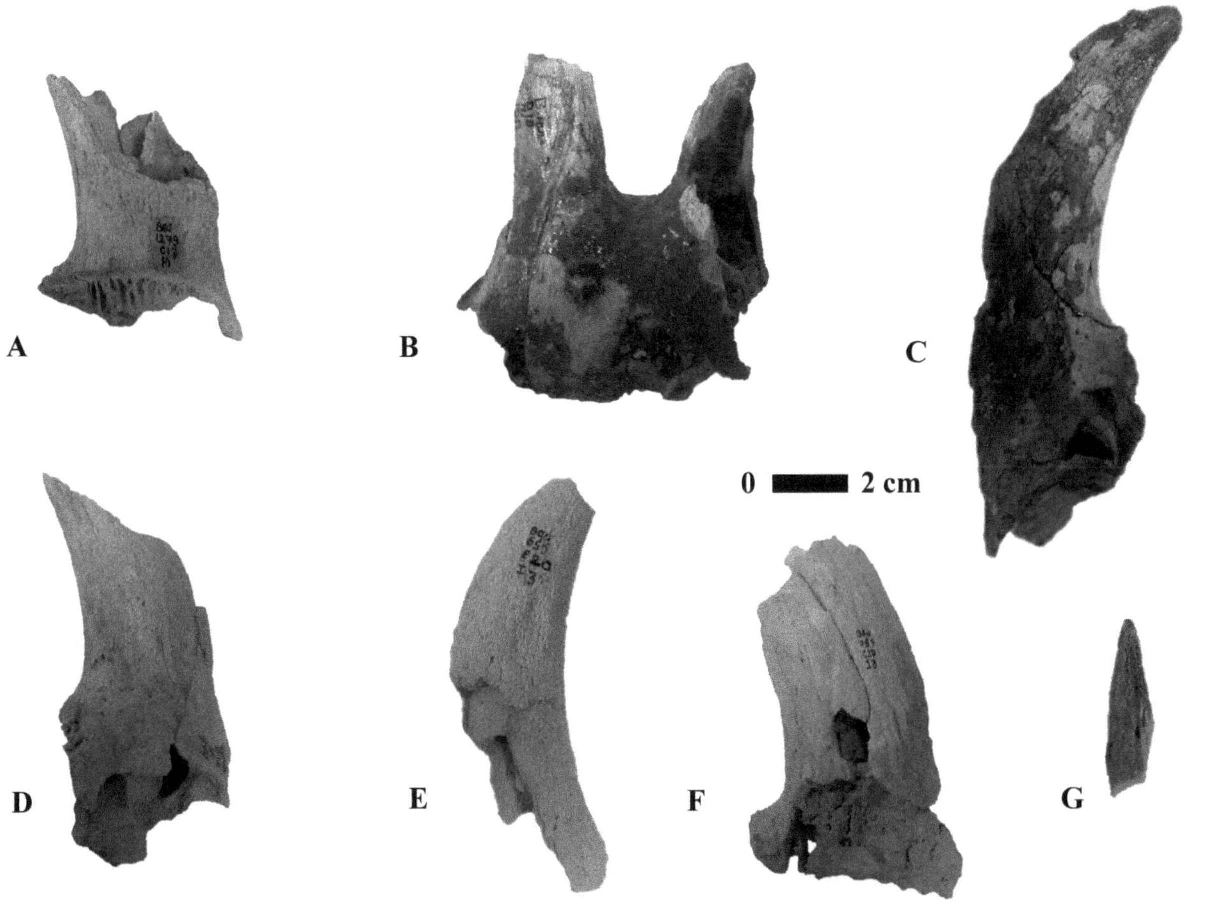

Figure 5.11. ***Hemitragus cedrensis*. A, left horn core C17 H1 1279, medial face. B, frontal bone with horn cores D18 H1 130, anterior face. C, right horn core with upper orbit C18 H1 203, medial face. D, right horn core B16 H1 779, lateral face. E, fragmentary horn core C20 I3 655, profile view. F, right horn core C20 I3 797, lateral face. G, apex of fragmented horn core B16 H1 716, anterior face. Photos P. Fernandez.** ***Hemitragus cedrensis. A, cheville gauche C17 H1 1279, face médiale. B, os frontale avec chevilles osseuses D18 H1 130, face antérieure. C, cheville droite avec orbite supérieure C18 H1 203, face médiale. D, cheville droite B16 H1 779, face latérale. E, cheville fragmentée C20 I3 655, vue de profil. F, cheville droite C20 I3 797, face latérale. G, apex cheville osseuse fragmentée B16 H1 716, face antérieure. Clichés P. Fernandez.***

were used intensively: the bones are very much broken up, and the breakage patterns show that the bones were fresh when broken. They also show many cut marks, even on bones which would not have provided much meat. The positions and orientations of the cut marks on those bones suggest that flesh and muscles were being removed in order to facilitate breakage for the extraction of marrow (Fernandez 2001). Patterns of burning suggest that bones were subsequently burned as fuel, although some heating may have taken place before breakage, too, since this has been suggested to make marrow removal easier.

The second potential source of fat is other, larger and fatter, animals. Table 5.3 shows tahr as being the most numerous species among the remains, but that actually does not mean that it is what the hominins ate the most. The next two species, horse and aurochs, account for almost as large a percentage of the remains (33.7% combined total, versus 34.6% for the tahr), but since they are much larger animals they would have provided much more meat, and would have had more fat on them than the smaller tahrs. In fact, even though horse and aurochs do not dominate the H1 faunal list in terms of number of remains, they probably did still dominate the diet of the hominins, although not as much as in the lower layers where they account for a large majority of the remains (a combined total of 74.2% of remains in layer I, for instance).

The horse present in layer H is *Equus cf. taubachensis* (Figure 5.12), while in lower layers we found *Equus mosbachensis* (Fernandez 2001, 2006). This change is probably due to both chronologic and environmental reasons, since layer H is younger and mainly reflects a milder climate than the lower layers. *E. taubachensis* was a smaller animal than *E. mosbachensis*, but they are both considered to be large horses (Fernandez 2001), well within the size range of modern riding horses, with *E. mosbachensis* at the upper end of that range (but not as big as a draught horse). There would still have been plenty of meat on *E. taubachensis*.

Thirteen individual horses were identified and aged in layer H1. Their age ranges include everything from juveniles to old adults, with the majority falling within the range of prime-aged adults (Fernandez 2001, Fernandez & Legendre 2003). Fernandez (2001) considers that they

Figure 5.12. *Equus* cf. *taubachensis*. A, right upper tooth row D18 H1 132, with from left to right M2/, M1/ and P4/. B, right lower tooth row B15 H1 686, with from left to right M/1, P/4, P/3 and P/2; A and B occlusal face. Photos P. Fernandez. *Equus* cf. *taubachensis. A, rangée dentaire supérieure droite D18 H1 132, avec de gauche à droite M2/, M1/ et P4/. B, rangée dentaire inférieure droite B15 H1 686, avec de gauche à droite M/1, P/4, P/3 et P/2; A et B face occlusale. Clichés P. Fernandez.*

were obtained through mass hunting of family groups, and that hunting took place during the period from July to September (plus or minus one month). However, there were also two horses found in layer H2, one of which would have died in July to September, but the other in November (plus or minus one month) (Fernandez & Legendre 2003). Fernandez's (2001) Table 156 shows that one of these was a juvenile, the other was a prime adult, but I have been unable to determine which of the two died in November.

Aurochs specimens come from prime-aged and older adults, obtained through selective hunting. Their bones show cut marks related to the removal of meat, the possible removal of tendons, and butchery marks as they were cut into pieces; the bones are highly broken for marrow removal.

Roe deer remains come only from the combustion levels of H1 (Fernandez 2001), and probably for that reason show a higher frequency of burning than the bones of other taxa, where Fernandez' sample includes remains from less-burned or non-burned levels of H, too. They include both juveniles and adults, and Fernandez (2001) attributes this to ambush hunting, which he still considers a type of selective hunting. Some of the animals must have been killed during the birthing period (broadly, March to July, but it may be possible to narrow that down to May to July), but others are attributed to a wider range of time, from June to March. This means that there is a slight possibility that all animals were obtained within that narrow window of time from June to July, but it is more likely to represent two different times of occupation of the site. In fact, it is most likely that the layer includes multiple occupations, widely spaced in time and at different seasons of the year.

The roe deer remains show some differences from the other taxa. They were much less intensively exploited for marrow (only the metapodia and tibias were fractured), and they show few cutmarks. The next most common deer, the red deer, also shows few cut marks, but these bones have a lot of percussion marks (Fernandez 2001). They include adults and a few juveniles, and the season of death is unknown.

Among the less common taxa, there is not much information to be gleaned, although Fernandez (2001) does mention that the Irish elk bones were fractured for marrow removal, and the chamois sample consists of two juveniles killed during the winter or spring (November to May). They were probably the unlucky victims of opportunistic encounter hunting. Also, the mere presence of the reindeer tells us that part of layer H dates to a cold period.

Looking again at the times of death of the taxa, we see that hominins must have been present at the site in the summer, when most of the horses (and perhaps some of the roe deer) were killed. They must also have been present at other times, too, though: one horse in H2 was killed in November, the tahr were killed somewhere between November and February, the chamois some time between November and May, and some of the roe deer must have been killed any time between June and March.

In the site report for 1993-1994-1995 (Lebel 1996), Fernandez makes some interesting remarks about the state of the bones, subdivided by level. In the lowest level that had been excavated by then, the cobble level under the combustion level, he found that the bones showed cutmarks that were both oblique and longitudinal (fillet cutting), with butchery activities concentrated in zones C17 and C18. There had been intense marrow recovery, and the bones were very burned. In the combustion level itself, 19% of the bone fragments showed cutmarks (due to meat removal before breakage), with 80% of the cutmarks longitudinal or oblique and 20% transversal, near the articulations, due to cutting the animals up into pieces (Fernandez in Lebel 1996). The topmost layer, above the so-called "*plancher d'effondrement*" described above, had many fewer bone fragments, with signs of both hominin-induced breakage (for marrow), and gelifraction. The bones had been leached by water passing through the sediments.

Tables 5.4, 5.5 and 5.6 give the results of my attempts to allot identified faunal remains to smaller segments (tranches) within H1, and also to add up identified remains in H2. The numbers must not be taken as absolutely representing what was there: they are only indications of potential trends. Still, we see that tahr tends to be most important in the upper half of H1. The largest number of tahr specimens is in tranche 50 (8 pieces, representing at least 4 individuals, possibly as many as 7, based on repetition of body parts and on age at death, allowing a margin of plus or minus one year or so for overlaps). Roe deer, on the other hand, have their peak number of pieces in tranche 70, with 19 pieces (I did not try to calculate MNI for roe deer or aurochs).

The lowest segments were dominated by horse and aurochs (as had been the case for layers I and J), but it is

Table 5.4. Numbers of identified pieces for the four most abundant species, by tranche in H1 and for H2 all together. See "From The Chest: How to "un-lump..." for details. Data extracted from Fernandez (2001): Tableaux 2 and 52 for tahr; Tableaux 140 and 156 for horse; Tableaux 81, 102 and 104 for aurochs; Tableaux 62 and 176 for roe deer. ***Nombres de pièces identifiées pour les quatre espèces les plus abondantes, par tranche en H1 et pour H2 au complet. Voir "From The Chest: How to "un-lump..." pour plus de détails. Données extraites de Fernandez (2001): Tableaux 2 et 52 pour le tahr; Tableaux 140 et 156 pour le cheval; Tableaux 81, 102 et 104 pour l'aurochs; Tableaux 62 et 176 pour le chevreuil***

Tranche	tahr	horse	aurochs	roe deer	Total for tranche
10	6	22	4	8	40
20	4	14	9	11	38
30	4	10	5	10	29
40	2	9	2	4	17
50	8	10	5	11	34
60	3	18	10	10	41
70	2	20	14	19	55
80	3	14	8	9	34
90	1	19	12	3	35
100	2	13	5	0	20
H2	2	18	3	1	24
Total per species	**37**	**167**	**77**	**86**	**367**

Table 5.5. Distribution (percent) of pieces of each species among the tranches. ***Distribution (pourcentage) des pièces de chaque espèce parmi les tranches***

Tranche	tahr	horse	aurochs	roe deer
10	16.2	13.2	5.2	9.3
20	10.8	8.4	11.7	12.8
30	10.8	6.0	6.5	11.6
40	5.4	5.4	2.6	4.7
50	21.6	6.0	6.5	12.8
60	8.1	10.8	13.0	11.6
70	5.4	12.0	18.2	22.1
80	8.1	8.4	10.4	10.5
90	2.7	11.4	15.6	3.5
100	5.4	7.8	6.5	0.0
H2	5.4	10.8	3.9	1.2
Total %	**100**	**100**	**100**	**100**

noteworthy that the largest number of horse specimens is in tranche 10, at the very top of H1 (22 pieces, representing at least 6 individuals). Horse is also the most common type of animal (although with a relatively small number of pieces) in tranche 40. However, with those exceptions, horse seems to be more important towards the bottom of the profile, and especially in H2. Aurochs tends to mirror the horse values, but diverges at the very top and the very bottom, in both tranche 10 and tranche 100 and in H2: in all of these cases, horse increases relative to its neighbouring value (i.e., higher in 10 than in 20; higher in 100 than in 90, higher in H2 than in 100), but aurochs decreases. Despite that, the distribution of pieces of horse is strongly significantly correlated (at the 0.01 level) with that of pieces of aurochs (see Table 5.7). This is the behaviour we have seen throughout the pente du remplissage deposits so far: complementary use of horse and aurochs, likely at different seasons of the year.

Table 5.6. Percent of each species in the assemblage from each tranche. ***Pourcentage de chaque espèce dans les assemblages de chaque tranche***

Tranche	tahr	horse	aurochs	roe deer	Total %
10	15.0	55.0	10.0	20.0	100
20	10.5	36.8	23.7	28.9	100
30	13.8	34.5	17.2	34.5	100
40	11.8	52.9	11.8	23.5	100
50	23.5	29.4	14.7	32.4	100
60	7.3	43.9	24.4	24.4	100
70	3.6	36.4	25.5	34.5	100
80	8.8	41.2	23.5	26.5	100
90	2.9	54.3	34.3	8.6	100
100	10.0	65.0	25.0	0.0	100
H2	8.3	75.0	12.5	4.2	100

Table 5.7. Correlations (Kendall's tau_b) between values for pieces of the four most common faunal species distributed among the H1 tranches and in H2. * = Correlation is significant at the 0.05 level (2-tailed). ** = Correlation is significant at the 0.01 level (2-tailed). N = 11 in all cases. ***Corrélations (Kendall's tau_b) entre les valeurs pour la distribution des pièces des espèces fauniques les plus abondantes dans les tranches pour H1 et dans H2. * = Corrélation est significative au niveau de 0.05 (2-tailed). ** = Corrélation est significative au niveau de 0.01 (2-tailed). N = 11 dans chaque cas***

		tahr	horse	aurochs	roe deer
tahr	correlation coefficient	1.000	.098	-.020	.535*
	sig. (two-tailed)	.	.689	.936	.030
horse	correlation coefficient	.098	1.000	.642**	.229
	sig. (two-tailed)	.689	.	.007	.343
aurochs	correlation coefficient	-.020	.642**	1.000	.495*
	sig. (two-tailed)	.936	.007	.	.040
roe deer	correlation coefficient	.535*	.229	.495*	1.000
	sig. (two-tailed)	.030	.343	.040	.

There are also significant correlations between tahr and roe deer, and between aurochs and roe deer. These may be due to a variety of different factors. All three of these species may have been hunted at least partly in the winter. Roe deer and tahr both become more abundant after the bottom-most tranches, which might be due to climate change allowing those two species to become more common in the area. In addition, roe deer and tahr are considerably less imposing animals than aurochs, so there may have been more similarities in hunting practices used to obtain tahr and roe deer, versus the way an aurochs would be approached.

The small number of pieces in tranche 40 stands out. This is likely due to a combination of factors: a fairly high number of lithic pieces in that tranche of numbers (as we will see below, tranche 40 has the third-highest number of lithics among the tranches), and a very high degree of breakage and burning: tranche 40 is in the very heart of the combustion level. It is probable that a larger than usual proportion of the bones had been broken and burned to the point of no longer being identifiable. Horse bones and teeth (being larger than those of roe deer and tahr) may have retained more identifiable characteristics under those circumstances. The low proportion of aurochs bones in that tranche, on the other hand, suggests that aurochs really were not common in that assemblage, since their bones are even larger than those of horses.

5.6. Lithics

In order to make the results for layer H more easily comparable to those of other layers, I will use the same formats for the H lithics as I did for the lithics in earlier chapters, presenting the entire assemblage together in terms of typology and of raw materials, in Tables 5.8 (typology),

Table 5.8. Lithic assemblage of layers H, H1 and H2, summarised by typological category. ***Assemblage lithique sommaire des couches H, H1 et H2, par catégorie typologique***

Category	Typology list (n)	Typology list (%)	Raw materials list (n)	Raw materials list (%)
Liste Bordes	288	3.2	431	6.2
Debris	57	0.6	47	0.7
Eclats	1290	14.4	1538	22.0
Lames	94	1.0	54	0.8
Tamis/Collectif	7007	78.0	4481	64.0
Nucléus	244	2.7	222	3.2
Galets	0	0	10	0.1
Cortex	1	<0.1	1	<0.1
non-identifié	n/a	n/a	220	3.1
Total	**8981**	**100**	**7004**	**100**

5.9 (typology by raw material source, numbers), 5.10 (same, in percentages), and 5.11 (raw materials by geographic source region). However, I will also analyse the results by tranche, below. Note that whereas the tranches include only material from the richest zones of H1 (bands 15 to 18), the full assemblage has the advantage of also including material attributed to H1 from outside of that area, whose stratigraphic relationships to the levels in the burned area are not really clear.

Looking at Table 5.8, it is immediately apparent that the sample size in layer H is much larger than in the previously discussed layers, and also that, yet again, the numbers in the typologically-described list do not match up with the numbers on the raw materials list. This is, as usual, mainly due to a large difference in the number of "tamis" (sieve) pieces. There is a larger percentage than usual (i.e., compared to the lower layers) of tamis pieces in H, which would tend to make other values a bit lower: compared to K1-J and to I, H has relatively low percentages of items from the Liste Bordes, éclats (flakes), and lames (blades). Its percentage of nucléus (cores) is intermediate between those other two layers.

Tables 5.9 and 5.10 show that Layer H has about the same percentage of unidentifiable pieces as layer I did (58.2 vs. 58.4%), but it is in fact much less patinated than I, because 15.6% of the pieces in H fall into raw material categories that were burned, which also makes them unidentifiable. H has relatively high values for material from Faraud etc., but low numbers of pieces from Murs (11.1%, vs. 13.1% in layer I and 27.9% in K1-J), and intermediate values for the other main source areas.

Finally, Table 5.11, which recalculates the values excluding unidentifiable and unknown pieces, shows that layer H contains material from almost entirely the same source regions as the other layers, except that three of the very minor source areas represented in lower layers are absent, while H contains the first piece identified from the Far SE (Rocher de Guérin, near St. Michel l'Observatoire, about 30 km from the Bau). This is piece number C16 H1 833, a double sidescraper, convex on one side and concave on the other, made on a blade. Figure 5.13 shows the locations of the source areas used in layer H.

The fact that the tables show that 20 pieces of local material are present in the H assemblage is actually misleading: all of these are cobbles (or in some cases, I have them noted as "so-called" cobbles) of the local limestone, not flint. The number of local pieces is thus too high if we are considering flint, and too low if we are considering the presence of cobbles. Most of the cobbles were not given catalogue numbers, so this is just the 20 that were numbered, for some reason. That leaves us with the same question as in the other chapters: why no use of local flint? And still no answer.

Table 5.11 shows that, overall, layer H contains an unusually low amount of material from the SW (Murs area), and an unusually high amount from the nearer SW. Rather than try to explain that result, however, let us move on to the data by tranche, so that we can see the interesting patterns that lumping it all together obscures. To start, Table 5.12 gives some summary data about the tranches, and Tables 5.13 and 5.14 divide them up by typological category (number and percent). As usual, there are differences in the numbers of pieces identified typologically and those for which the raw material was identified. The total weight shown relates to the typological list, and that is used to calculate the weight per piece. The total weight is highest in the tranches with the biggest assemblages, which makes sense. The largest weight/piece is found in tranches with high proportions of cores, which also makes sense since they are generally larger objects than flakes or scrapers.

Table 5.12 shows that there is a consistently important percentage of patinated pieces, but that the most patinated

Table 5.9. Number of pieces by typology and source area in layers H, H1 and H2. ***Nombre de pièces par catégories typologiques et groupes de sources des matières premières dans les couches H, H1 et H2***

H-H2 n	Liste Bordes	Debris	Eclats	Lames	TamisColl	Nucléus	Galets	Cortex	non-id	Total
Unidentifiable	211	10	709	23	2954	72	0	0	94	4073
Unknown	5	0	10	0	12	0	0	0	3	30
Faraud etc.	62	21	265	10	577	33	0	1	40	1009
Méthamis etc.	38	4	174	7	255	38	0	0	24	540
Les Sautarels etc.	2	0	1	0	11	0	0	0	1	15
Sault	28	0	90	1	305	6	0	0	18	448
Murs	76	12	262	12	314	72	0	0	30	778
Local	0	0	1	0	2	0	10	0	7	20
St. Jean de Sault	0	0	0	0	0	0	0	0	0	0
Roussillon	0	0	0	0	0	0	0	0	0	0
Murs-Bezaure Olig.	3	0	23	1	46	1	0	0	1	75
Ravin de la Treille	1	0	1	0	4	0	0	0	1	7
St. Trinit	4	0	2	0	0	0	0	0	1	7
Durance	0	0	0	0	0	0	0	0	0	0
Tertiary Calavon	0	0	0	0	0	0	0	0	0	0
Tertiary Murs VdeV	0	0	0	0	1	0	0	0	0	1
Mormoiron	0	0	0	0	0	0	0	0	0	0
R de Guérin	1	0	0	0	0	0	0	0	0	1
N.Aurel	0	0	0	0	0	0	0	0	0	0
Total	**431**	**47**	**1538**	**54**	**4481**	**222**	**10**	**1**	**220**	**7004**

tranches are at the bottom of H1 (56.8% in tranche 90 and 60.4% in tranche 100). Patination of lithics results from their exposure to carbonate-bearing water within the calcareous sediments. The lowest levels of H1 were the levels at which we found a series of calcareous slabs, which had channeled water flow, which means that the water had to flow through a smaller area, in between and underneath the slabs. And that is exactly where the lithics had to be, too. Thus we can conclude that tranches 90 and 100 match up to the slab layers (granted, as they should, being at the bottom). If it was just that those levels are lower down, therefore somehow received more water, or received water with a higher load of dissolved carbonate, we could expect high patination in layer H2, too, but its values are back down to normal.

Table 5.12 also shows some variability in the percentage of burned pieces. Those are minimum numbers, calculated by adding up all the pieces attributed to types which include being burned in their definitions. Other pieces which were still identifiably some other type would be noted down as that type, whether they showed traces of burning or not. That includes fully patinated pieces. Thus it is likely that a much higher percentage of pieces than these numbers were actually burned. However, the variations in the numbers should be related to overall burning: higher in burned layers, lower in others. We see in Table 5.12 that the highest values run from tranche 20 through 70, with the highest value of all (27.0%) being in tranche 40. These tranches therefore cover the most burned part of the layer, probably including the cobbles towards the bottom.

In terms of typological characteristics, Table 5.8 showed an overall percentage of pieces identified according to the Bordes list that was low in H, relative to the lower layers, and Table 5.14 shows that this is broadly true,

Table 5.10. Percentages of pieces by typology and source area in layers H, H1 and H2. ***Pourcentages des pièces par catégories typologiques et groupes de sources des matières premières dans les couches H, H1 et H2***

H-H2 %	Liste Bordes	Debris	Eclats	Lames	TamisColl	Nucléus	Galets	Cortex	non-id	Total
Unidentifiable	49.0	21.3	46.1	42.6	65.9	32.4	0	0	42.7	58.2
Unknown	1.2	0	0.7	0	0.3	0	0	0	1.4	0.4
Faraud etc.	14.4	44.7	17.2	18.5	12.9	14.9	0	100.0	18.2	14.4
Méthamis etc.	8.8	8.5	11.3	13.0	5.7	17.1	0	0	10.9	7.7
Les Sautarels etc.	0.5	0	<0.1	0	0.2	0	0	0	0.5	0.2
Sault	6.5	0	5.9	1.9	6.8	2.7	0	0	8.2	6.4
Murs	17.6	25.5	17.0	22.2	7.0	32.4	0	0	13.6	11.1
Local	0	0	<0.1	0	<0.1	0	100.0	0	3.2	0.3
St. Jean de Sault	0	0	0	0	0	0	0	0	0	0
Roussillon	0	0	0	0	0	0	0	0	0	0
Murs-Bezaure Olig.	0.7	0	1.5	1.9	1.0	0.5	0	0	0.5	1.1
Ravin de la Treille	0.2	0	<0.1	0	<0.1	0	0	0	0.5	0.1
St. Trinit	0.9	0	0.1	0	0	0	0	0	0.5	0.1
Durance	0	0	0	0	0	0	0	0	0	0
Tertiary Calavon	0	0	0	0	0	0	0	0	0	0
Tertiary Murs VdeV	0	0	0	0	<0.1	0	0	0	0	<0.1
Mormoiron	0	0	0	0	0	0	0	0	0	0
R de Guérin	0.2	0	0	0	0	0	0	0	0	<0.1
N.Aurel	0	0	0	0	0	0	0	0	0	0
Total	**100**	**100**	**100**	**100**	**100**	**100**	**100**	**100**	**100**	**100**

Table 5.11. Percentages of identified pieces (excluding unidentifiable and unknown source pieces) from each geographic region for layers K2, K1-J, I and H. ***Pourcentages des pièces identifiées (excluant pièces indéterminables ou de sources inconnues) de chaque région géographique pour les couches K2, K1-J, I et H***

Geographic area	K2 %	K1-J %	I-I4 %	H-H2 n	H-H2 %
Local	0.2	3.4	1.3	20	0.7
NE	19.4	20.4	12.3	455	15.7
E	0	0.3	0.1	0	0
Far SE	0	0	0	1	<0.1
Far S	0	0	0.1	0	0
SSW	0	0	0.1	0	0
SW	44.3	54.1	35.0	869	30.0
Near SW	23.0	15.0	29.1	1009	34.8
WSW	12.4	6.6	21.9	540	18.6
NW	0.7	0.1	0.1	7	0.2
other	0	0	0	0	0
Total	**100**	**100**	**100**	**2901**	**100**

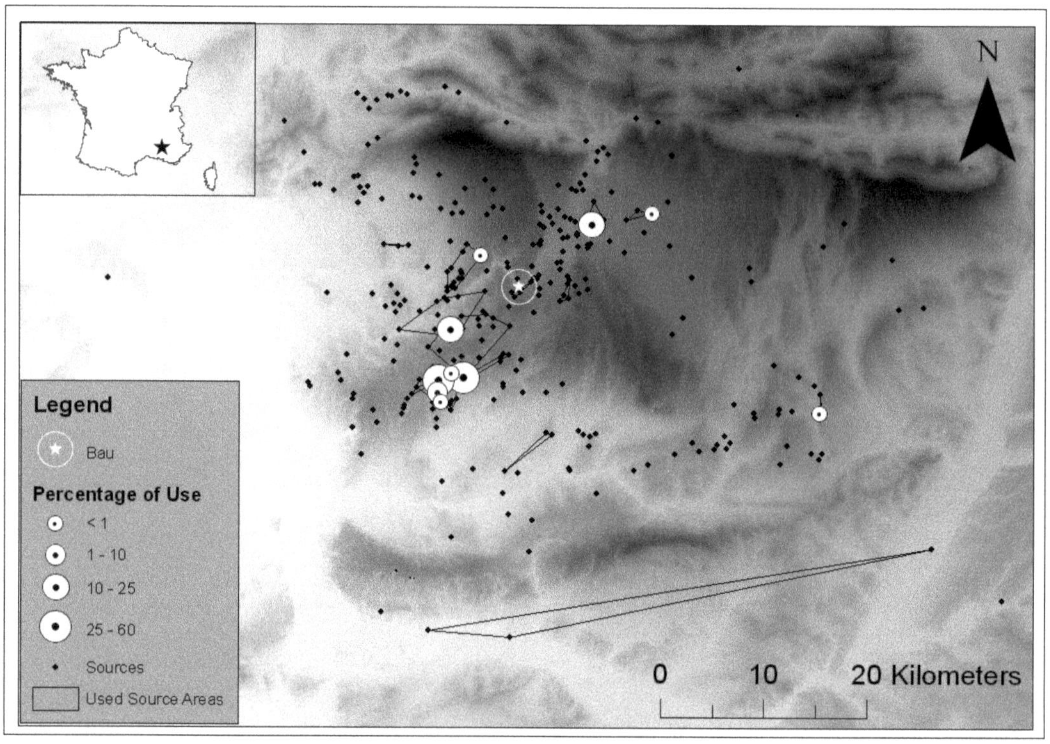

Figure 5.13. Map of the raw material source areas used for layers H through H2. ***Carte des aires de sources de matière première utilisées pour les couches H jusqu'à H2.***

Table 5.12. Summary data for H1 by tranches, and H2. ***Données sommaires pour H1 par tranches, et H2***

Tranche	Typology list (n)	Total weight (g)	Weight / piece (g)	Raw materials list (n)	Patinated (%)	Burned (%)
10	852	6012.6	7.1	904	43.1	12.7
20	1077	7457.8	6.9	735	39.6	21.2
30	1492	8456.0	5.7	655	37.4	16.6
40	1059	6740.5	6.4	684	40.2	27.0
50	936	7165.4	7.7	597	39.7	16.2
60	735	6380.8	8.7	640	42.2	22.0
70	491	4505.6	9.2	486	46.7	17.7
80	419	2315.2	5.5	509	49.9	12.4
90	436	1921.0	4.4	426	56.8	11.3
100	359	1419.1	4.0	361	60.4	13.3
H2	483	3287.7	6.8	413	44.3	6.1

with the exception of tranche 60 (at 4.8% it is almost as high as in layer I, and higher than K1-J). The numbers are particularly low in the lower tranches, to a lowest value of 2.1% in tranche 90, but back up to 4.1% in H2. These values are influenced by the percentages of tamis in each tranche, however, so even though they are not entirely controlled by that, we should probably not put too much weight on them. The tranche values also confirm that H1 is consistently low in percentages of flakes, with only tranche 70 (at 20.8%) in what would be considered normal territory, based on layers K1-J and I, and with the lowest percentage in tranche 100, at 8.4%. That tranche also has low percentages in the other categories (Liste Bordes, blades, cores, even debris), but a particularly high value for tamis. Layer H2, on the other hand, has an unusually high percentage of flakes (30.0%).

Turning to the raw materials, Table 5.15 gives the percentages of pieces in each tranche from each of the geographic areas (for simplicity's sake, excluding potential source areas which are not represented in these tranche assemblages). The percentages are calculated

Table 5.13. Typological data for H1 by tranche, and H2: numbers of pieces. ***Données typologiques pour H1 par tranche, et H2: nombres de pièces***

Tranche n.	Liste Bordes	Debris	Eclats	Lames	TamisColl	Nucléus
10	31	5	108	7	674	27
20	37	5	156	10	843	26
30	49	6	171	16	1218	32
40	34	5	144	10	841	25
50	29	11	114	13	739	30
60	35	8	120	10	533	29
70	13	3	102	5	342	26
80	14	3	59	8	326	9
90	9	2	48	7	361	9
100	10	1	30	1	311	6
H2	20	2	145	5	299	12

Table 5.14. Typological data for H1 by tranche, and H2: percentages of pieces. ***Données typologiques pour H1 par tranche, et H2: pourcentages des pièces***

Tranche %	Liste Bordes	Debris	Eclats	Lames	TamisColl	Nucléus
10	3.6	0.6	12.7	0.8	79.1	3.2
20	3.4	0.5	14.5	0.9	78.3	2.4
30	3.3	0.4	11.5	1.1	81.6	2.1
40	3.2	0.5	13.6	0.9	79.4	2.4
50	3.1	1.2	12.2	1.4	79.0	3.2
60	4.8	1.1	16.3	1.4	72.5	3.9
70	2.6	0.6	20.8	1.0	69.7	5.3
80	3.3	0.7	14.1	1.9	77.8	2.1
90	2.1	0.5	11.0	1.6	82.8	2.1
100	2.8	0.3	8.4	0.3	86.6	1.7
H2	4.1	0.4	30.0	1.0	61.9	2.5

based on pieces from identified sources, excluding pieces from unidentifiable and unknown sources.

As was mentioned above, the Local material is limestone cobbles, not flint, and does not even represent all of the cobbles. It might be indicative that the highest number is in tranche 70, which might mean that that corresponds to the cobble layer in the stratigraphy, but then, maybe not. It is safer just to disregard that column.

For the other regions, dividing the assemblage up into tranches turns out really to be worthwhile. In Table 5.11, where the overall H assemblage was contrasted with those from K2, K1-J and I, it looked like H had an intermediate use of the NE region's sources. Table 5.15 shows that in fact the NE% starts out low at the bottom, in H2 and in tranche 100 (even lower than layer I's value of 12.3%), but basically increases as we go up, so that by tranche 10 it has reached a percentage which is higher than in any of the other layers. Tranche 30 stands out from its neighbours in having a fairly low percentage of NE material (and an unusually high – for H1 – percentage of material from the SW), but the trend is still clear. In contrast, the WSW region, which also seemed to have an intermediate value in the full H assemblage, starts out high at the bottom, with H2, 100 and 90 all showing values higher than the previously highest value, in I (21.9%), then decreasing upwards. If we had just looked at K1-J, then I, then H, we would think that use of the NE sources declined from K1-J to I, but then went back up a bit in H, while use of the WSW increased from K1-J to I, then went back down a bit in H. In fact, what happened was that the decline in use of the NE sources carried on from I into the lowest parts of H, and then it started to increase, reaching greater abundance by the top of H1. Meanwhile use of the WSW increased in I, continued to increase in the lower parts of H, and only then declined through the rest of H. Thus the slabs level within H shows a continuation of the pattern of behaviour seen in layer I. The combustion level in H1 is where the behaviour really changes.

For the other main areas, the SW and the Near SW, we again see interesting patterns. H as a whole seemed to have an unusually low percentage of material from the SW, which follows a pattern of decline from K1-J through I. The results by tranches, despite some fluctuations (notably the high value in tranche 30), show that this pattern does continue, with three of the top four tranches having the three lowest values, as low as 23.0% in tranche 40. Tranche 40, on the other hand, has the highest value for material from the Near SW, and in general (with the exception of tranche 30, again) the upper tranches have higher values than do the lower tranches. In fact, the values for the SW show a statistically significant negative correlation with the values for the Near SW: increased use of the SW is correlated with decreased values for the Near SW (see Table 5.16). On the other hand, increased use of the Near SW is positively correlated with use of the NE. Use of the NE, in addition to being positively correlated with the Near SW, is negatively correlated with the SW, and is also negatively correlated with the WSW. The WSW has no significant correlation with use of the SW, but is negatively correlated with the NE and the Near SW.

Getting back to typology, briefly, there are a couple of points worth mentioning. Dividing the samples in each tranche up by both typological and raw material category, as was done for the overall assemblage in Tables 5.9 and

Table 5.15. Percentages of identified pieces (excluding unidentifiable and unknown source pieces) from represented geographic regions for tranches in H1, and for H2. ***Pourcentages des pièces identifiées (excluant les pièces indéterminables ou de sources inconnues) des régions géographiques représentées dans les tranches de H1 et dans H2***

Tranche %	Local	NE	Far SE	SW	Near SW	WSW	NW
10	0.5	21.3	0.0	28.1	36.7	13.4	0.0
20	1.0	20.1	0.0	26.9	36.0	15.9	0.0
30	0.3	13.0	0.0	41.6	29.8	14.9	0.3
40	0.8	20.6	0.0	23.0	37.9	17.7	0.0
50	0.7	15.7	0.0	32.2	36.7	14.3	0.3
60	1.2	13.4	0.0	30.8	34.4	19.4	0.8
70	1.7	16.0	0.0	32.6	32.0	17.7	0.0
80	1.0	12.8	0.5	33.3	31.3	20.0	1.5
90	1.4	14.2	0.0	28.4	29.1	27.0	0.0
100	0.0	11.5	0.0	36.5	28.1	24.0	0.0
H2	0.0	10.7	0.0	32.7	30.7	25.9	0.0

Table 5.16. Correlations (Kendall's tau_b) between values for use of source areas in the H1 tranches and H2. * = Correlation is significant at the 0.05 level (2-tailed). N = 11 in all cases. ***Corrélations (Kendall's tau_b) entre les valeurs pour l'utilisation des aires de sources dans les tranches pour H1 et dans H2. * = Corrélation est significative au niveau de 0.05 (2-tailed). N = 11 dans chaque cas***

		NE	SW	NearSW	WSW
NE	correlation coefficient	1.000	-.564*	.587*	-.514*
	sig. (two-tailed)	.	.016	.012	.029
SW	correlation coefficient	-.564*	1.000	-.587*	.147
	sig. (two-tailed)	.016	.	.012	.532
nearSW	correlation coefficient	.587*	-.587*	1.000	-.537*
	sig. (two-tailed)	.012	.012	.	.023
WSW	correlation coefficient	-.514*	.147	-.537*	1.000
	sig. (two-tailed)	.029	.532	.023	.

5.10, results in numbers which are really too small to be meaningful. However, the pattern of use of material from different areas does seem to be broadly the same across the tranches as it is in the overall assemblage. In particular, there are very few cores from the NE: six cores, equalling 2.7% of the overall cores assemblage, are spread across five tranches. In contrast, 14.9% of the overall core assemblage comes from the Near SW, 17.1% from the WSW, and 32.9% from source areas to the SW, which actually beats the percentage of cores of unidentifiable raw material (32.4%). This means that even in the tranches where material from the NE is particularly common, it was not being brought in as unworked or partly prepared raw material: it was showing up mainly as flakes or tools (see Figure 5.14 for an example).

On the other hand, material from the Near SW, which Table 5.11 shows as being, overall, the most common source region, dominates because of the high number of pieces in the sieves. Among the four main source regions, the Near SW is first in the sieves, second in the liste Bordes pieces and in flakes, and third in cores. Despite being third in terms of cores, it is likely that the cores from the Near SW were brought in as raw material and reduced further at the Bau, to generate the abundant small pieces in the sieves. We even found one piece from that region that consisted entirely of cortex (Table 5.9).

Figure 5.14. A tool in raw material type AE, from the NE source region. Note the mollusc shell showing as a dark oval in the centre of the lower part of the tool. ***Un outil confectionné dans le type de matière première AE, d'une source dans la région NE. Noter la coquille de mollusque de forme ovale foncée, au centre de la partie inférieure de l'outil.***

Before we put all this together, there is one final piece of the puzzle to consider: the results from Wilson and Browne (2014) concerning which factors influenced the choice of which raw material sources to use. As was explained in earlier chapters, our models look at the relative weights (importance) of ten variables in explaining the quantity (number of pieces) of raw material from different source areas in the assemblage, based on the characteristics of the sources that were used, as well as the characteristics of the sources that were not used. Table 5.17 gives the results for layer H as a whole, as presented in Wilson & Browne (2014), and Table 5.18 converts the Wald Chi-Square values to percentages, which allows us to contrast the weights of factors between models which are not otherwise directly comparable. In Tables 5.17 and 5.18 we see that in order of importance, the biggest factor is AOSISA (Area of the Sources in the Source Area: the total amount of the source area that actually has flint available within it, rather than the empty spaces in between sources in the source area). This is followed by Quality (how good is the raw material?) and then by Difficulty (Calories per kilometre expended travelling from the source area to the Bau). Difficulty has a negative Beta value (Table 5.17), so greater difficulty makes the source area less likely to be used. This is followed by Large Rocks and then the Extent of the source chosen to represent the source area.

In Wilson & Browne (2014) we grouped the variables into two sets of five: those having to do with the raw material itself (Quality, and the rock sizes), and those having to do with the terrain at the source and between it and the Bau. The Wald Chi-Square values, converted to percentages, for Layer H come to a total of 39.6% for the Raw Material variables, and 60.3% for the Terrain variables. This is the second highest total for Terrain of any layer at the Bau, with only layer G (to follow in chapter 6) having a higher Terrain value, and a lower Raw Material value. This continues a pattern of increasing importance of Terrain variables as we move upward through the deposits at the Bau. So, in layer H, although raw material quality still matters a great deal, and the availability of large rocks at the source is important, too, the most important considerations have to do with the effort necessary to get and transport the rocks: do the sources cover a lot of ground, and is it easy to get from there to here?

5.7. Conclusion – or, Tying it all together

In layer H, we have more detailed information about stratigraphic changes than usual. We have evidence of change in climate, through both micromorphological and palynological analyses. We have evidence of hiatuses in occupation. We also have evidence of change in prey species through time, and seasonal use of different prey. We have evidence of different choices about lithic raw materials, and from that, use of the surrounding landscape. Do these different lines of evidence converge at all? Yes, they do.

First of all, let's look at the patterns of raw material use. The same source areas are used through time, but their

Table 5.17. Generalised linear model for raw material source choices in H-H2. P-values in bold font are statistically significant. Sample: 2827 pieces from 10 source areas

Omnibus test: likelihood ratio chi-square = 779.580, df = 10, p = <0.001

Variable	Beta	S.E.	Wald Chi-Square	P-value	AIC
Quality	1.561	0.3902	16.003	**<0.001**	180.873
Extent	1.848	0.5345	11.959	**0.001**	
Difficulty (Cal/km)	-0.093	0.0236	15.537	**<0.001**	
Small rocks	-6.722	2.3825	7.961	**0.005**	
Medium rocks	-3.657	1.4379	6.468	**0.011**	
Large rocks	2.344	0.6546	12.826	**<0.001**	
Very large rocks	0.517	0.7170	0.520	0.471	
Calories	-0.001	0.0006	1.755	0.185	
AOSA	-1.7 e-7	6.5 e-8	7.171	**0.007**	
AOSISA	8.9 e-5	1.6 e-5	30.183	**<0.001**	
(Intercept)	1.412	3.9890	0.125	0.723	

Table 5.18. Wald Chi-Square values as percentages, H-H2 assemblage

Variable	Wald Chi-Square	Wald Chi-Square as %
Quality	16.003	14.5
Extent	11.959	10.8
Difficulty (Cal/km)	15.537	14.1
Small rocks	7.961	7.2
Medium rocks	6.468	5.9
Large rocks	12.826	11.6
Very large rocks	0.520	0.5
Calories	1.755	1.6
AOSA	7.171	6.5
AOSISA	30.183	27.3
(Intercept)	0.125	0.1
Total	**110.508**	**100**

relative proportions change. They seem to break down into three groups, or strategies:

Strategy #1: Concentrate on the SW and the Near SW. The SW has abundant, good quality flint, and when travelling to the Bau from the SW you traverse the length of the area I am calling the Near SW, so all of those outcrops are reasonably easily available to you, too.

Strategy #2: Still using some material from the SW, also use material from the WSW. This cuts down on the material from the Near SW because when you travel to the Bau from the WSW you intersect only the northern edge of the Near SW area.

Strategy #3: More use of the NE and the Near SW, less use of the SW and of the WSW. The increased use of the NE seems to shift the centre of gravity of the utilised area, so that travel towards the SW is reduced, increasing the relative proportion of Near SW material.

These strategies may well have been influenced by topography (see Figure 5.13). From the SW to the Bau the topography is, certainly, irregular, but overall it is mainly uphill through the SW and the Near SW and then along the top plateau until you get to the gorge, very near the Bau. Between the SW and the WSW, however, there is a ridge running E-W, so that if you are in the WSW and heading for the Bau you would want to stay on the north side of that ridge, thus missing out on most of the Near SW area. From the NE to the Bau, on the other hand, you can travel through fairly flat land (flat for this region!), across the plateau near Sault and through the wide, rolling landscape of the valley between Sault and Monieux, before finally arriving at the gorge, or carrying on across the top of the hills around the gorge to reach the Near SW, if you choose to do that. In terms of elevation, for example, Murs (in the SW) is at about 500 m asl, while the bottom of the gorges de la Nesque at Méthamis (in the WSW) is at 316 m. Faraud, the closest outcrop of the Near SW to the Bau, is at 679 m. The plateau above the Bau itself varies in elevation between about 700 and 800 m, as does the area to the NE, until finally starting to rise gradually on the Plateau d'Albion (still only at 862 m asl south of Revest-du-Bion, more than 17 km from the Bau as the crow flies). In terms of difficulty, the straight-line route from the gorge at Méthamis to the Bau has a difficulty of 91.8 Cal/km over

8.5 km, while the straight-line path from the Hippodrôme at Sault to the Bau has a considerably lower difficulty of 68.9 Cal/km over a distance of 9.4 km.

None of these strategies was used exclusively in any tranche in H1, or in H2, but some tranche assemblages show stronger tendencies one way or another, and from the bottom to the top we see a decline in use of strategy number 2 (with declining emphasis on the WSW), and an increase in strategy number 3 (with increasing emphasis on the NE). The faunal assemblages also change through time, with horse decreasing in importance as we move upwards, and then increasing again in the uppermost tranche. As we saw above, there are some statistically significant correlations between the presence of different faunal types. There is also one statistically significant correlation between lithics and fauna: use of the WSW is positively correlated with abundance of horses in the tranches (Table 5.19). This does not necessarily mean that the hominins went to the WSW to hunt horses. I personally would not want to lug a horse carcass all the way up from Méthamis to the Bau, and in any case it is very likely that horses were present on the plateaus close to the Bau. Certainly they are known to have been there in the later Pleistocene: there are natural traps, pitfalls due to karst holes in the limestone, which have been excavated and found to contain numerous horses, for instance near St. Jean de Sault (Pelletier et al. 2020). Perhaps, though, the correlation is not between the WSW and horses, per se, but rather between using raw material sources in the WSW and hunting near the Bau *in the summer*.

Of all of the main source regions used in the layer H assemblage, the WSW area is the most difficult of access. Just as we learned that tahr have restricted mobility in the winter, and stay in a smaller territory, perhaps we can imagine the same to be true for the hominins. In the winter, they could stay within the upland territory, where the topography is a bit easier. They would hunt tahr – who would be in these upland areas because that is their preferred habitat. They could hunt aurochs, who would be on the plateaus and in the broad valley of Monieux, in the open areas. Aurochs are at their most formidable in the later summer (Van Vuure 2002), but would be easier to hunt in other seasons, particularly if they also reduced their territory when their mobility declined due to snow in winter or mud in the rainy spring and autumn. Deer would be available in the forested areas on the slopes, and are particularly easy to track in winter snow.

Table 5.19. Correlations (Kendall's tau_b) between values for use of the WSW source area in the H1 tranches and H2 and abundance of pieces of horse. * = Correlation is significant at the 0.05 level (2-tailed). N = 11 in all cases.
Corrélations (Kendall's tau_b) entre les valeurs pour l'utilisation de l'aire de source dans l'WSW dans les tranches pour H1 et dans H2 et l'abondance des pièces de cheval. * = Corrélation est significative au niveau de 0.05 (2-tailed). N = 11 dans chaque cas

		horse	WSW
horse	correlation coefficient	1.000	.477*
	sig. (two-tailed)	.	.042
WSW	correlation coefficient	.477*	1.000
	sig. (two-tailed)	.042	.

In the summer, on the other hand, in addition to hunting in the local area for deer and horses, it would be easier for the hominins to move farther afield, and travel through other areas, including the WSW and the SW, and also including even more distant areas. The one piece of flint from the Far SE was in tranche 80, which also has a fairly high value for the WSW, and has the highest proportion of multiple tools of any tranche. By that I mean tools which have been retouched on more than one edge, forming two or more identifiable tool segments, such as the double-sided scraper in this case. Multiple tools are generally thought to reflect curation of tools, keeping them and re-using them over time and often across distance, as they formed part of a personal toolkit. A high proportion of multiple tools suggests that the hominins were highly mobile, moving across the landscape and bringing their tools along with them. That may have been easier in the summer, rather than the winter. The fact that the piece from the Far SE ended up in a tranche with a fairly high proportion of material from the WSW – the opposite direction from the Bau – also suggests that the hominins at that time were highly mobile.

Climate also changed over the course of layer H, from cold to warm to cold again. In the metre-thick deposits we called H, we may well have captured the end of the Riss III glacial period (MIS 6), the subsequent interglacial (MIS 5e) and the very beginning of the next glacial period, the Würm I in Alpine terminology. In the lower parts of H, including H2 and the slabs level of H1, we are in the MIS 6, and the hominins using the Bau carried on much as they had in previous layers, hunting horses in the summer and aurochs at other times. They continued to increase their use of raw material sources in the WSW, although they did use other areas, too. The climate was cold, which may have contributed to their tendency to use the Bau more during the summer than at other times of year: both because of increased mobility in the summer, and perhaps because they spent their winters at lower altitudes, where the climate was not as harsh.

Then the climate changed. It got milder and wetter. The landscape remained mainly open, but was broken up by more abundant and larger patches of forest. Deer and tahr became more common near the Bau. Hominins spent more time at the site during the winter. They restricted their territory and emphasised ease of access of raw materials. They used fire, possibly to help them extract the marrow they needed to complement the meat in their diet, and no doubt for heat and cooking. Repeated visits over thousands of years resulted in the thick combustion layer of H1. The milder climate reduced the amount of cryoclastic debris and windblown sand and dust, making the burned residues the main component of the sediments. The repeated use of much the same part of the rock shelter for these fires (just the right distance inside the overhang, just the right distance from the rock wall to allow it to reflect the light, and absorb and re-radiate the heat) allowed those residues to build up into a thick, greasy, black layer – and an archaeological bonanza.

Then it got cold again. The Bau became too cold for substantial winter use, and reverted to a summer stop-over. The hominins using it were, once again, concentrating on summertime hunting of horses. However, this time there is no complementary increase in use of aurochs (perhaps even less use of the site outside of the summer months?), and they did not go back to using a lot of raw material from the WSW. Instead, the emphasis on ease of access to raw materials continued, and more material from the NE was brought to the site. It was, however, as in lower levels, brought in more as part of a finished tool-kit than as raw material for later reduction, thus corroborating the idea that the Bau was just one place to stop on a round through the territory, not a home base of any sort. Above H1, the breccia composed of cryoclastic debris, and the scarcity of archaeological material, suggest that hard times are coming. In chapter 6 we will move on to layer G, and see what happened next.

5.8. Sommaire du chapitre

Les couches décrites dans les premiers chapitres de ce livre comprenaient des sediments calcaire blanchâtres en grande majorité tombés du plafond et des parois de l'abri. Avec la couche H nous rencontrons pour la première fois quelque chose de vraiment différent. Dans l'épaisseur que nous avons appelée H (de 605 à 705 cm sous zØ), se trouve une couche noire, épaisse d'entre 30 et 45 cm, très brûlée, graisseuse, remplie de fragments d'os et de silex. Cette couche, H1, est plus épaisse près de la paroi rocheuse et continue sur 4 à 5 mètres vers l'est. Nous n'avons pas trouvé sa limite vers le sud, puisque nous n'y avons pas fouillé: la couche est bien protegée sous 2 mètres de brèche dense et fortement cimentée. Pour information: en 1994 une équipe de 7 personnes a mis deux semaines en travaillant avec des marteaux-piqueurs pour ouvrir 4 mètres carrés jusqu'au niveau de la couche H1. Vers le nord, la couche noire disparaît à la bande 18. A l'extérieur de cette zone très brûlée et très riche, la couche H2 correspond à la partie inférieure de H; elle est moins brûlée et moins riche. Sous H1, par contre, nous passons directement à la couche I. La couche H1 a été datée par thermoluminescence (entre 169 ± 17 et 191 ± 15 mille ans), et par sa faune (MIS 6 et/ou MIS 5e).

La stratigraphie à l'intérieur de la couche H1 est complexe. Au plus bas elle est la continuation de la couche I sous-jacente: sediments calcaires peu ou non-brûlés, et au moins deux niveaux superposés de grandes dalles rocheuses tombées du plafond. Entre ces dalles (verticalement et horizontalement) se trouvent des sediments riches en vestiges archéologiques, que nous avons fouillés en deux décapages. Sur les dalles il y avait des galets roulés, amenés du lit de la rivière, entremêlés dans les dépôts archéologiques brulés. Au dessus, nous avons fouillé en 3 ou 4 décapages totalisant environ 20 cm des sediments très brûlés, graisseux, riches en os et en silex. Il y avait également une grande roche cassée ayant servi d'enclume, sur lequel des os avaient été cassés pour en extraire la moelle. Par dessus tout ça il y avait une dalle fragmentée et des plaquettes calcaires, le tout plus ou moins cimenté ou induré, formant un plancher. Sur ce plancher il y avait un peu de matériel archéologique, qui n'était pas brûlé. La partie supérieure de la couche H correspondait à de la brèche calcaire, presque complètement stérile archéologiquement.

L'aire de combustion était sans structure évidente, mis à part sa couleur noire et la nature brûlée de son contenu. Sa génèse est liée à une succession de feux posés directement à terre, le plus souvent à une distance d'environ 1,5 mètres de la paroi rocheuse. Les dépôts brûlés pouvaient atteindre leur épaisseur maximale (jusqu'à 40 cm) près de la paroi. L'étude micromorphologique des sediments a montré qu'ils avaient été piétinés (moins intensément près de la paroi) et compactés, et qu'ils contenaient beaucoup de résidus brûlés à très haute température. Ils témoignent également de variations climatiques importantes sur de longue durée. Les résultats micromorphologiques sont corroborées en grande partie par la palynologie: la partie inférieure de la couche était plutôt associée à un climat froid (période glaciaire), par la suite un rechauffement est visible, avec plus ou moins d'humidité selon le niveau, et on observe enfin une augmentation de la proportion de pollens d'arbres. En haut de la couche le climat redevient froid et rigoreux et la végétation plus steppique.

Etant donné ces différences stratigraphiques dans l'épaisseur de la couche H, l'hypothèse de détecter aussi des variations du matériel archéoologique a été faite. Pour la vérifier, j'ai divisé le matériel de la couche H1 en 10 tranches superposées (nommées 10, 20 . . . jusqu'à 100 au plus bas), et j'ai attribué la faune et les pièces lithiques aux différentes tranches. Même si chaque tranche est susceptible parfois d'inclure des restes de plusieurs occupations ou seulement une partie d'un seule occupation, elles nous permettre de suivre les grandes tendances de changements à travers les dépôts.

L'importance de cette division en tranches est visible par exemple en comparant les proportions d'espèces fauniques de chaque tranche (Tableau 5.3) rapportées au nombre d'espèces totale de l'assemblage (Tableaux 5.4, 5.5 et 5.6). Par exemple le tahr est important seulement dans les tranches supérieures, et c'est le cheval qui domine les tranches inférieures. En revanche, le chevreuil est moins abondant dans les tranches inférieures. Ces différences résultent d'une combinaison de facteurs: choix des hominidés dans leur stratégie de chasse, changements climatiques qui ont modifié l'habitat autour du site (plus ou moins ouvert, fermé…), l'évolution des espèces dans le paysage, et le choix des hominidés d'utiliser le Bau en période estivale (tranches inférieures avec abondance du cheval) ou hivernale (tranches correspondant à l'aire de combustion et à la chasse du tahr). Par ailleurs, complètement au-dessus des tranches, quand le climat redevient froid, la proportion du cheval est encore importante et la frequentation du site plus probablement estivale.

Les origines géographiques des matières premières lithiques nous informent sur le territoire utilisé par les hominidés, et les tendances peuvent changer à travers le temps. Bien que les roches du sud-ouest (Murs) aient toujours été favorisées de même que celles des sources du sud-ouest moins éloignées ("Near SW"), dans les tranches inférieures on voit une proportion notable de matériaux du ouest-sud-ouest (WSW, près de Méthamis). Ces derniers diminuent vers les tranches centrales de la couche H1, puis vers le sommet les matières premières des sources dans le nord-est apparaissent en plus grandes proportions.

En résumé, dans les niveaux inférieurs le climat était froid, le site a plutôt été fréquenté en été avec une chasse importante de chevaux, les matières premières comprennaient des quantités notables de roches du WSW, et une pièce lithique du sud-est plus éloigné a été retrouvée. On ne suggère pas que les préhistoriques chassaient les chevaux près de Méthamis pour les ramener au Bau – ce serait un non sens, d'autant plus qu'il devait y avoir des chevaux aux alentours immédiats du site. Non, ce que nous voyons ici est le résultat d'une grande mobilité des hominidés pendant les étés durant cette période glaciaire pendant lequel le paysage était ouvert, peu boisé, et que les hominidés utilisaient un grand territoire.

Plus tard, tout change. Le site est utilisé plus souvent pendant l'hiver, le climat est plus tempéré, le paysage plus boisé. Les hominidés chassent plus de chevreuil et le tahr, et le cheval et l'aurochs sont moins abondants. Ils se servent du feu... beaucoup plus! L'utilisation des matières premières provenant du WSW a diminué, et les matières premières du nord-est sont devenues plus importantes. Les déplacements sur le territoire semblent se concentrer plus près du Bau, entre les gisements du NE et du sud-ouest plus proches et le tahr est chassé dans les gorges. Cela limite la distance et la difficulté de trajets dans un paysage boisé et enneigé en hiver. Le Bau devient presqu'un lieu résidentiel et non un campement de passage.

Et cela change encore. Le climat est plus froid, le paysage est plus ouvert et steppique, la chasse des chevaux est plus importante, et le Bau est plutôt utilisé en période estivale... mais cette fois-ci, avec une utilisation moindre du matériel du WSW. En effet, les roches du NE restent abondantes, mais toujours (comme auparavant) importées au Bau plus en forme d'outils finis qu'en matière première brute. Le Bau représente un lieu de passage, un campement temporaire lors de périples à travers le territoire pour des hominidés mobiles qui connaissent la qualité des ressources minérales mais qui favorisent des gîtes d'accês facile. Enfin, au sommet de la couche H, la brèche cryoclastique annonce l'arrivée d'une période glaciaire difficile que nous aborderons dans le chapitre 6.

Appendix 5.1. H1 tranche divisions by zone

The ten subdivisions of H1 (tranches) are based on the maximum number attributed to a lithic piece in each zone, divided by ten, rounded up or down to the nearest whole number. I recognise that I used the same subdivisions for C15 (maximum number 1072) and C16 (maximum number 1076) for my own convenience, even though the C16 subdivisions could or should have been offset by 1 from the C15 ones. The faunal numbers were subsequently allotted to the same tranches; in a few cases, higher numbers existed among the faunal specimens, and they were at first put into a tranche of their own (called 100+), but there were very few of these so they have been added in to the 100 tranche.

The tranches were only calculated for the richest zones of H1, where the burned layer was, and did not include any zones beyond band 18. These richest zones are also the ones where no pieces were ever labelled H2. The H2 tranche thus consists of all pieces attributed to layer H2, all of which came from outside of the zones in tranches 10 through 100.

The following table lists the piece numbers allotted to each tranche of H1, in each zone. The number ranges are inclusive, e.g., B15 H1 90 would be in tranche 10.

	A15	A18	B15	B16	B17	B18	C15	C16	C17	C18	D17	D18
10	1-22	1-5	1-90	1-80	1-54	1-49	1-107	1-107	1-138	1-109	1-75	1-100
20	23-44	6-10	91-180	81-160	55-108	50-98	108-215	108-215	139-276	110-218	76-150	101-200
30	45-66	11-15	181-270	161-240	109-162	99-147	216-322	216-322	277-414	219-327	151-225	201-300
40	67-88	16-20	271-360	241-320	163-216	148-196	323-429	323-429	415-552	328-436	226-300	301-400
50	89-110	21-25	361-450	321-400	217-270	197-245	430-536	430-536	553-690	437-545	301-375	401-500
60	111-132	26-30	451-540	401-480	271-324	246-294	537-643	537-643	691-828	546-654	376-450	501-600
70	133-154	31-35	541-630	481-560	325-378	295-343	644-750	644-750	829-966	655-763	451-525	601-700
80	155-176	36-40	631-720	561-640	379-432	344-392	751-857	751-857	967-1104	764-872	526-600	701-800
90	177-198	41-45	721-810	641-720	433-486	393-441	858-964	858-964	1105-1242	873-981	601-675	801-900
100	199-221	46	811-892	721-796	487-539	442-491	965-1072	965-1076	1243-1380	982-1092	676-751	901-1002

6

Layers G and F

6.1. Introduction

The final section of the pente du remplissage to deal with is the uppermost one: the mainly brecciated deposits that form the top of the slope underneath the upper plateau part of the site. There is much less information available about this section of the site, both because there was very little archaeological material in it, and because it was so strongly cemented. It required serious effort with jackhammers to get through it, and the only reason for doing that was to make it possible to excavate the rich archaeological layers underneath. That means that we really treated these deposits as something to get through as quickly as possible, rather than something to study in their own right. Nonetheless, there were pockets of archaeological materials within the breccia, and they constitute the findings from layer G.

This chapter also includes layer F. Whereas layer G comes from the breccia at the top of the pente du remplissage, below the Tranchée Moulin and the upper plateau, layer F is the floor of that tranchée. This chapter therefore covers the link and the transition between the pente and the Tranchée Moulin. Layer F was not very well preserved, but it was a distinct level of concentrated material, exposed both at the front (on top of G) and at different places along the Tranchée Moulin. Since there is very little information about layer F available, as with G, the two layers fit well together into one chapter.

In moving from layer G into F, we also move from the part of the site where layers were named based on arbitrarily assigned one-metre depth attributions (e.g., layer H is from 605 to 705 cm below zØ) to the part of the site where layers were named based on archaeological and/or sedimentological characteristics. The bottom of layer G is the -605 cm level, but the top is really "where the slope starts", which was around 480 cm below zØ. Layer F is the concentration of apparently unburned material mixed with plaquettes and loose to indurated finer sediments, which is located below a sterile layer about 35 or 40 cm thick, and above the pente du remplissage. Its z values run from about 400 or 405 cm below zØ, down to about 480 cm below zØ, although it was not equally concentrated all through that thickness.

Layer G was dug in the bands A to D, 14 to 18, but some squares did not really exist, because of the bulges of the rock wall (see Figure 1.11, and remember that the wall is irregular vertically as well as laterally). I have information about the lithics from the squares A17, B15-18, and C15-18, but not from the test pits dug in 2006. For layer F, I have information about lithics from the squares A17, C11, C13-16, D9, D14-16, E8, E16, F16 and F17, G15-17, H17 and I17. Many of these do not appear on Figure 1.11 because that shows the most important excavated areas, and very little material was found in these zones. However, the Tranchée Moulin extends from about band 5 to band 14, with the trench getting wider starting around band 13. The bands 14 to 17 were, when we started, at the same level as the floor of the Tranchée Moulin, but they have been excavated since then in bands A through C or D, depending on the band. The plateau part of the site includes a large central mass, about 2 to 4 m thick, which is separated from the rock shelter wall on the west side by the Tranchée Moulin. There has always been a sort of very irregular floor, at about the same level as the Tranchée Moulin, that extends across the site in front of that central mass. Layer F was excavated from that floor in bands E-I, 15-17, in the places where it was exposed, as well as within the Tranchée Moulin.

6.2. From The Chest: Dealing with breccia

Figure 1.11 shows a big unexcavated area in the middle of the site. It is true that the figure does not show every square that we excavated – most of the Tranchée Moulin is not labelled, for instance. However, it is also true that most of that central area was not excavated, because it consists of a thick mass of breccia. We fought our way through thick layers of breccia in the pente du remplissage, and in the Tranchée L. They were enough to keep us busy.

Breccia is a rock composed of large angular fragments of a pre-existing rock (in this case, pieces of the rock shelter's ceiling and walls), that have been cemented back together – in this case, by re-precipitated calcium carbonate that infiltrated the deposits while dissolved in water that had dripped down from the ceiling. The breccia at the Bau is not just a little bit stuck together: it is fully cemented. It has no "give". When you hit it with a hammer and chisel, the shock waves you create do not travel more than a few cm before they have been absorbed, so you have to work to break off every small fragment. The site is so rich not only because people used it over and over for a long time, but because the remains they left behind were protected by this breccia. But to get at those remains, you have to deal with the breccia. How?

We started with hammers and chisels. We did not get very far. I am pretty sure I have had a small bone chip in my right hand since 1988 . . . Anyway, it soon became clear that we needed extra help, and by 1990 Serge had arranged for funding to rent the necessary equipment: jackhammers. For several seasons we started each year with a week or two, or three, of jackhammering, usually with a small team of five to seven people, before the main team arrived to excavate. We had an electric jackhammer and a generator to run it, but the real work was done with pneumatic hammers. That required an air compressor, which would be towed down our rough forest track to be placed on the plateau above the site. Two sets of air hoses at least 40 m long hung over the cliff to reach the site and feed two jackhammers at a time. We always had a heavy hammer (25 or 35 kg), and two or three smaller ones, usually 8 kg, but we did have a 2.5 kg hammer sometimes (for "fine" work!). We also had a supply of different bits for the hammers: we generally preferred the pointed picks over the chisel-ended ones.

We would set the air compressor going in the morning on the way to the site, go down and put on our hardhats, goggles and earplugs, and work until it ran out of gas. I suppose we disturbed the peace of nature in the gorge, with the jackhammer noise reverberating back and forth, and the hiss of the air hoses (those things always leak a bit). When the compressor ran out of gas and stopped – oh! The quiet! Our shoulders relaxed, we took our earplugs out, we said "Aaahhhh . . .", and we took a five-minute break while someone ran up to the top to put more gas in. The hoses would come back to life and hiss, and away we go again! All day, every day, maybe six days a week, maybe ten days straight before a day off.

We used buckets and trowels to collect the debris from the jackhammers, one or two people crouched in front of the working bit, scraping rocks into their buckets and passing them back to someone behind, getting an empty one back. The bucket line would pass the buckets along until they reached someone at the top end of a large tube (Figure 1.10 shows the bottom end of such a tube), who would dump them into the tube. The rocks would slide down the tube to end up farther from the site, but would pile up and need to be spread out from time to time. We traded jobs around; everybody did everything, eventually. I remember being at the top end of the tube in the pente du remplissage in 1994, perched on a narrow ledge just wide enough for my two feet, a sheer drop of about three metres to the rocks of the excavated area below: dump bucket, hand back empty with one arm, get full bucket with other arm, dump, hand back and get, dump, hand back and get, dump. . . If anything interrupted that flow, if no-one passed me anything for 20 seconds or so, I would fall asleep. And wake up with a start, to receive – thankfully! – another bucket. It was hard work, but it was worth it, and to this day, if I hear a jackhammer I lift my head like a hunting dog scenting prey. A jackhammer! I wonder if I can convince whoever has it to let me use it?

We were able to use this technique in the pente du remplissage because we could work from the top near the edge, and work our way back. It is much easier to break breccia that is open on at least one side. It did get harder and harder to work as we got into the rear corner of our desired area. We were able to use jackhammers in the Tranchée L, too, because we got lucky. Little did we know when we decided to excavate band L that the breccia in the upper part of the deposit formed in two main domes, one centred on band H and the other farther east, probably over band P. Band L runs down between those two domes, and although it has plenty of breccia, it is thinner there than it is on either side.

We were also lucky because the breccia we excavated was usually sterile: it contained little or no archaeological material in most cases. There are exceptions: part of layer 4 was brecciated, as we will see in a later chapter, and in the pente, band D was consistently the most brecciated, even in the archaeologically-rich layers. But for the most part, we could feel free to have at her with the jackhammers. When we did find some material, quickly coordinate and collect, and we are off again. That was how layer G was dug.

Digging breccia by hand is actually less effective than using jackhammers. It is slower, but that would be fine if the results were better. They are not. Digging cemented flint with a hammer and chisel is all right, and can be fun. You learn to "feel" the breccia, so that when a piece of flint has been uncovered but is still stuck in there on one side, you know to hit some spot a few centimetres from the piece, and the flint will pop out of its cement. You can do that with a jackhammer, too. Digging bones, though? Not fun. What are the softest, most fragile things in this mass of breccia? The bones. If anything in here is going to break, what will it be? The bones. Even if you do get fragments of them out, chances are you have lost any identifying features, and good luck looking for cut marks!

The outer layers of the bones stay stuck to the breccia, and you get fragments of the interior. We tried to avoid destroying anything, of course, but at times we decided it was better to sacrifice the few in order to get to the many. That is the problem with breccia.

6.3. The sediments, and evolution of the site in G and F

Layer G never provided much archaeological material, but, ironically, it was the first layer to provide *any* archaeological material below the upper part of the site, and the reason it did was because it was so highly cemented. In my personal notes from August 29th 1987, towards the end of our first excavation season, I note that "en cherchant des planchers stalagmitiques susceptibles à être datés, j'ai remarqué des silex collés en dessous d'un plancher épais, signalant donc un niveau archéologique bien plus ancien que ceux de la tranchée Moulin (dont F est le plus ancien), puisque localisé sous une couche de cailloutis et un plancher stalagmitique épais. Une couche aussi ancienne n'avait jamais était signalée auparavant à ce site." I was looking for a stalagmite which could be dated to give the base, oldest, age of the site; at the time, we thought the upper platform was the entire site. The flint pieces that I found were in zone A18, in what we later named layer G. Figure 6.1 is a photograph from 1988, before we started digging the pente du remplissage, showing what might be the same stalagmitic layer near the rock wall. Figure 6.2 shows the broader area in 1993, and Figure 6.3 shows the pente as it looks now, with a wooden structure protecting the deposits where we excavated the richest area of layer H1. For orientation, notice a large bloc of breccia propped up against the wall in all three pictures (under the letter B in Figure 6.2, where it is harder to see).

My search for a datable stalagmitic layer did pay off: Blackwell et al. (1991) report that they obtained a $^{230}Th/^{234}U$ date from a stalagmite in G of 141 $^{+10}/_{-9}$ ka, which would put it at the end of the Riss III, in MIS 6. This is younger than the TL dates obtained for layer H, but older than the biochronostratigraphic age of MIS 5e that Fernandez (2001) suggests for H1. That makes the exact timing of these deposits uncertain, but somewhere in the transition from the end of the Middle Pleistocene to the beginning of the Late Pleistocene. The dating of layer F is even more uncertain, but stratigraphically we know it is younger, and its sediments imply that it formed during a glacial period (Würm I?).

Figure 6.2 clearly shows how variable the deposits are. We see large blocks, some of which are ceiling fragments and others of which are stalagmitic floors. We also see interweaving layers of finer material, creating an uneven appearance. The vertical white marks are from the jackhammer bit. In Figure 6.3, taken twenty years later, we see that the walls of breccia that we carved out are still standing. The variation in grain size is still visible, as is the variation in the amount of interstitial matrix. It should also be evident that the deposits nearest the rock

Figure 6.1. A stalagmitic layer and the top part of layer G in 1988. ***Un plancher stalagmitique et la partie supérieure de la couche G en 1988.***

Figure 6.2. The deposits of layer G (centre of photo), with the Tranchée Moulin on top, during excavation in 1993. Lower layers are covered in a tarp for protection. ***Les dépôts de la couche G (au centre de la photo), avec la Tranchée Moulin en haut, pendant les fouilles de 1993. Les couches inférieures sont couvertes de bâches pour leur protection.***

wall tend to not be cemented: there is a tendency for them to slump onto the roof of the wooden structure. This was a pattern that we found many times. For instance, layer I1 was a fine-grained, unconsolidated layer that was found only near the rock wall.

The variability in these deposits suggests that they formed over a long period of time, under a variety of climatic conditions: cryoclastic debris means that it was cold enough (and wet enough) for frost to form in the roof and walls and break chunks off, while stalagmitic deposits and cement in the breccia mean that later it was warm enough to have water flowing through the limestone, dissolving calcium carbonate and redepositing it in the sediments. There are also thin clay-rich layers. The clay may have come from dissolution of the limestone, leaving behind only the small insoluble fraction of the rock. Fine grains may also have been blown into the site by winds, as loess. In the first scenario, it must have been warm enough to have water flowing through the deposits for a long time. In the second, it must have been cold enough to allow wind to pick up dust from bare (unvegetated) ground outside the cave, and blow it in. I did a quick microscope study of a sample collected at z = -550 cm in August 1997, and compared it to the insolubles extracted from the site's

Figure 6.3. Recent photograph of the lower part of the site. Layer G consists of the deposits above the wooden structure, and the Tranchée Moulin is above that, along the wall of the rock shelter. ***Photographie récente de la partie inférieure du site. La couche G se trouve en haut de la structure en bois, et la Tranchée Moulin plus haut, le long de la paroi rocheuse.***

limestone. They both contained clays and angular to sub-angular quartz grains, mainly silt-sized but up to medium sand-sized. The sample from the clay layer in G also contained some rounded quartz grains of fine sand size. As a provisional conclusion, based on only one sample, it is likely that the fine-grained layers in G mainly derive from dissolution of the rock shelter's limestone, but that there might also be a small addition of windblown material, i.e., loess.

In 1988, soon after we started digging in the pente du remplissage, I recorded the sediment types exposed along the line B/C. Table 6.1 gives a summary of the deposits as seen at that time. The thicknesses are often given as ranges, because the layers varied laterally or from front to back of the site, sometimes pinching out entirely.

By 1995 we had dug into the deposits far enough that my personal stratigraphic notes from that year mention that at the section wall forming the north side of band 18, layer G starts (at its contact with H) as a series of thin layers of fine-grained loose sediments separating thicker layers of breccia, or plaquettes, or large rocks. The thicker layers may represent colder time periods, separated by the thinner layers formed during warmer periods. The layer or ensemble above that follows a similar pattern except for a larger proportion of large or very large rocks and slabs within it. This layer can be cemented, or washed clear of interstitial sediments, and above that is another thin, fine-grained layer.

This set of breccia layers carried on upwards until we get into layer F. The archaeological layer F consisted of unburned sediments containing cryoclastic rock fragments and finer interstitial material, which could be loose, consolidated, or cemented, depending on the location. My notes from 1987, zone C9, describe it as "couche à cailloux calcaires de l'abri, de 1 ou 2 jusqu'à 20 ou 30 cm de long, normalement 5 à 8 cm. Matrice marron-jaunâtre assez fine (sables fins ou très fins). Cette couche fait au moins 40 cm d'épaisseur."

In 1993, while digging down towards H1 in the squares D14 through D17, layer F was noted as forming an approximately horizontal grey layer, about 25 cm thick, in the eastern section wall (Lebel, 1993). It contained "quelques lits noirs lenticulaires de 2 à 3 centimètres d'épaisseur" (Lebel, 1993), as it had also done in the Tranchée Moulin, and it pinched out towards the north and the west.

6.4. Pollen and Fauna

J. Argant (Argant, 2000) took a series of samples for pollen analysis down the section wall in C15 in 1999. Eight of these fall within the levels of layers F and G, and of these, two were analysed (number 2 in layer F, and number 6 in layer G). Unfortunately, neither of these contained enough pollen for statistical analysis, which as Argant (2000) says, restricts the degree of interpretation possible. Table 6.2 gives the numbers of pollen grains of various taxa, extracted from her Tableau 4. Sample #6 contained mainly pine pollen (both fresh and old, worn pollen) and monolete spores, as well as a very few grains of fir, juniper, walnut, and non-arboreal pollen including grasses. The presence of abundant old pine pollen is further evidence of water passing through the deposits of layer G. Sample #2 had even fewer pollen, again including pine and spores, but it did contain a greater proportion of other arboreal pollen (fir, oak, hazel, linden, juniper, buckthorn and cedar) and a greater proportion and diversity of non-arboreal pollen.

Table 6.1. Summary of stratigraphic notes (1988) concerning the part of the slope of the deposits relevant to layers G and F, at the beginning of our excavations in this part of the site. "Petit" means 1-3 cm, "moyen" = 3-7 cm, and "gros" = 7-10 cm. "Blocs" are larger than 10 cm. ***Notes stratigraphiques sommaires (1988) concernant la partie de la pente du remplissage au niveau des couches G et F, au début de nos travaux dans cette partie du site. "Petit" = 1-3 cm, "moyen" = 3-7 cm, et "gros" = 7-10 cm. Les "blocs" sont plus grands que 10 cm***

Thickness in cm	Description
5 à 6	brèche à cailloux petits à moyens, compacte et bien consolidée
12 à 50	brèche poreuse bien cimentée à compacte, cailloux petits à gros, et blocs
20	petit cailloutis compact, induré, à blocs plus ou moins horizontaux intercalés
60-70 (limite F/G à 15 cm sous la couche supérieure)	cailloutis un peu cimenté, poreux, cailloux petits à gros et quelques blocs anguleux, souvent aplatis, orientation à tendance horizontale ou parfois inclinée, voire verticale
11-12	plancher stalagmitique, pendage au Nord
12-13 --> 1	terre caillouteuse non-indurée
3 --> 0	terre à granules et quelques petits cailloux, indurée, presque concrétionnée
3	plancher stalagmitique, pendage au Nord
10-12 --> 30-33	cailloutis non-induré, cailloux petits à moyens, anguleux, aplatis, sans orientation préférée; plus gros et plus horizontaux vers l'avant de l'abri
15	plancher stalagmitique, pendage ~30° au Nord

Table 6.2. Numbers of pollen grains and spores in samples from layers F and G, after Argant (2000), Tableau 4. ***Nombres de graines de pollen et spores dans les échantillons des couches F et G, extrait d'Argant (2000), Tableau 4***

Taxon	sample #2 (layer F)	sample #6 (layer G)
Pinus	11	47
Pinus grande taille	-	4
Pinus usés	-	86
Abies	2	2
conif. indet.	-	1
Quercus	9	-
Corylus	4	-
Tilia	3	-
Juniperus	2	1
Juglans	-	1
Frangula	1	-
Cedrus	1	-
POACEAE	3	2
Plantago	-	1
APIACEAE	1	-
LAMIACEAE	3	-
ROSACEAE	-	1
SCROFULARIACEAE	1	-
Rumex	1	-
Filipendula	1	1
indet.	1	-
spores monolètes	8	49
amas	7	3

In her report from 2006 (J. Argant, 2006), after further work, Argant describes six pollen zones within the C15 samples. Layers G and F fall into three of those zones, presented in Table 6.3. The lowest of these zones, BAU-4, includes part of layer H and the lower half of G, and shows an improvement in climate, with increasing temperature. BAU-5 is based on one sample from layer G, but seems to show that climate got colder again, and then BAU-6, from the upper part of G and in F, shows a temperate climate, warmer and wetter, but with the heliophile taxa remaining preponderant, so the landscape was still open. She says that this is still far from the climatic optimum (J. Argant, 2006).

Information on the faunal content of layers F and G is sparse. Lebel (1987) reports having found one cervical vertebra of a horse in layer F, along with numerous teeth; it is not clear whether the teeth could all be from the same animal, or are even of the same species. Fernandez (2001) lists one tooth of a horse (*Equus sp.*) from layer F, and reports that the animal was approximately 2 years old. He does not list any skeletal material from any species in either layer F or layer G, but does report that both layers contain at least one tooth of ibex (he does not specify how many), and one tooth each of an adult roe deer (in layer G), and red deer (in layer F). It is likely that more faunal material than that was found, but much of it was probably unidentifiable. I have no other precise information. Lebel (1987), for instance, lists how many lithics, bones and teeth were found that season by square, but does not differentiate them into layers. This sparse information gives the Minimum Number of Individuals (MNI) figures shown in Table 6.4. About all we can say from this is that hominins obtained animals and brought them to the site, and that the landscape probably had some trees (to suit the deer, although they are flexible about their habitat). That corroborates the pollen information.

6.5. Lithics

In keeping with the scarcity of information about every other aspect of these two layers, the lithic assemblages are small, and there are major differences between the

Table 6.3. Pollen zones including layers F and G, after J. Argant (2006). ***Zones polliniques concernant les couches F et G, d'après J. Argant (2006)***

zone pollinique	z et couches	caractéristiques	interprétation
BAU-6	430-510cm F et Gsup.	AP jusqu'à 80%, feuillus autour de 40%, herbacées steppiques disparaissent presque, semblable à BAU-4.	conditions climatiques tempérées, pas encore optimum car *Pinus* reste abondant, vers le haut chêne diminue et herbacées augmentent.
BAU-5	530 cm dans couche G	nette régression des feuillus au profit du pin, spores de fougères dominent, faible nombre de taxons.	péjoration climatique ou problème de conservation? Difficile à dire avec un seul échantillon.
BAU-4	... moitié inf. de G; couche H	... progression de feuillus jusqu'à 40%: chêne surtout, noisetier, buis puis frêne, tilleul, châtaignier, noyer, filaire, platane. Taxons steppiques régressent considérablement.	... [amélioration] climatique se confirme et se précise, augmentation considérable de température.

typological lists and the raw material lists. Remember that these two lists were established at two different times. The raw materials were recorded on a yearly basis as we excavated, from 1987 to 1999, and although I tried to record as many as possible, I did miss some. That list does not include any pieces excavated in 2000 or 2006. The typological list was developed in 2013 and 2014 as some students and I went through every piece in storage at the government repository in Vaison-la-Romaine. I then spent several months in 2015 and 2016 trying to match up the two lists. For layers with large assemblages, the fact that a small proportion does not match is not statistically important. For layers with small assemblages, several different causes of difference can coalesce to make a major impact.

Starting with layer G, the raw materials list includes 224 pieces, but the typology list only has 41 pieces, 39 of which are on the raw materials list. In fact, one of those is only on the raw materials list, but when I first saw it I noted that it was a "mauvaise qualité-rognon/nucleus", so I am including it with the cores here. The only explanation I have for the lack of typological pieces is that the other pieces either were not returned to Vaison-la-Romaine after being studied in Canada, or they were separated from the rest of the assemblage in Vaison, and we did not see them. Another issue is that in Wilson & Browne (2014), the GLM for layer G was calculated on the basis of a sample size of 240 identifiable pieces, more than are now included in that layer. The extra pieces are from square B16, and were recorded during excavation as being from layer "G?". Since that was the only information we had when working on the GLMs, we included them as G. However, they had subsequently been re-defined as coming from layer H1, so they appear on the typological list with that layer, and were included with the assemblage from H1 in chapter 5 (from the top tranche, as might be expected given the initial uncertainty of their attribution).

Table 6.4. Minimum Number of Individual animals, based on faunal remains in layers G and F, after Lebel (1987), Fernandez (2001, 2006), Fernandez in Lebel (2006) and A. Argant (2006). ***Nombre Minimal d'Individus dans la faune des couches G et F, d'après Lebel (1987), Fernandez (2001, 2006), Fernandez in Lebel (2006), et A. Argant (2006)***

Species	Common name	layer G	layer F
Equus sp.	horse	n/a	1
Capra cf. *caucasica*	ibex	1	1
Capreolus capreolus	roe deer	1	n/a
Cervus elaphus	red deer	n/a	1

Layer F is even worse: the raw materials list has 299 pieces, while we were able to record the typology for 59 pieces, but only 10 pieces have both kinds of information. One additional typologically identified piece (a sidescraper) matches a number on the raw materials list, but that number corresponds to two petrographically identified pieces (one from Murs, one from Sault), and I do not know which one this piece is. This is all complicated by the fact that much of layer F was excavated in 1987, and in that year, for some foolish reason, I did not record the catalogue numbers of the pieces, just their zones and layers. That is not a problem when it comes to analysing the proportions of raw materials within a layer, but it makes it practically impossible to match up the typological list with the raw material list. Also, we just did not see as many pieces in 2013 and 2014 from layer F as I had seen years earlier. This problem will reappear in the next chapter, too, concerning the other layers in the Tranchée Moulin. Either the material is missing, or it is in storage at Vaison but not with the rest of the Bau material. Furthermore, 270 of the 299 pieces in F were unidentifiable as to their source, mainly because they were highly patinated, although despite the layer's unburned appearance, that also includes 38 burned pieces. Nine of the 10 pieces on both lists are patinated, and one is burned. That leaves only 29 pieces with identified sources (but no typological identification), which made it impossible to use this assemblage in a GLM, and one sidescraper which comes from either of two diametrically opposed sources.

Table 6.5 shows the breakdown of typologically identified pieces in both layers. Layer G has only a few flakes and one

Table 6.5. Lithic assemblages of layers G and F, summarised by typological category. ***Assemblages lithiques sommaires des couches G et F, par catégorie typologique***

Category	G: Typology list (n)	G: Typology list (%)	F: Typology list (n)	F: Typology list (%)
Liste Bordes	2	4.9	5	8.5
Debris	0	0.0	1	1.7
Eclats	3	7.3	12	20.3
Lames	1	2.4	2	3.4
Tamis/Collectif	33	80.5	37	62.7
Nucléus	2	4.9	2	3.4
Total	**41**	**100**	**59**	**100**

blade, and only two tools: one double concave sidescraper, and another which is both a sidescraper and an atypical couteau à dos, in raw material type Y, from the Near SW (Figure 6.4). That piece shows an interesting pattern of concentric rings, due to a large fossil. Layer G contains a fairly large proportion of sieve material, consisting of small flakes and chips. This is independent of the way that we collected the material, because for the typological list we sorted through the collections, looking for anything that could be further described. The two cores include one of poor quality, and the other is a fragment. It is easy to understand why they would have been abandoned at the Bau. This seems therefore to be an assemblage that results from *in situ* knapping, with most of the products carried away.

Layer F's typological assemblage has a greater proportion of flakes, blades and tools (all sidescrapers, one of which also has a notch), and a lower proportion of sieved material. It may reflect a lower proportion of knapping at the site, and a bit more abandonment of tool-kit components, than layer G, but there is not much to say about such a small assemblage.

Tables 6.6 and 6.7 break down the typology and source area data for layer G, as was done in previous chapters; with F's small assemblage, there is no point breaking it down in this way. We see that G has a relatively small unidentifiable proportion, mainly patinated, but also including 8 pieces (3.6%) in burned raw material types, although we found no traces of hearths in the excavated area. The identifiable assemblage is dominated by materials from Murs and Sault, with Faraud a distant third, and Méthamis barely used.

Figure 6.4. A tool from layer G, number B16 G 2, in raw material type Y, from the Near SW. Note the fossil in the centre of the piece. ***Un outil de la couche G, numéro B16 G 2, réalisé dans la matière première type Y de la région sud-ouest proche, "Near SW". Notez le fossile au centre de la pièce.***

Table 6.8 presents the proportions of raw materials from different source regions (also shown in Figure 6.5), comparing layer G to other layers considered earlier. G is very different from those layers. It continues the increase in material from the NE that we saw in the upper tranches in layer H1, and the decrease in material from the WSW. In fact, it exaggerates them both, with the WSW nearly disappearing. On the other hand, it has a fairly high amount of material from the SW, and a minor amount from the Near SW. These results are probably influenced by a large number of pieces of small size due to reduction of cores. The one nodule/core for which I do have raw material data did come from the SW.

It would be interesting to compare the regional results by weight of accumulated material, rather than number of pieces, but the only weight data is for the much smaller typological assemblage, not the raw materials assemblage. Even a quick glance at the results in Tables 6.6 and 6.7 shows that this would not be representative, because there are drastic differences between these two assemblages: for instance, only 2 of 78 pieces from Sault are on the typological list, and only 5 of 87 from Murs. This stems from the combination of how the material presented itself at the dig, and the absence of the complete assemblage when we did the typological study. As we dug, we occasionally found pockets of lithics in small areas within the breccia. We treated these as collectives, so all of them were taken together and assigned a single number. For example, C18 G 1 consisted of 67 pieces: 12 patinated, but 10 from Murs (type Q) and 45 from Sault (type R). C18 G 4 included 42 pieces: 2 patinated, 12 type Q and 28 type R. C18 G 7 had 56 pieces: 55 of type P, also from Murs, and 1 type AD, of unknown source. This lends credence to the idea that we were finding the remains of flint-knapping episodes. It also means that the absence of just those three collectives from the typological assemblage led to the loss of 165 pieces: more than half the total known.

There is not much to say about the sources of raw material in layer F's assemblage, with only 29 of 299 pieces identifiable, so it is not included in Table 6.8. However, those pieces include 10 from the Near SW (34.5% of identifiable pieces), 5 from the WSW (17.2%), 5 from the NE (17.2%), and 9 from the SW (31.0%): much more like layer H than layer G.

The GLM that Wilson & Browne (2014) presented for layer G is given below as Table 6.9. The small assemblage size was not ideal for this type of analysis, and one factor (Small Rocks) had to be removed. The combination of the other factors did give a useful result, however, even though when judged by a cut-off of $p = 0.05$, the only factor that showed a significant result was AOSISA (Area Of the Sources In the Source Area), which influences how easy it is to find raw material. Table 6.10 gives the Wald Chi-Square values converted to percentages. Layer G continued the trend toward greater importance of Terrain variables, as we go up through the layers in the lower part of the site. In fact, layer G has the most extreme set of

Table 6.6. Number of pieces by typology and source area in layer G. ***Nombre de pièces par catégories typologiques et groupes de sources des matières premières dans la couche G***

G n	Liste Bordes	Debris	Eclats	Lames	TamisColl	Nucléus	non-id	Total
Unidentifiable	0	0	1	0	14	0	17	32
Unknown	0	0	0	0	1	0	1	2
Faraud etc.	1	0	2	0	10	0	5	18
Méthamis etc.	1	0	0	0	0	0	0	1
Les Sautarels etc.	0	0	0	0	0	0	0	0
Sault	0	0	0	0	2	0	76	78
Murs	0	0	0	0	4	1	82	87
Local	0	0	0	0	0	0	0	0
St. Jean de Sault	0	0	0	0	0	0	0	0
Roussillon	0	0	0	0	0	0	0	0
Murs-Bezaure Olig.	0	0	0	0	2	0	2	4
Ravin de la Treille	0	0	0	0	0	0	0	0
St. Trinit	0	0	0	0	0	0	0	0
Durance	0	0	0	0	0	0	0	0
Tertiary Calavon	0	0	0	0	0	0	0	0
Tertiary Murs VdeV	0	0	0	0	0	0	0	0
Mormoiron	0	0	0	0	0	0	0	0
R de Guérin	0	0	0	0	0	0	0	0
N.Aurel	0	0	0	0	0	0	0	0
not identified	0	0	0	1	0	1	0	2
Total	**2**	**0**	**3**	**1**	**33**	**2**	**183**	**224**

values in that direction, of all the layers we analysed: Raw Material variables total 28.2%, and Terrain variables total 62.9%. This certainly implies that the main consideration was ease of acquisition of raw material, not its size or quality.

6.6. Conclusion

During the time span of formation of layer G, while at least 125 cm of sedimentary deposits were accumulating, it seems that hominins used the site only sporadically and for very short periods, in contrast to the repeated, intense occupations of the lower layers. Climate changed from relatively temperate to colder, and back to warmer and wetter, possibly fluctuating many times. These deposits fit somewhere into the transition from Middle to Late Pleistocene, possibly including parts of the Riss III (MIS 6), the subsequent interglacial (MIS 5e), and perhaps the beginning of the next glacial (Würm I, MIS 5d). Judging from the sparse archaeological remains, very small groups of hominins came through on occasion, bringing with them a meal (at least part of a deer, or an ibex, although other remains may just not have been preserved) and some flint to knap. We have no signs of them using fire, except that a few flint pieces were burned – but of course it is possible that the fire was somewhere else in the site. We have very little knowledge of their toolkit, probably because most of it was brought in then carried away again when they left. It seems that the Bau was used as a bivouac, a campsite for a night, and only rarely used even for that.

By the time of formation of layer F, a more temperate climate was established, although it was still fairly cool and with only patchy woods. The Bau was once again used more often, or for somewhat longer periods; we still have very little evidence of fire, but we do find more of the lithic *chaîne opératoire* present, and the archaeological material is concentrated into an identifiable layer, rather than scattered through separate pockets in the breccia. There is evidence of use of ibex, red deer and horse, but perhaps only one animal of each type. This was still just an occasional campsite, but that was going to change.

Table 6.7. Percentages of pieces by typology and source area in layer G. ***Pourcentages des pièces par catégories typologiques et groupes de sources des matières premières dans la couche G***

G %	Liste Bordes	Debris	Eclats	Lames	TamisColl	Nucléus	non-id	Total
Unidentifiable	0	0	33.3	0	42.4	0	9.3	14.3
Unknown	0	0	0	0	3.0	0	0.5	0.9
Faraud etc.	50.0	0	66.7	0	30.3	0	2.7	8.0
Méthamis etc.	50.0	0	0	0	0	0	0	0.4
Les Sautarels etc.	0	0	0	0	0	0	0	0
Sault	0	0	0	0	6.1	0	41.5	34.8
Murs	0	0	0	0	12.1	50.0	44.8	38.8
Local	0	0	0	0	0	0	0	0
St. Jean de Sault	0	0	0	0	0	0	0	0
Roussillon	0	0	0	0	0	0	0	0
Murs-Bezaure Olig.	0	0	0	0	6.1	0	1.1	1.8
Ravin de la Treille	0	0	0	0	0	0	0	0
St. Trinit	0	0	0	0	0	0	0	0
Durance	0	0	0	0	0	0	0	0
Tertiary Calavon	0	0	0	0	0	0	0	0
Tertiary Murs VdeV	0	0	0	0	0	0	0	0
Mormoiron	0	0	0	0	0	0	0	0
R de Guérin	0	0	0	0	0	0	0	0
N.Aurel	0	0	0	0	0	0	0	0
not identified	0	0	0	100	0	50.0	0	0.9
Total	**100**	**0**	**100**	**100**	**100**	**100**	**100**	**100**

Table 6.8. Percentages of identified pieces (excluding unidentifiable and unknown source pieces) from each geographic region for layers K2, K1-J, I, H and G. ***Pourcentages des pièces identifiées (excluant les pièces indéterminables ou de sources inconnues) de chaque région géographique pour les couches K2, K1-J, I, H et G***

Geographic area	K2 %	K1-J %	I-I4 %	H-H2 %	G n	G %
Local	0.2	3.4	1.3	0.7	0	0
NE	19.4	20.4	12.3	15.7	78	41.5
E	0	0.3	0.1	0	0	0
Far SE	0	0	0	<0.1	0	0
Far S	0	0	0.1	0	0	0
SSW	0	0	0.1	0	0	0
SW	44.3	54.1	35.0	30.0	91	48.4
Near SW	23.0	15.0	29.1	34.8	18	9.6
WSW	12.4	6.6	21.9	18.6	1	0.5
NW	0.7	0.1	0.1	0.2	0	0
Total	**100**	**100**	**100**	**100**	**188**	**100**

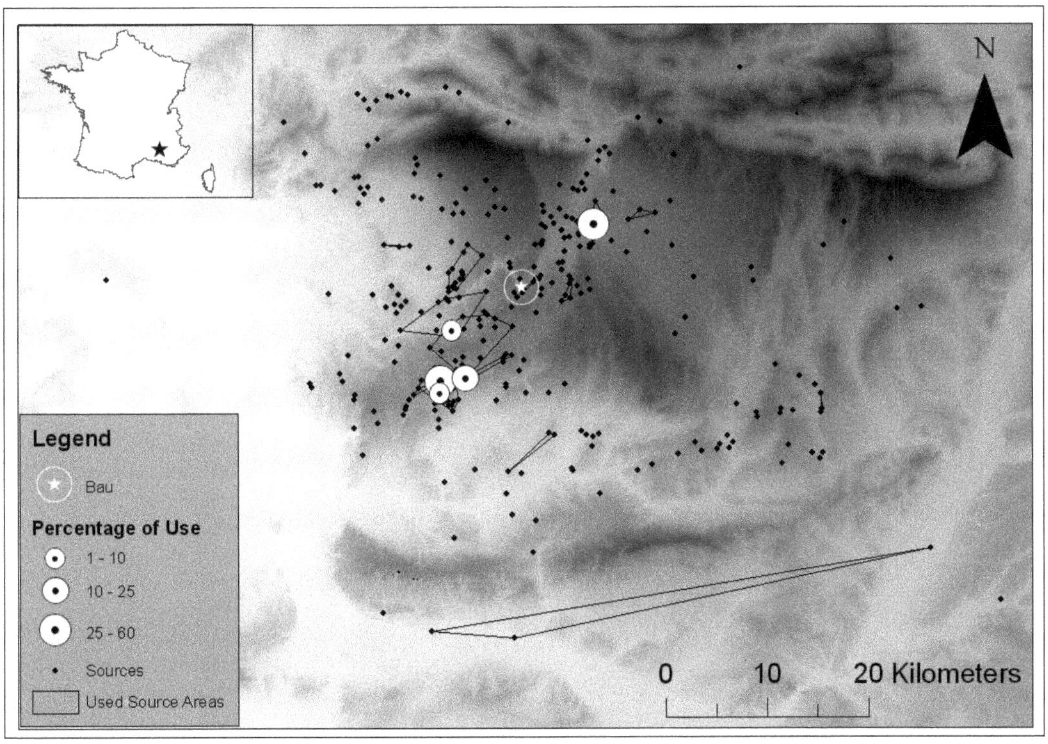

Figure 6.5. Map of the raw material source areas used for layer G. ***Carte des aires de sources des matières premières utilisées pour la couche G.***

Table 6.9. Generalised linear model for raw material source choices in G. P-values in bold font are statistically significant. Sample: 240 pieces from 5 source areas

Omnibus test: likelihood ratio chi-square = 379.152, df = 9, p = <0.001

Variable	Beta	S.E.	Wald Chi-Square	P-value	AIC
Quality	5.514	4.7487	1.348	0.246	76.589
Extent	8.484	6.6590	1.623	0.203	
Difficulty (Cal/km)	-0.189	0.1227	2.371	0.124	
Medium rocks	-5.363	5.3435	1.007	0.316	
Large rocks	19.928	15.6939	1.612	0.204	
Very large rocks	4.444	6.2795	0.501	0.479	
Calories	-0.006	0.0046	1.525	0.217	
AOSA	3.9 e-8	2.2 e-7	0.030	0.861	
AOSISA	7.6 e-5	3.6 e-5	4.421	**0.035**	
(Intercept)	-71.491	60.2640	1.407	0.236	

* Had to remove Small rocks because of Hessian matrix singularity

6.7. Sommaire du chapitre

Ce chapitre fait le lien entre la partie inférieure du site (que nous appelons la pente du remplissage), avec la couche G, et la partie supérieure, avec la couche F de la Tranchée Moulin. La couche G se présente en petites concentrations parsemées dans plus d'un mètre de brèche cryoclastique calcaire. En haut de cette brèche, le niveau sur lequel on marche est la couche F, qui correspond en quelque sorte au plancher des dépôts supérieurs et de la Tranchée Moulin. Cette dernière n'était pas très bien conservée, mais était représentée par des concentrations de matériel archéologique localisées en avant de la Tranchée Moulin (en haut de la couche G), ainsi qu'à plusieurs endroits à l'intérieur de la tranchée.

La couche G s'étend du début de la pente, à environ 480 cm sous zØ, jusqu'à la limite arbitraire de -605 cm. La

Table 6.10. Wald Chi-Square values as percentages, G assemblage

Variable	Wald Chi-Square	Wald Chi-Square as %
Quality	1.348	8.5
Extent	1.623	10.2
Difficulty (Cal/km)	2.371	15.0
Medium rocks	1.007	6.3
Large rocks	1.612	10.2
Very large rocks	0.501	3.2
Calories	1.525	9.6
AOSA	0.030	0.2
AOSISA	4.421	27.9
(Intercept)	1.407	8.9
Total	15.845	100

couche F est située en haut de la pente, entre environ 400 ou 405 cm sous zØ et environ -480 cm. Elle est définie comme étant une couche constituée de vestiges archéologiques peu ou pas brûlés mélangés avec des plaquettes cryoclastiques et des sédiments à grain fin: elle n'est pas définie arbitrairement par ses valeurs de z. Dans toute la partie supérieure du site nous avons identifié les couches par leurs caractéristiques sédimentologiques et archéologiques, et non pas en fonction des valeurs de z, à la différence des couches de la pente du remplissage.

La caractéristique marquante de la couche G est sa nature bréchique. Très dense, très dure, elle nous a obligé à travailler très dur. Mais elle a également protégé les vestiges, ainsi qu'un plancher stalagmitique qui a été daté d'environ 141 000 ans. L'âge de la couche F est incertain, mais les sédiments laissent supposer qu'elle correspond à une période toujours glaciaire mais plus tempérée que pendant la formation de la couche G.

L'étude palynologique (voir Tableaux 6.2 et 6.3) montre que le climat a dû varier pendant la période d'accumulation des couches G et F, sans jamais atteindre un optimum climatique. On n'a que très peu d'informations sur la faune, mais les restes d'au moins un chevreuil dans la couche G, du bouquetin dans les couches G et F, ainsi qu'un cheval et un cerf dans la couche F étaient présents. Quant à l'assemblage lithique, les informations sont encore plus réduites. Lors des fouilles de 1987, pour je ne sais pas quelle raison stupide, dans ma liste des matières premières j'ai seulement noté les couches et zones mais pas les numéros de catalogue des pièces lithiques. C'est le numéro de pièce qui sert à l'identifier uniquement, alors en 2013-14 quand j'ai étudié la typologie des pièces je n'ai pas pu rejoindre l'information typologique aux données pétrographiques. De plus, le matériel entreposé à Vaison-la-Romaine semblait manquer bon nombre de pièces de la Tranchée Moulin, ainsi que de la couche G. De fait, malgré les quelques centaines de pièces des couches G et F figurant sur la liste pétrographique, la liste typologique ne comprend que quelques dizaines de pièces provenant de ces deux couches. Ceci limite sérieusement ce que je peux dire sur leur typologie, et encore plus les liens entre typologie et pétrographie. Pire encore, pour la couche F il n'y a de toute façon presque rien à dire sur la pétrographie des pièces puisque la grande majorité d'entre elles sont si fortement patinées (ou parfois brûlées, même si la couche elle-même ne paraît pas carbonisée) que je n'ai pas pu identifier les origines de leur matières premières. Heureusement, on peut tirer quelques conclusions utiles pour la couche G.

Typologiquement, ce que nous avons dans la couche G correspond à la taille sur place (beaucoup de débitage), dont les produits recherchés ont été sortis du site. On peut facilement imaginer que les deux nucléus y trouvés ont été abandonnés délibérément, puisque l'un est de mauvaise qualité et l'autre n'est qu'un fragment de nucléus cassé. Les matières premières viennent presque aussi souvent de Sault (au NE du site) que de Murs (au SW). Les petites concentrations que l'on trouve dans cette couche comprennent souvent quelques dizaines de pièces, probablement liées à des épisodes différés de taille et d'occupation. Souvent ces concentrations comprennent une majorité de pièces provenant soit de Sault ou bien de Murs, bien qu'il existe également des mélanges. On peut imaginer que les préhistoriques étaient de passage, peut-être en petits groupes, pour une nuit ou deux, et que parfois ils venaient de la direction de Sault, parfois de Murs. Selon nos études concernant le choix des matières premières, dans la couche G les hominidés utilisaient surtout des gisements faciles d'accès, privilégiant ces derniers plus que ceux avec de grands rognons de bonne qualité.

La couche G correspond probablement à une longue période qui inclut la transition du Pléistocène moyen jusqu'au Pléistocène supérieur. Les conditions climatiques glacaires la plupart du temps rigoureuses, ont cependant variées avec des épisodes plus ou moins froid. Les hominidés n'ont utilisé le site que sporadiquement en bivouac. Ils ne nous ont pas laissé beaucoup d'outils mais des rejets et débitage produits par leurs activités de taille, ainsi que quelques restes de leur nourriture (les os d'animaux). S'ils utilisaient le feu, nous n'avons que quelques silex brûlés qui en témoignent.

Pendant l'accumulation de la couche F, en revanche, le climat était plus tempéré mais toujours froid, et si le site est toujours interprété comme un campement de courte durée, il a probablement été occupé pendant des périodes plus longues. En effet, nous avons la chaîne opératoire lithique quasi-complète, et le matériel archéologique est concentré cette fois-ci en une couche identifiable, au lieu de poches parsemées dans les sediments. Dans les chapitres à venir, nous verrons que le Bau redevient un lieu d'occupation important pour les préhistoriques.

7

The Tranchée Moulin

7.1. Introduction

The Tranchée Moulin is the part of the site that was originally dug by Franki Moulin (Moulin 1904), starting in 1901. As was explained in chapter 1, it was further studied (as were the debris piles from Moulin's excavation) by a number of people in the 1950s and 1960s, culminating in Henry de Lumley's brief visit in the mid-1960s, which resulted in publication in his compendium of Palaeolithic sites in southern France (de Lumley 1971).

The material amassed by the early investigators was dispersed among a variety of different collectors and museums, in Sault, Carpentras, Avignon and farther afield. An important quantity of it was collected by M. Louis Gauthier of Sainte Cecile-les-Vignes, and is now housed in the Musée Louis Gauthier in that town. I was able to examine it in 2014, thanks to the willingness and cooperation of M. Vincent Faure and Mme. Solenne Ricou-Fétisson. Also in 2014, thanks to M. Thierry Tabourdeau and the Groupe archéologique de Carpentras, I was able to examine some collections from a now-defunct museum in Carpentras. In both cases the material is necessarily without context, other than the fact that it must have come from the Tranchée Moulin, but it is typical of the Bau and can serve as a good example of what one finds there: plenty of Levallois flakes, blades and points, along with sidescrapers and Mousterian points, with a notable proportion of pieces whose bulbs have been removed. The raw material is (to me) very familiar and in the usual condition: often patinated and/or burned. Seeing these collections gives me a feeling which is a cross between the strange pleasure of meeting old friends for the first time, and a pang of regret for these orphaned lithics, who have lost their assemblages. Figure 7.1 gives an example of one of the display panels from the old museum in Carpentras, showing a variety of Levallois points.

Since the early work was described in chapter 1, this chapter will mainly concentrate on the work done by the excavations directed by Serge Lebel, starting in 1987. That work was greatly influenced by the condition of the site when we arrived. The existence of the Tranchée Moulin meant that decades'-worth of passers-by knew where to look for archaeological material. The Bau is not very easy to get to, but people do hike, and many people are interested in archaeology, so a lot of people must have come by and visited, and many of them must have scraped away at the deposits for a few minutes, to see if they could find anything. The bulk of the deposits exposed in the Tranchée Moulin consists of very strongly cemented breccia, with layer F (see chapter 6) as its floor, and two other layers (E and D) appearing towards the bottom of the breccia mass. Layer C is incorporated within the breccia, higher up in the mass. The breccia would have deterred the idly curious, but layers E and D are black and soft: easy for fingers or simple tools to dig in. By 1987, when we started excavating there, the breccia wall of the Tranchée Moulin had deep linear indentations within it, due to people scraping away at the soft layers until they were out of arm's reach, leaving only the breccia behind.

We planned at first to straighten out the wall of the trench, but the breccia was too hard for us in our first year, so we started digging in band L instead, as will be explained in chapter 8. By the time we had the equipment to deal with the breccia, we had made major discoveries in two other sectors of the site, and the work necessary to "regularise" the Tranchée Moulin seemed like too large an investment of time for too little return. We therefore dug in only a few squares along the trench, where there was some prospect of remnant archaeological layers. We did devote considerable energy to excavating a few square metres at the back of the Tranchée Moulin, at first in order to understand the stratigraphy of the upper deposits, and later in a series of attempts to see what connections (if any) we could make

Figure 7.1. A display panel from a defunct museum in Carpentras, showing Levallois points from the "Période Paléolithique, Epoque de Moustier". The labels at the top of the panel read (on the left): "Monieux, Vaucluse", and (on the right): "La Nesque, Baume de l'Aubesier". The central label at the top identifies the tools as coming from the "Collection Mistral, à Sault (Vaucluse)". ***Un panneau d'un ancien musée à Carpentras, avec des pointes levallois de la "Période Paléolithique, Epoque de Moustier". Les étiquettes en haut du panneau (à gauche): "Monieux, Vaucluse", et (à droite): "La Nesque, Baume de l'Aubesier". L'étiquette centrale en haut sert à identifier la source des pièces: "Collection Mistral, à Sault (Vaucluse)".***

between the layers in the Tranchée Moulin and those of the tranchée L, 4 or 5 metres to the east.

The rather spotty nature of our excavations in the Tranchée Moulin explains why the trench does not really appear on the map of the most important excavated areas in the site (see Figure 1.11). However, it extends from band 5 to about band 14, and its western wall is the rock shelter's rock wall. Since it curves, it takes in bands from A or B towards the front, to G in the back. The width of the trench varies from about one to three metres, with an extra "nook" carved out in the back by our efforts in F/G 5 and 6. Figure 7.2 is a recent picture of the trench, after several years with no dig. It is messy, but the indented nature of the breccia wall is clear. The view is towards the back of the site, and the area we excavated with jackhammers, in F and G, is tucked around the corner in the back, to the left of the number 32 painted on the wall.

We have always suspected that Moulin chose to dig along the rock wall because the sediments were unbrecciated there, at least in spots. We ourselves found, for instance, that the sediments were looser next to the wall in several areas, such as layer I1 in squares B19-B21, and in layer G in A14. However, we did also find remnants of breccia stuck to the rock wall here and there in the trench, some containing a lithic or a bone or two. An example is visible in Figure 7.2, next to the number 28 on the rock shelter wall. Some of those remnants matched up across the trench with the layers on the other side; others were higher in the stratigraphy. Moulin (1904) divided the deposits into at least four layers, and implies that the archaeological material came from only one of them ("la couche industrielle" [Moulin, 1904, p. 14]), while de Lumley (1971) provides a stratigraphic sketch including several layers, some labelled from A to D, others unlabelled. We could not be sure of which layers he meant, so we labelled the layers that were obvious in the lower part of the breccia as D, E and F, leaving room earlier in the alphabet for other layers higher up. This means that our layer names do not (necessarily) match any previous layer names, but the earlier names do not seem to have been used much, in any case. This also means that we were able to create a layer C when we found a lot of material higher up in the breccia. A couple of aurochs teeth may have also been attributed to a layer B: they appear in Fernandez (2001) as coming from layer B3. However, nothing else was reported from layer B, and it is possible that when they were labelled the layer name was inverted, and should have been 3B, which exists in the tranchée L. Since Fernandez (2001) does not mention what square they came from, I have to exclude them from further consideration. Thus, this discussion of the Tranchée Moulin will concentrate on the stratigraphy and material relevant to our layers C, D and E. Layer F was dealt with in chapter 6.

The layers were defined based on archaeological concentrations, not by arbitrary z values. Roughly speaking, though, layer E was found between z values of approximately 355 and 365 cm below zØ, D anywhere between about 330 and 350 cm below zØ, and C was more spread out and could be found between about 170 and as much as 330 cm below zØ, but was usually higher. There was always some distance between each layer, but how much separation there was, and the actual heights of the layers, varied along the trench. Note that layer C is being included in this chapter because it is in the Tranchée Moulin. It is actually higher than at least a couple of the layers in the tranchée L, which will be dealt with later, so in stratigraphic terms it is out of sequence here, but it makes more sense to have it in this chapter.

We have no direct chronological indicators for these layers. Henry de Lumley (1971) attributed the deposits to the Würm II of the Alpine glacial system, which would equate to MIS 4 and 3, around 75 to 35 ka. That would put these

Figure 7.2. The Tranchée Moulin, looking towards the rear of the site. Note the indentations in the breccia, where generations of passers-by scraped away the soft archaeological layers. ***La Tranchée Moulin, vue vers le fond de l'abri. Notez les trous dans la brèche, où pendant plusieurs décennies les promeneurs ont gratté dans les sédiments archéologiques les plus meubles.***

deposits squarely within the time period of existence and then disappearance of the Neandertals. However, the only relevant radiometric dates we have are for the stalagmitic layer in G, which was 141 $^{+10}/_{-9}$ ka (Blackwell et al. 1991), and a date of 141.85 +/- 5.3 ka (Lebel 2006) for a speleothem from square I5, between the Tranchée Moulin and the tranchée L. Both of these dates fall at the end of MIS 6 (Riss III). The speleothem extends underneath layer D in the Tranchée Moulin, and is located stratigraphically about 2 m higher than layer G. This age is therefore problematic, since if the date for layer G is correct, and the law of superposition still holds true, anything higher up in the stratigraphy should be younger, not exactly the same age. The solution to this is most definitely not that the two stalagmitic layers were once connected, because the main sedimentary layers within the upper part of the site are roughly horizontal, and even if we could imagine that the deposits sloped steeply from the back of the cave to the front, how did layer E manage to get underneath the speleothem in the back of the cave, but well above it in the front? I have no solution to offer. All I can say is that the thick deposits of the Tranchée Moulin most likely date to the late Riss and/or early to mid-Würm, and the uppermost layers may be younger still.

7.2. From the Chest: T or t?

The attentive reader will have noticed that I refer to the Tranchée Moulin with a capital T, but the tranchée L with a small t. You may well be bothered by that. I would be! I should pick one or the other and be consistent. However, I can not make myself do so. For some reason, the "Tranchée Moulin" seems to me to be a name, but the "tranchée L" is a description as well as a name. I just wanted to tell readers that I have noticed, and I apologise for the inconsistency, but that is the way it is.

7.3. Sediments and the evolution of the site

The overall picture of the stratigraphy of the Tranchée Moulin is presented in a simplified way in Figure 7.3, based on a detailed sketch in my personal notes from 1995. This figure of course ignores most of the lateral variations. Layer F (see chapter 6) consisted of loose to cemented fine and coarser sediments, with mainly unburnt archaeological material within it. Above this is a 30 to 40 cm-thick of series of layers of breccia, showing variations in grain size of the clasts as you go up. The degree of cementation also varies, but in general these are very hard breccias.

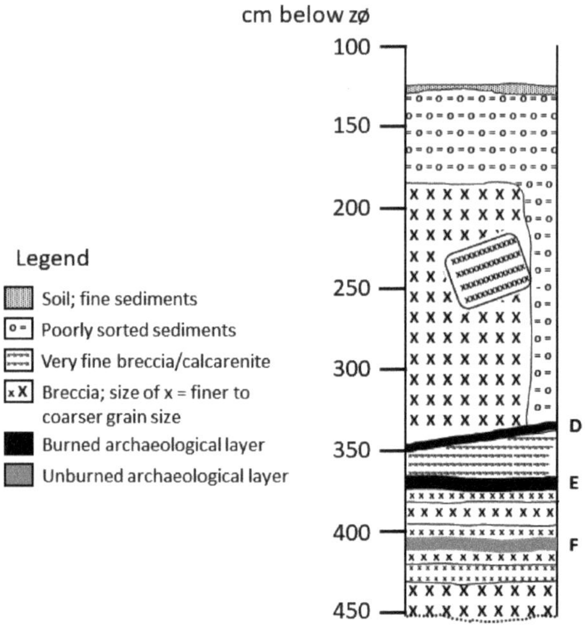

Figure 7.3. Simplified representative stratigraphic column for the Tranchée Moulin. ***Colonne stratigraphique simplifiée représentative de la Tranchée Moulin.***

Layer E is a highly burned archaeological layer, very black where it was freshly exposed, and becoming a dull grey when weathered. It was about 4 cm thick near the front of the trench, thickening to 10 or 12 cm at the rear, and was close to horizontal, but with a slight dip towards the north (outside) of the site (Lebel 1987). Other than the bone fragments and lithics, the sediment itself was fine-grained, and felt damp and plastic when excavated. It seemed to include a lot of minute charcoal fragments, which gave it the black colour. Layer E also seemed to pinch out in some areas, and the distance between it and layer F varied as well. In the area we excavated under the breccia at the back of the Tranchée Moulin, layer E was thick and rich in zone F6, but thinned and disappeared towards the east (zone H6) and south (zone G5) of the site (Lebel 1990).

Above layer E is a thick slab (henceforth referred to as "the dalle") of very fine-grained breccia – perhaps better described as a calcarenite, since the grains are mainly sand-sized. This layer is very strongly cemented, and while its base as seen in the Tranchée Moulin is roughly horizontal, its upper surface has a distinct slope towards the front of the site, making it considerably thinner there. Our work attempting to make the link between the Tranchée Moulin and the tranchée L showed that the top surface of this dalle also includes stalagmitic deposits (i.e., what was sampled and dated from zone I5), and it has a steep slope towards the west, so it rises as much as 90 cm in about 4 m, from the Tranchée Moulin towards the zone I; then it changes direction and slopes eastward into the tranchée L. Its thickness varies from as much as 35 cm at the rear of the Tranchée Moulin, to 10 or 12 cm at the front; its thickness in the area between the two trenches is unknown, but in the tranchée L it shows up again as being about 20 cm thick. The dalle also becomes coarser and less well cemented towards the front of the site, so that it does not stand out from the other deposits as clearly.

The dalle is archaeologically sterile, and it represents a lengthy period of time in which fine sediments accumulated, got cemented together, and then seem to have been eroded into a smooth surface on top, while accumulating at least three layers of stalagmite (Lebel 2006) at the rear of the site. This provided a clear surface upon which subsequent hominins could set up camp. Figure 7.4 shows zone F6 while I was in the process of drawing the stratigraphic section of its north wall (which is not visible, because it is oriented away from this east-facing photograph). The section drawing is almost legible, on the board on the floor of the square, where it is sitting on remnants of layer E. The dalle is the thick hard layer on the left, which was cut off along the band 7/6 boundary line to make F6 accessible: it is what the tools are sitting on. Layer D would have been on top of the dalle.

One other interesting thing about the dalle is that in the rear of the Tranchée Moulin it is cracked or faulted, with one side extending down into layer E and a matching dislocation in the sediments above it. This must have resulted from a sudden stress many thousands of years ago.

Layer D was found directly on top of the dalle. It was a black to dark grey, fine-grained layer containing abundant lithics and bone fragments, and seemed to contain many tiny charcoal fragments. It was badly affected by the groping hands of clandestine visitors, but the gaps where it "should" have been range from a few cm thick to as much as 25 cm. We were able to excavate remnants of it in places along the Tranchée Moulin, and found it in our new excavation area in the rear of the trench, but it disappeared at the limit between zones G6 and G7. It also seemed to be more diffuse towards the northern end of the trench (Lebel 1987). Like layer E, it was damp to the touch and plastic when first excavated.

Above layer D is a very thick layer of breccia. In our 1988 excavation in zones F 5, 6 and 7, for example, we had to get through more than 2 m of breccia to reach layer D (Lebel 1988). Most of this was archaeologically sterile, but along the length of the trench, at various heights, we did notice concentrations of archaeological material: a series of objects 20 cm above layer D in bands 11 to 14; tiny flakes and bones 90 cm above layer D, approximately where two human teeth had been found by earlier excavators; and 1 m higher than layer D in A11 and A12, in the remnants of breccia stuck to the rock shelter wall (Lebel 1987, 1988). In the squares that we excavated, we found a complicated mixture of sedimentary deposits with and without archaeological material. The deposits with archaeological material make up layer C; the rest make an interesting story of depositional and post-depositional processes at work.

The long-exposed trench wall is composed of breccia which appears to be consistently fine- to medium-grained

Figure 7.4. Zones F7 (unexcavated) and F6 (excavated, on the right) in 1988, with remnants of layer E exposed, giving the dark colour. See text for further details. ***Les zones F7 (non-fouillée) et F6 (fouillée, à droite) en 1988, avec des traces de la couche E visible, de couleur noire. Voir texte pour plus de détails.***

and dense, unlike the breccia lower down in the site, with its occasional large slabs and its matrix-free pockets. The irregularly bulbous appearance of it does suggest that there are variations in degree of cementation, but overall it seems rather uniform. Our excavations in zones F and G 5-6 showed clearly, however, that the mass is not uniform. In that area, above layer D there was a brecciated layer or deposit about 60 to 80 cm thick, broken up into unstable chunks and with some variations in grain size and cementation. I cannot say whether these chunks are part of a single, broken, layer, or represent breccias deposited at different times under different conditions. We dug the band F (5, 6 and part of 7) in 1988, and Lebel (1988) describes this as follows: "Avant d'arriver à la couche D, des brèches calcaires instables, couvraient ce niveau anthropique et un concrétionnement avait cimenté les objets archéologiques." In 1990 we used jackhammers to excavate G5 and 6, and in those squares the breccia above layer D was more firmly cemented and did not show as many variations.

Above that initial brecciated layer there is a block of a different nature: a combination of very fine-grained breccia and thin stalagmitic layers, about 25 to 30 cm thick and easily distinguished from the breccia around it. Figure 7.5 shows the eastern section wall of G6, with the block in it, tilted on a steep angle; it has had a sample removed from the centre of it. This block matches some breccia that we found stuck to the rock shelter wall nearby, as if it had broken off of it. In fact, the shape of the breccia seems to match the contours of the rock wall in this area. The block also has a thin stalagmitic layer along its top surface, suggesting that it was an exposed surface for some length of time.

On top of the stalagmitic layer and the lower breccia there is another thick layer (40 to 90 cm) of strongly cemented, medium-grained breccia. Finally, when we excavated this area, the space between the breccia and the rock shelter wall was filled with loose cryoclastic sediments, which also extended above and covered the top of the breccia. Over that was a very thin layer of modern soil.

My interpretation of this was developed at the dig in consultation with other excavators who noticed interesting aspects and pointed them out to me. However, if it is wrong, I accept the blame. Remember that layer D is very soft and plastic, and is located directly on top of the fine-grained, strongly cemented dalle, which is itself located on top of the soft, plastic layer E. Cryoclastic sediments accumulated on top of layer D during a cold period, then were cemented into breccia during a subsequent warm period, which also resulted in the accumulation of stalagmitic layers mixed with fine sediments (tufa) against the rock wall. The next

Figure 7.5. Eastern section wall of zone G6, showing breccia and the sampled block of different material incorporated in the deposits. The gap between the breccia and the rock shelter wall on the right was previously filled with loose sediments. ***Coupe est de la zone G6, avec de la brèche incorporant un bloc (échantillonné) de nature distincte. L'espace entre la brèche et la paroi rocheuse était remplie de sédiments meubles avant la fouille.***

cold period caused more cryoclastic debris to accumulate on top of this layer, and these sediments in turn were cemented together, fusing the entire lot into a solid, heavy mass. Water percolating through the site did not have many outlets, since it could not flow through densely cemented, impermeable, breccia. Some of it would have flowed across the surface, but some must have had to flow down the rock wall until it reached layer D, which channeled it outward. This further weakened both layer D and the connection of the breccia to the rock wall. This weakening may even have been exacerbated by repeated freezing (swelling) and subsequent thawing of the wet layers (D and E), as opposed to the solidity of the impermeable breccia layers. Eventually, the weight of the mass of breccia was too much for the connections and the strength of D to take, and they gave way. The mass fell onto the dalle – causing the faults noted there, as parts of it broke down into layer E. Layer D was squeezed out of some areas, causing it to pinch out, as noted at the junction of squares G6 and G7. The dislocation within the mass resulted in some of the breccia being broken into chunks, and this caused parts of layer C to be noticeably offset from each other. For instance, Lebel (1990) notes that "Un décrochement séparait en deux compartiments la couche anthropique et brèchique C-1, épaisse de 15 à 20 centimètres. Le déplacement d'une trentaine de centimètres s'est fait dans le sens horizontal et vertical. Le décalage d'une même couche doit être mis en relation avec des processus post-dépôts importants."

When this had settled down again, the resulting gap between the breccia and the wall was gradually filled with more cryoclastic debris, probably during the last glacial period. It has not been cemented into breccia. The clasts are angular and often vertical, which may result from cryoturbation, i.e. frost action after they were deposited. Lebel (1990) reports that these deposits also contained archaeological material, but does not specify what types, nor what layer they would have been attributed to.

Layer C was defined in 1988, and included four sub-layers (C1 through C4) as well as layer C itself. C1 was the richest of the sub-layers. C itself was named based on a concentration of archaeological material in F6, which Lebel (1988) describes as "une concentration ou densité importante de vestiges au milieu de la coupe (couche C), juste sous la brèche . . . Beaucoup d'ossements présentaient différents stades de combustion et des silex étaient brûlés." Layer C1 was underneath C. C2 was on top of C3, and both were located in the north-eastern quarter of zone F6. C4 was farther back, towards the rear of the zone, and "couvrait les plaquettes bréchoïdes du fond" (Lebel 1988). The material was very broken and showed signs of having been transported by water: bone fragments were rounded and worn down. There were many small lithic flakes with mechanical breakage. When we excavated the band G in 1990 we found material in similar condition, and attributed it to layer C1 (according to my personal notes from July 23rd and 24th 1990). There is no chance that these remains represent a single, coherent occupation. It is more likely that they cover a very long period of time, and many of them have been considerably reworked before ending up at the back of the site. However, given their small numbers, any kind of statistical analysis requires lumping them all together, so in this book they will be presented as a single layer C.

7.4. Pollen and fauna

There has been no pollen study done for deposits in the Tranchée Moulin itself, so the only relevant data come from the study of the speleothem in I5, reported by J. Argant (2006). She compares it to pollen zone BAU-6, which she described at the top of the sequence taken in zone C15, and which included layer F and part of layer G. The I5 speleothem has a higher proportion of arboreal pollen, but the principal taxa present are the same, and appear in similar proportions, except for juniper, "qui apparaît ici beaucoup plus nettement." There are few herbaceous plants, and she suggests that this might mean that pollen grains were arriving at the site differently, due to wind, being washed in, or the presence of humans. Nonetheless, the results are similar to those found for BAU-6: "Ce spéléothème ne se distingue en tous cas pas fondamentalement des échantillons de la zone pollinique BAU-6 et pourrait à la rigueur s'inscrire dans leur continuité" (J. Argant 2006).

Moulin (1904) reported finding pieces of charcoal which were identified as being *Amelanchier*, a shrub or small tree that "craint autant les froids rigoureux que la grande chaleur" (p. 19). He says that the charcoal was found in "les couches inférieures" (p. 18). Its presence implies a mild, temperate climate.

Together, these results give us two data points for the entire trench (and beyond). The fact that they corroborate each other in suggesting that the climate was fairly mild does not tell us that it was always mild, nor give us any information on what else might have happened over the long time period of deposition that resulted in the accumulation of these sediments.

Moulin (1904) also provided a list of fauna found in the Tranchée Moulin, which was repeated in de Lumley (1971) with a few differences. The differences are not explained, but may have been due to updated identifications. The layer or level at which specimens were found was not specified, but de Lumley (1971) does usually provide indications of abundance, as shown in Table 7.1. During our excavations, abundant faunal remains were found; for instance, Lebel (1990) reports that in G5 and G6 we found 2511 bones, with layers C1 and E especially rich, and in 1988 zone F6 provided 2100 bones and teeth, from all layers combined (Lebel 1988). However, apparently almost none of these were in an identifiable state. Digging through the tables in Fernandez (2001) yields very few mentions of layers C and E (summarized in Table 7.2), plus the aforementioned layer B3 (two aurochs teeth), and nothing at all from layer D.

Three of the four taxa that de Lumley (1971) listed as abundant or very abundant appear on Fernandez' (2001) list: horse, aurochs, and ibex. The only one missing is the rabbit. Chamois and red deer appear in both tables, but de Lumley lists them as rare. Other rare or very rare species from de Lumley's list are absent from Fernandez', but Fernandez does add fallow deer. It is perhaps more interesting that, judging by the information from Fernandez (2001), the hominins of layer E hunted larger fauna (horse, red deer and aurochs) than those of layer C (ibex, chamois and fallow deer). Horse and aurochs require large, open grasslands, while red deer need at least some forest, implying that layer E accumulated during the kind of cool period that we have become familiar with in the data about the lower layers. Ibex and chamois, on the other hand, are more associated with craggy, mountainous areas, while fallow deer are highly adaptable but do require at least a small amount of forest cover for shelter. The difference in fauna between layers E and C thus may be due to climate and landscape differences, or it may be due to choices about where and how to hunt. Or it may just be an artifact of the small sample size.

7.5. Human remains

Two human teeth were found in the Tranchée Moulin, the first by Moulin himself (Moulin 1904), and the second by B. Mary in 1964 (as reported in de Lumley [1971]) "en nettoyant le front de taille supérieur de Franki Moulin" (p. 385). Moulin describes the tooth that he found in detail: it was a deciduous upper second molar with resorbed roots, so it had arrived at "son terme de durée" (Moulin 1904, p. 16), and fallen naturally from the child's mouth. The child would have been about 10 or 11 years old. The tooth showed no pathology, and only a little use-wear, as would be normal. Unfortunately, that tooth has since been lost.

The tooth that B. Mary found in 1964 was studied by M.-A. de Lumley (1973). It is a left lower second premolar, well preserved and only lightly worn on its occlusal surface. Compared to modern human teeth, it is of moderate length (front to back), but very wide (cheek to tongue), which puts it in company with other Neandertal teeth such as La Quina 9 and Spy II (de Lumley 1973).

Table 7.1. Faunal list reported in de Lumley (1971). ***Liste faunique selon de Lumley (1971)***

Taxon	common name	abundance
Equus caballus	horse	very abundant
Sus scrofa	boar	rare
Cervus elaphus	red deer	rare
Cervus alces	elk	very rare
Cervus capreolus	roe deer	very rare
Cervus canadensis	wapiti (Canadian elk)	?
Capra ibex	ibex	very abundant
Rupicapra	chamois	rare
Bos primigenius	aurochs	abundant
Canis lupus	grey wolf	fairly rare
Ursus arctos	brown bear	rare
Felis leo spelaea	cave lion	very rare
Felis pardus	panther	very rare
Castor fiber	Eurasian beaver	rare
Oryctolagus cuniculus	European rabbit	abundant
Vitrea pseudohydatina	land snail	?
Vitrea mentonica	land snail	?

Table 7.2. Faunal results for layers C and E, from Fernandez (2001): Minimum Number of Individuals (MNI) for each species. ***Liste faunique pour les couches C et E, d'après Fernandez (2001): Nombre Minimal des Individus (MNI) de chaque espèce***

Taxon	common name	MNI in layer C	MNI in layer E
Capra sp.	ibex	1	n/a
Rupicapra rupicapra	chamois	1	n/a
Bos primigenius	aurochs	n/a	1
Equus sp.	horse	n/a	1
Cervus elaphus	red deer	n/a	1
Dama dama	fallow deer	1	n/a

Moulin reports that the tooth he found came from "la base du plancher stalagmitique supérieur" (Moulin 1904, p. 16). His description of the stratigraphy is hard to match up to the reality as I know it, but in any case this is above the "couche industrielle", so it is probably approximately layer C. The stratigraphic sketch in de Lumley (1971) and in de Lumley (1973) shows its location as being well within the upper part of the breccia, which again sounds like our layer C. The tooth found in 1964 is shown on de Lumley's sketch as coming from a lower level, but probably still within the breccia above our layer D, so perhaps it also is from the equivalent of our layer C.

7.6. From the Chest: Early archaeologists

Franki Moulin's report on the Bau de l'Aubesier (Moulin, 1904) is very well done, by any standards. He discusses the sediments, the fauna, the human remains and the lithics. (In fact, I now realise that the format in the chapters of this book is very similar!) I have been told that his was the first excavation, at least in southern France, to include a sedimentological study. Still, he was a pioneer in the field; he and his contemporaries did not proceed in the same way as modern archaeologists do. We take note of the locations of the remains, and keep everything we can: even tiny chips of rock, and unidentifiable splinters of bone. They did not bother with that. (I have also been told that Moulin used dynamite to get through the breccia.) We tell ourselves that we should keep things because future archaeologists may have new techniques and will be able to glean information that we can not even dream of. The early archaeologists seemed to believe that they already knew how to study the sites perfectly well. A case in point is the way lithics were handled. We analyse lithic assemblages for their signs of technology and raw material provenance, as well as typology. We often base our interpretations around the concept of the *chaîne opératoire*, and for that we need to know what stages in the reduction of the raw material are present in the site, and in what proportions. Do we have nodules, cores, blanks, debris, retouch flakes and tools, or, for instance, mainly tools and a few chips from occasional retouch? For this, we need to know what is in the whole assemblage, from smallest chip to largest chunk. Early archaeologists were guided more by the concept of the *fossile directeur*: the individual tool type that lets you know which "industry" you are dealing with. If there are handaxes, it is Acheulian; if there is a laurel leaf, it is Solutrean; if there are Mousterian points it must be Mousterian. The rest of the pieces were uninformative and therefore generally not worth bothering with. The combination of lack of information about the positions within the site from which each piece came, and the lack of entire categories of material from the assemblages, means that it is very difficult to study the old collections in anything resembling a modern way.

7.7. Lithics

The early analyses of the lithic industry from the Tranchée Moulin (Moulin 1904; de Lumley 1971) agree that it represents a progressive Mousterian industry, but those analyses do not distinguish between assemblages by layer, and it is not clear whether they refer to a single layer or a combination of two or more layers. The collections have since become fragmented, and no doubt some parts have been lost. Even if this were not the case, a modern study of them would not be able to answer the questions that modern archaeologists ask, since these are not representative samples of their industries. Early archaeologists, even ones as scientifically inclined as Franki Moulin, asked different questions than we do (see From the Chest: Early archaeologists). They kept the pieces that they considered meaningful for those questions, and discarded the rest. The fact that several people could spend weeks or months at the Bau in the mid-twentieth century sorting through what Moulin left behind shows that he left a lot. Those later investigators established collections of tools from the Bau, which make for impressive displays – but they are *collections*. Collectors choose interesting pieces. These are not full assemblages. Understanding these industries therefore requires fresh excavation and analysis. Unfortunately, our efforts have not helped very much.

The lithics excavated starting in 1987 were shipped back to Canada for typological analysis by Serge Lebel, then shipped back to France. Lebel (2000) provides some typological data for several layers at the Bau, but he does not provide much detail, and he does not include layers D or E. In 2013 and 2014, as already mentioned elsewhere, I got access to all of the material stored at Vaison la Romaine, but when it had all been seen and added up, I realised that some must still be missing. The information Lebel (2000) provides for layer C, for instance, includes typological attributions for 72 pieces, and he says C had a total of 716 pieces. We saw 272. When I recorded the raw materials during the excavations, I recorded a total of 860 pieces in layer C. This discrepancy extends to the other layers, too. For D, our typological data include 77 pieces, but I have raw material information for 147. Layer E is even worse: the raw material list includes 1998 pieces, but our typological study only found 171. Given this incomplete dataset, it is impossible to say much about the typology of any of these layers. What data we have are given in Table 7.3.

These data are obviously incomplete. Not a single retouched piece in layer D? That does not sound like Moulin's "couche industrielle". The results for layers D and E must therefore be considered indicative, at best. For layer C, we do have Lebel's (2000) list of identified pieces,

Table 7.3. Lithic assemblage of layers C, D and E, summarised by typological category. ***Assemblage lithique sommaire des couches C, D et E, par catégorie typologique***

	Layer C	Layer C	Layer D	Layer D	Layer E	Layer E
Category	Typology (n)	Typology (%)	Typology (n)	Typology (%)	Typology (n)	Typology (%)
Liste Bordes	14	5.1	0	0	14	8.2
Debris	0	0	0	0	0	0
Eclats	5	1.8	15	19.5	13	7.6
Lames	3	1.1	1	1.3	8	4.7
Tamis/Collectif	248	91.2	61	79.2	129	75.4
Nucléus	2	0.7	0	0	7	4.1
Total	**272**	**100**	**77**	**100**	**171**	**100**

which differs substantially from the list we established (Table 7.4 shows both). Most of the same categories are represented, but the numbers and relative proportions are very different, especially concerning blades, sidescrapers, and endscrapers. Finally, our data for layer E are also listed in Table 7.4 in the same order as Lebel's list, with some of the blades appearing twice since they also serve as supports for retouched tools. It is also worth noting that 3 of the blades are Levallois. Taken all together, about all we can say is that these do indeed look like progressive Mousterian industries. For illustration, Figure 7.6 shows a tool from layer E, number E6 E 28, which is a burned and patinated piece combining a sidescraper and an atypical endscraper formed by bifacial retouch. Figure 7.7 shows a heart-shaped retouched piece from layer C, number F6 C 41, which manages to be a notch and an atypical endscraper, and has had its bulb removed.

In previous chapters I have broken down the typological list by raw material source areas, but that is not possible for the Tranchée Moulin: there are too few matches between the two lists. For layer E, 89 pieces match, of which 75 are tamis, and of those 89, only 17 pieces are in identifiable raw materials. For layer D, there are no matches at all. For layer C, 6 pieces match, of which only 3 are identifiable. Part of this is due to my regrettable decision in 1987 to not record the catalogue numbers of the pieces; this accounts for 60 pieces in layer D not being able to be matched by number to the typology list, and 412 pieces in layer E. The rest is due to the mismatch between what I observed at the excavations, and what we found in storage years later.

Still, there are respectable numbers of lithics on the raw materials list for each layer, and we could hope they would give us some information about source usage. Table 7.5 gives the break down by numbers of pieces, and Table 7.6 gives the percentages of each assemblage from various areas. Unfortunately, the great majority in each case is unidentifiable. Layer C is best, with a total of "only" 76.1% unidentifiable and of unknown source combined. Layers D and E each exceed 90% for that total. Those numbers include burned pieces, which account for 14.7% of layer E's assemblage (294 pieces), 14.3% of layer D's assemblage (21 pieces), and 21.9% of the assemblage from layer C (188 pieces). Layers E and D, being relatively continuous, more-or-less horizontal, accumulations of ash, charcoal, bones and lithics, seem to have been living floors, which preserved traces of fires and all other aspects of use of the site, thus explaining the presence of burned lithics. Layer C, on the other hand, is an accumulation of reworked material scattered throughout at least 30 to 50 cm of apparently unburned deposits, much of it obviously rolled by water. It is likely that traces of fire would have been washed away or destroyed while that happened, but the lithics tell us fire was indeed used, as do the burned bones mentioned above.

Considering the identifiable pieces, for layer D only eleven pieces can be identified to a source area. There really is no point in trying to extract meaningful information from that. Table 7.7 recalculates the results for layers E and C, as percentages of identifiable pieces, and contrasts the results for those of the lower layers. For both layers, but especially in C, use of sources in the NE seems fairly high, there is a substantial proportion from the Near SW, and sources in the SW are less common than in lower layers. This resembles the strategy described for the upper sections of layer H1, with the "centre of gravity" of the territory

Table 7.4. Typologically identified pieces in layers C and E, according to Lebel (2000), and our study done in 2013 and 2014 (shown by *). ***Pièces identifiées typologiquement des couches C et E, selon Lebel (2000), et selon notre étude entreprise en 2013 et 2014 (marquée *)***

Typology	Lebel (2000) layer C n	Lebel (2000) layer C %	* layer C n	* layer C %	* layer E n	* layer E %
Eclats levallois	10	13.9	2	9.1	1	3.0
Lames et lamelles	19	26.4	3	13.6	12	36.4
Pointes levallois	1	1.4	1	4.5	2	6.1
Racloirs simples	12	16.7	5	22.7	7	21.2
Autres racloirs	4	5.5	4	18.2	2	6.1
Grattoirs	12	16.7	2	9.1	1	3.0
Autres types Paléo. sup.	6	8.3	0	0	0	0
Denticulés	2	2.8	0	0	0	0
Encoches	1	1.4	3	13.6	1	3.0
Nucléus levallois	3	4.2	0	0	2	6.1
Nucléus non-levallois	2	2.8	2	9.1	5	15.2
Total	**72**	**100**	**22**	**100**	**33**	**100**

Figure 7.6. Piece combining sidescraper and atypical endscraper from layer E, in burned and patinated flint. ***Pièce combinant racloir et grattoir atypique de la couche E, dans du silex brulé et patiné.***

Figure 7.7. Retouched piece from layer C, F6 C 41: a notch and atypical endscraper which has had its bulb removed. ***Pièce retouchée de la couche C, F6 C 41: encoche et grattoir atypique, avec son bulbe enlevé.***

Table 7.5. Number of pieces by source area in layers E, D and C. ***Nombre de pièces par groupes de sources des matières premières dans les couches E, D et C***

source area	layer E	layer D	layer C
Unidentifiable	1800	136	652
Unknown	1	0	3
Faraud etc.	59	3	63
Méthamis etc.	25	1	23
Les Sautarels etc.	0	0	3
Sault	60	7	81
Murs	42	0	28
Local	0	0	1
St. Jean de Sault	0	0	0
Roussillon	0	0	0
Murs-Bezaure Olig.	8	0	5
Ravin de la Treille	2	0	0
St. Trinit	1	0	0
Durance	0	0	0
Tertiary Calavon	0	0	1
Tertiary Murs VdeV	0	0	0
Mormoiron	0	0	0
R de Guérin	0	0	0
N.Aurel	0	0	0
Total	**1998**	**147**	**860**

closer to the Bau, rather than greater use of the SW. On the other hand, layer C contains the first piece we have seen so far from the Oligocene sources along the Calavon river to the south of the Bau, type AH. This identification was made based on several criteria, including that the rock contains a type of fossil called a charophyte, which is a freshwater algae. In the Vaucluse and surrounding region, the only flint liable to contain charophytes is of Oligocene age, when the area had freshwater lakes and hypersaline lagoons. (Most of the flint at the Bau is of Cretaceous age, and formed in entirely marine conditions.) Figures 7.8 and 7.9 show the distribution of the raw material sources and abundance of use for layers E and C, respectively, while Figure 7.10 shows the specimen of type AH, and Figure 7.11 shows an example of a charophyte in Oligocene flint, seen in thin section.

The identifiable assemblages for layers E and C were large enough, and included enough different source areas, that Wilson & Browne (2014) were able to use them for Generalised Linear Models, shown in Tables 7.8 and 7.9. The Wald Chi-square results from each, converted to percentages, are in Table 7.10. These models gave very different results from each other. In layer E, the Terrain variables (Extent, Difficulty, Calories, AOSISA and AOSA) account for the majority of the weight in the model (56.1%), with the Raw Material variables (Quality, and the rock sizes) adding up to only 33.5%. Of the Raw Material variables, the most important is Large rocks, which is the size of nodule that is most suitable for tool-making. Thus ease of access and of use seems to be the priority. In contrast, in layer C it is the Raw Material variables that dominate (63.7%) over the Terrain variables (34.4%), with Large rocks, Quality, and Medium rocks appearing as three of the four most important variables (with AOSISA in second place). This emphasis on raw material characteristics may mean that the hominins prioritised the quality of their tool kits, but without a clear picture of the industry, we cannot be sure.

Table 7.6. Percentages of pieces by source area in layers E, D and C. ***Pourcentages des pièces par groupes de sources des matières premières dans les couches E, D et C***

source area	layer E	layer D	layer C
Unidentifiable	90.1	92.5	75.8
Unknown	0.1	0.0	0.3
Faraud etc.	3.0	2.0	7.3
Méthamis etc.	1.3	0.7	2.7
Les Sautarels etc.	0.0	0.0	0.3
Sault	3.0	4.8	9.4
Murs	2.1	0.0	3.3
Local	0.0	0.0	0.1
St. Jean de Sault	0.0	0.0	0.0
Roussillon	0.0	0.0	0.0
Murs-Bezaure Olig.	0.4	0.0	0.6
Ravin de la Treille	0.1	0.0	0.0
St. Trinit	0.1	0.0	0.0
Durance	0.0	0.0	0.0
Tertiary Calavon	0.0	0.0	0.1
Tertiary Murs VdeV	0.0	0.0	0.0
Mormoiron	0.0	0.0	0.0
R de Guérin	0.0	0.0	0.0
N.Aurel	0.0	0.0	0.0
Total	**100**	**100**	**100**

7.8. Conclusion

During the time period of accumulation of the deposits in the Tranchée Moulin, climate changed several times: sediments were deposited then cemented together, then more were deposited, and so on. At times the site was abandoned for long periods. At other times Neandertals used it, laying fires, bringing in the animals they had hunted, making and using stone tools . . . but the details of this remain elusive. The archaeological layers were excavated by different teams, over a period of a century; their finds have been dispersed, the collections fragmented. The remains themselves – at least those excavated most recently – are in poor condition: lithics highly patinated and burned, bones highly broken and weathered.

Layer E is the first substantial and well-circumscribed layer (counting upwards in the sequence) since layer H, after more than two metres-worth of breccia that included the patches of material that make up layer G, and the rather spotty and ill-defined layer F on top of that. Layer E is several centimetres thick (or more) and extends at least ten metres along the trench. It is approximately horizontal, and probably was a living floor, used many times. It has also had a lot of water pass through it, since it is squeezed between two layers of breccia, and that (plus extensive use of fire by the hominins) accounts for the poor condition of the archaeological material. The identifiable lithics and bones may not be telling the whole story, but if anything, layer E seems to resemble to top part of layer H1 (see chapter 5), in its distribution of raw material sources, and in the fauna used (horse, aurochs, red deer). It seems like a continuation of the story of the lower levels.

Layer D is frankly enigmatic. It definitely was a layer: we excavated it in a few square metres, and it is easy to

Table 7.7. Percentages of identified pieces (excluding unidentifiable and unknown source pieces) from each geographic region for layers K2, K1-J, I, H, G, E and C. ***Pourcentages des pièces identifiées (excluant les pièces indéterminables ou de sources inconnues) de chaque région géographique pour les couches K2, K1-J, I, H, G, E et C***

Geographic area	K2 %	K1-J %	I-I4 %	H-H2 %	G %	E %	C %
Local	0.2	3.4	1.3	0.7	0	0	0.5
NE	19.4	20.4	12.3	15.7	41.5	31.0	39.5
E	0	0.3	0.1	0	0	0	0
Far SE	0	0	0	<0.1	0	0	0
Far S	0	0	0.1	0	0	0	0
S	0	0	0	0	0	0	0.5
SSW	0	0	0.1	0	0	0	0
SW	44.3	54.1	35.0	30.0	48.4	25.4	17.6
Near SW	23.0	15.0	29.1	34.8	9.6	29.9	30.7
WSW	12.4	6.6	21.9	18.6	0.5	12.7	11.2
NW	0.7	0.1	0.1	0.2	0	1.0	0
Total	**100**	**100**	**100**	**100**	**100**	**100**	**100**

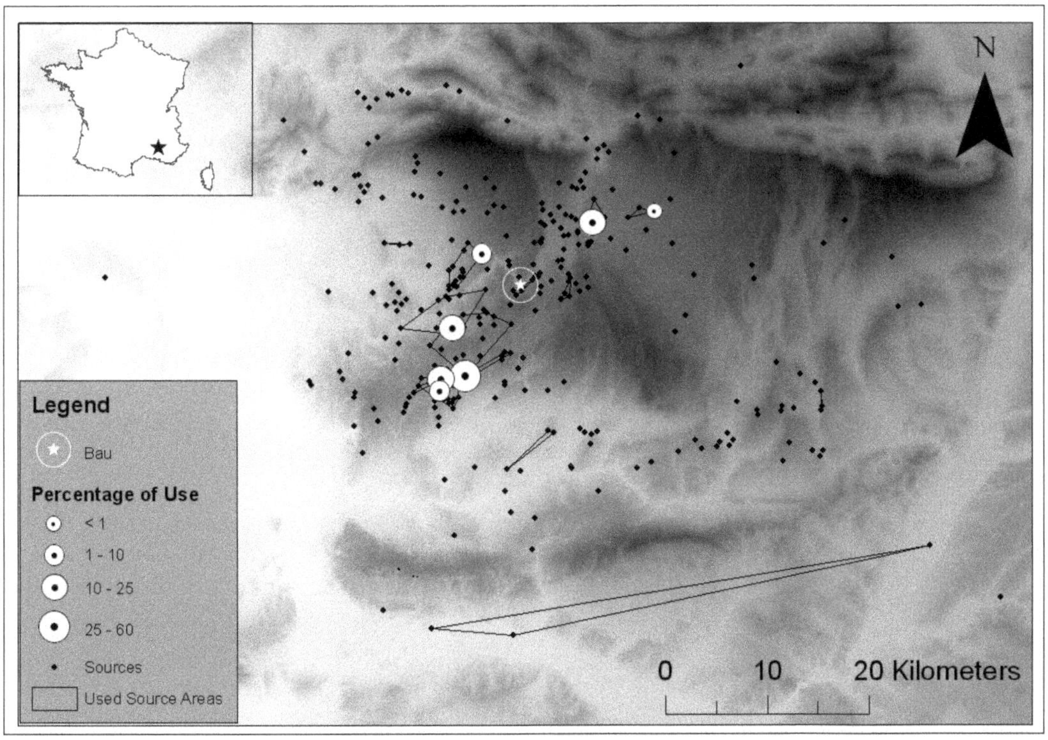

Figure 7.8. Map of the raw material source areas used for layer E. ***Carte des aires de sources de matière première utilisées pour la couche E.***

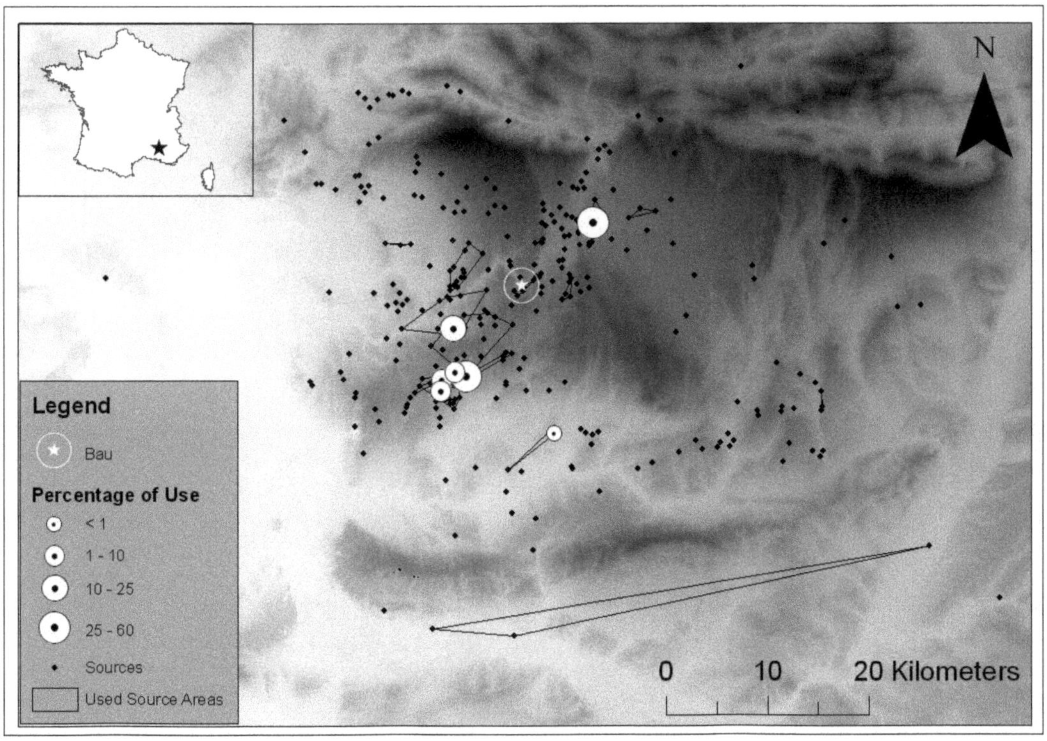

Figure 7.9. Map of the raw material source areas used for layer C. ***Carte des aires de sources de matière première utilisées pour la couche C.***

describe where it was (on top of the dalle), and what it was like (dark, fine-grained, with bone fragments and lithics). Beyond that, though, there remains so little data that there is nothing to say. Was this Moulin's "couche industrielle", or was that layer E? Or a combination? What tools were found here? What animals? We do not know.

Layer C, although not well-circumscribed, and not even really a layer, gives us much more information. It is very different from the lower layers. In the GLMs (Wilson & Browne, 2014) all of the layers up to and including E had higher total weights put on the Terrain variables than on the Raw Material variables. After E, that changed, and

Raw Material became more highly weighted than Terrain variables. Layer C is the first time, in this book, that we have met with that situation: remember that layer C is actually fairly high up in the stratigraphy, so we are dealing with it out of sequence, just because it is in the Tranchée Moulin. Out of sequence or not, though, it shows us that something has changed. The little information we have on fauna also differs from layer E and the lower layers, since it includes only smaller animals (fallow deer, ibex and chamois). For now, though, that is all we can say. We will move on to the tranchée L, and hope that the results for E, D and C will make more sense in the context of other results from the upper part of the site.

Figure 7.10. Specimen of raw material type AH, piece G6 C2 273 (a small flake with its bulb removed), containing charophytes (the small white ovals). ***Specimen du type de matière première AH, pièce G6 C2 273 (un petit éclat avec son bulbe enlevé), contenant des charophytes (petites formes ovales blanches).***

7.9. Sommaire du chapitre

Pendant longtemps, la seule partie de Bau qui ait été fouillée et étudiée était ce que nous appelons la Tranchée Moulin: elle était le site. Les fouilles ont été commencées par Franki Moulin à partir de 1901 et d'autres chercheurs y ont travaillé dans les années 1950 et 1960. Les vestiges ont été dispersés dans des collections et musées à Sault, Carpentras, Avignon et plus loin encore. Les collections de M. Louis Gauthier sont maintenant dans le Musée Louis Gauthier à Sainte Cécile-les-Vignes où je les ai examinées grâce à la bonne volonté de M. Vincent Faure et Mme. Solenne Ricou-Fétisson. C'est par l'obligeance de M. Thierry Tabourdeau et le Groupe archéologique de Carpentras que j'ai pu examiner les collections d'un ancien musée à Carpentras. Dans les deux cas, on trouve beaucoup de pièces lithiques typiques du Bau: éclats, lames et pointes levallois, racloirs, et pointes moustériennes; bon nombre de pièces ont eu leur bulbes enlevés, ce qui est également assez caractéristique du Bau. Les pièces sont souvent patinées et/ou brûlées.

Ce présent chapitre concerne surtout ce que nous connaissons de la tranchée à travers les fouilles menées par Serge Lebel (le travail antérieur a été décrit dans le chapitre 1). A notre arrivée en 1987, nous avons trouvé une longue tranchée avec la paroi rocheuse sur un côté, et un mur de brèche, très irrégulier, sur l'autre côté. Les couches plus meubles à l'intérieur de la brèche avait été grattées complètement y compris par les bras des passants curieux, laissant seulement la brèche la plus dure. Nous avons abandonné l'idée de rectifier cette coupe pour recommencer la fouille dans les règles de l'art: nous n'avions pas les outils nécessaires. Nous avons donc entrepris une seconde tranchée à 4 ou 5 mètres vers l'est, la tranchée L, qui sera le sujet du chapitre 8. Cependant, nous n'avons jamais complètement oublié la Tranchée Moulin, et nos travaux ont clarifié depuis plusieurs aspects de son contenu.

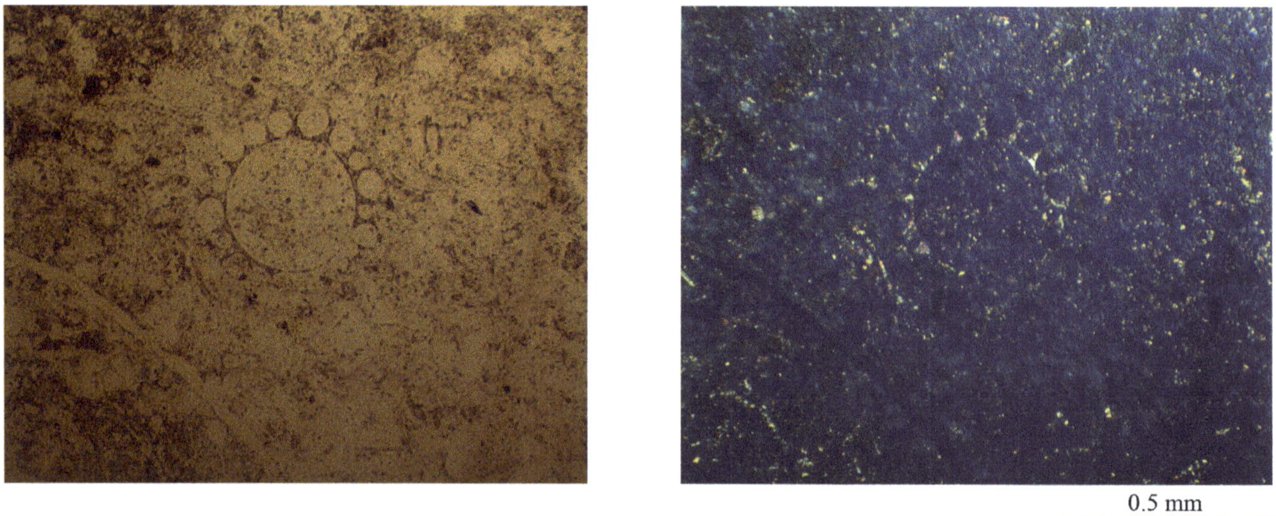

Figure 7.11. A charophyte (the round shape with smaller circles around it) in Oligocene flint from La Coquillade (south of Roussillon), sample number 169:270, in plane-polarised light on the left, and cross-polarised light on the right. ***Un charophyte (la forme ronde entourée de plus petites formes rondes) dans du silex oligocène de la Coquillade (au sud de Roussillon), échantillon numéro 169:270, vu en lumière polarisée plane à gauche et lumière polarisée croisée à droite.***

Table 7.8. Generalised linear model for raw material source choices in layer E. P-values in bold font are statistically significant. Sample: 180 pieces from 7 source areas

Omnibus test: likelihood ratio chi-square = 308.822, df = 9, p = <0.001

Variable	Beta	S.E.	Wald Chi-Square	P-value	AIC
Quality	0.683	0.3780	3.266	0.071	96.139
Extent	1.878	0.7422	6.401	**0.011**	
Difficulty (Cal/km)	-0.066	0.0334	3.901	**0.048**	
Medium rocks	-0.991	1.2747	0.604	0.437	
Large rocks	3.167	0.9691	10.682	**0.001**	
Very large rocks	-0.412	0.8912	0.214	0.644	
Calories	-0.001	0.0009	0.918	0.338	
AOSA	-1.2 e-7	7.6 e-8	2.487	0.115	
AOSISA	8.6 e-5	2.6 e-8	11.084	**0.001**	
(Intercept)	-10.749	5.0333	4.560	0.033	

* Had to remove Small rocks because of Hessian matrix singularity (see Wilson & Browne, 2014).

Table 7.9. Generalised linear model for raw material source choices in C. P-values in bold font are statistically significant. Sample: 151 pieces from 7 source areas

Omnibus test: likelihood ratio chi-square = 296.483, df = 10, p = <0.001

Variable	Beta	S.E.	Wald Chi-Square	P-value	AIC
Quality	2.162	0.8313	6.767	**0.009**	80.872
Extent	1.470	0.9195	2.555	0.110	
Difficulty (Cal/km)	-0.069	0.0498	1.911	0.167	
Small rocks	-8.163	4.3168	3.576	0.059	
Medium rocks	-5.663	2.4697	5.258	**0.022**	
Large rocks	5.032	1.4695	11.726	**0.001**	
Very large rocks	0.970	1.6614	0.341	0.559	
Calories	0.000	0.0011	0.145	0.703	
AOSA	-3.3 e-7	1.9 e-7	2.979	0.084	
AOSISA	0.000	4.0 e-5	7.380	**0.007**	
(Intercept)	-6.490	7.1582	0.822	0.365	

Nous avons identifié quatre couches, que nous avons nommées C, D, E, et F (du haut en bas); la couche F est décrite dans le chapitre précédent. La couche E se trouve entre environ 355 et 365 cm sous zØ, D entre environ 330 et 350 cm sous zØ, et C, est un peu plus éparpillée dans la brèche, entre 170 et 330 cm sous zØ (mais normalement plus haut que ce dernier). Nous n'avons pas de datations pour ces couches.

La couche E est très noire, fortement brûlée, avec un très léger pendage vers le nord. Elle se trouve sous une dalle de brèche à grains très fins, très dense, épaisse d'une dizaine de cm en avant mais jusqu'à 35 cm en arrière de la tranchée. Cette dalle semble être continue le long de la tranchée, et latéralement jusqu'à la tranchée L, servant de marqueur stratigraphique important. La dalle ne contient pas de vestiges archéologiques; sa surface, en revanche, aurait servi de plancher d'occupation pour les hominidés à venir. Dans la Tranchée Moulin la dalle a un pendage marqué vers le nord (vers l'avant du site), et latéralement elle monte jusqu'à la bande I puis redescend jusqu'à la tranchée L.

La couche D se situe directement sur la surface supérieure de la dalle, mais disparaît vers le nord et vers l'est entre les zones G6 et G7. Elle est de couleur noire, et contient

Table 7.10. Wald Chi-Square values as percentages, E and C assemblages

Variable	E: Wald Chi-Square	Wald Chi-Square as %	C: Wald Chi-Square	Wald Chi-Square as %
Quality	3.266	7.4	6.767	15.6
Extent	6.401	14.5	2.555	5.9
Difficulty (Cal/km)	3.901	8.8	1.911	4.4
Small rocks	n/a	n/a	3.576	8.2
Medium rocks	0.604	1.4	5.258	12.1
Large rocks	10.682	24.2	11.726	27.0
Very large rocks	0.214	0.5	0.341	0.8
Calories	0.918	2.1	0.145	0.3
AOSA	2.487	5.6	2.979	6.8
AOSISA	11.084	25.1	7.380	17.0
(Intercept)	4.560	10.3	0.822	1.9
Total	**44.117**	**100**	**43.46**	**100**

des sédiments à grains fins, ainsi que d'abondants vestiges lithiques et osseux. Elle est surmontée par une couche de brèche très épaisse; des concentrations ponctuelles de restes archéologiques à l'intérieur de cette brèche constituent la couche C (avec sous-couches C, et C1 à C4).

Il semblerait qu'après la formation de cette brèche dense et lourde, elle se soit séparée de la paroi rocheuse: peut-être parce que la couche D, très meuble et plastique, coincée entre la dalle et le poids de cette brèche, ne pouvait plus la soutenir. La masse de brèche est tombée sur la dalle en la fracturant. Cette brèche a emporté avec elle un autre bloc de brèche différente englobé dans la brèche principale (un morceau de cette dernière est toujours collé à la paroi). L'espace ainsi créé entre la brèche et la paroi a été rempli par la suite de sédiments meubles cryoclastiques.

Nous n'avons que très peu d'informations sur les conditions environnementales pendant l'accumulation des dépôts de la Tranchée Moulin: d'après les sédiments il a dû faire froid souvent, certes, mais avec des phases parfois plus tempérées et humides. Les quelques pollens et les restes fauniques ne nous aident pas avec cette question. Malgré l'abondance des restes osseux que nous avons trouvés, il n'y avait que très peu d'os identifiables dans les couche C et E, et rien du tout en D. On peut signaler simplement que les espèces identifiées dans la couche E correspondent à de grands animaux (cheval, cerf, aurochs), tandis que celles de la couche C étaient plus petites (bouquetin, chamois, daim).

Deux dents humaines ont été trouvées dans la Tranchée Moulin, la première par F. Moulin (maintenant perdue), et la deuxième en 1964 par B. Mary. Elles appartiennent respectivement à un enfant, et à un adulte néandertalien.

Nous avons trouvé plusieurs centaines de pièces lithiques dans les couches E, D et C, qui figurent sur la liste pétrographique. Malheureusement, un très grand nombre de ces pièces ne semblent pas être entreposées à Vaison-la-Romaine. La liste typologique est alors très différente de la liste pétrographique (voir Tableau 7.4), ainsi que de celle que S. Lebel avait publiée auparavant pour la couche C. On peut seulement dire que les assemblages sont d'un "Moustérien progressif".

L'étude des origines des matières premières est quant à elle entravée par l'état des pièces: très patinées et/ou brûlées, les matières premières sont indéterminables dans une grande majorité des cas. Pour le reste, une bonne proportion provient du NE du site, ainsi que du sud-ouest moins éloigné ("Near SW") et enfin du SW près de Murs. La couche C contient également un silex oligocène avec un charophyte (algue d'eau douce) fossile: c'est la première fois que l'on voit cette roche dans les industries du Bau.

En résumé, la couche E est une couche brûlée, facilement identifiable sur la fouille, se présentant en un niveau noir rempli de restes archéologiques. Par sa faune et son industrie elle s'inscrit dans la continuité des couches inférieures, avec néanmoins un changement perceptible dans les choix des matières premières dont les caractéristiques des roches (qualité, taille des rognons) propres aux gisements deviennent un critère de plus en plus important.

La couche D, également facile à identifier sur la fouille, ne nous a livré que très peu d'informations utilisables. Elle reste énigmatique. Par contre, bien qu'elle ne soit pas continue comme les deux autres, la couche C a livré plus d'informations. On y trouve les vestiges laissés par des hominidés qui chassaient des animaux de taille moyenne, disponibles dans ou près des gorges de la Nesque. Ces groupes ont favorisé des matières premières de bonne qualité avec un format optimal plutôt que de chercher des affleurements d'accès facile.

8

Introduction to the tranchée L, and layer 5

8.1. Introduction

This chapter moves us into the central part of the site, where we started digging in 1987. It tells the overall story of the central trench, the tranchée L – so called because it is centred on band L, although it grew into parts of bands from I to P (see Figure 1.11). Within the trench, we found several archaeological layers, the lowest of which, layer 5, is also included in this chapter. The other layers get their own chapters, and most of the stratigraphy of the trench will be described in the layer-appropriate chapter, since there are relationships between sediments and archaeological layers. However, some aspects of the excavation of the tranchée L are either separate from the layers, or need to be explained before we reach the layer in which they happened, so those stories start this chapter.

8.2. From the Chest: Roman or Arabic?

The layers in the tranchée L were given numeric names, rather than letter names, because we did not want to assume relationships between layers there and in the Tranchée Moulin: not until we could determine what those relationships really were. For several years we seemed to switch back and forth between Roman and Arabic numerals for this, with no firm decision on which was correct. Finally Serge came down on the side of Roman numerals, but by then I had decided that Arabic numerals were preferable. My reason for that is simple: which layer is I? Top of the upper platform, or in the pente du remplissage between H and J? I will therefore use Arabic numerals, but if I quote other people, who used Roman numerals, or use pictures of tools labelled with Roman numerals, I hope the reader will understand. In most cases the meanings should be clear, but if it seems likely to be confusing I will insert clarifications.

8.3. The story of the tranchée L

In 1987, the initial plan, to "fix up" the Tranchée Moulin and excavate there, was quickly found to be more difficult than expected: after a few days of setting up the grid system, and a few more days of cleaning out decades of accumulated rubbish from the site (including hundreds of empty wine bottles), we were finding ourselves without many places to dig. The breccia was too hard for the basic tools we had, and the soft parts of the Tranchée Moulin had already been scraped away by previous excavators and passers-by. What to do?

Serge Lebel had a co-director for the excavations that year (and in 1988): a French archaeologist, Robert Brandi, whose presence made it easier for Serge, as a Canadian, to get permission to excavate from the French authorities. Serge had a doctorate from France, and years of experience in French archaeological sites – he even had a French wife; but even still, having a "Franco-Canadian" project probably sounded better. The French authorities gave Serge a couple of years to prove himself before giving him such an important site by himself. Robert Brandi had many years of experience, especially with the technical side of excavating, and he and Serge decided to undertake a test excavation towards the centre of the site.

At that point, the front of the upper deposits was concealed behind a series of large boulders, several metres high and wide, which must have come from a collapse of the roof overhang many years ago. Figure 8.1 shows some of them. That picture was taken recently, from a vantage point standing on the breccia between the tranchée L and the Tranchée Moulin and looking east towards the tranchée L and beyond. In 1987, there was another rock of similar size, in what is now a gap between the breccia and the first rock. Robert Brandi had a portable winch, so he was able to winch that one out of its position, and we pushed it down onto the slope below the site. Removing that, and some

Figure 8.1. Large boulders at the front of the upper deposits; view is towards the east, with breccia in the foreground. ***Gros rochers qui couvrent le devant des dépôts supérieurs; vue vers l'est, avec de la brèche au premier plan.***

smaller rocks, left us with a way in to the deposits, and that is where the test excavation was located, in squares L12, L13 and L14 (Lebel, 1987).

Underneath the large rocks was a thin layer of relatively modern sediment or soil, loose and with rootlets and areas of various colours (shades of brown to grey), within which we found a few pottery fragments and lithics (Lebel, 1987). We named this layer 1. Underneath that we found sediments with less or no organic component, usually paler coloured, sometimes indurated, sometimes washed clean of matrix, sometimes cemented into breccia. All of these variations will be described in detail in future chapters concerning layers 2 and 3. Most importantly, as we got further into this trench, we realised that it was skirting the edge of a dome of breccia: the breccia thinned towards the front of the trench, and towards the east. Later we found out that a second dome is located further to the east, so that the breccia thickens again through band M and N – but band L has the least. It had breccia at least a metre thick in some areas, but that is less than in surrounding bands. We also found that the further we went inwards, as we lengthened the trench back towards the rear wall of the rock shelter, the more archaeological material we found, and the layer under the breccia, which we named layer 4, started to get a bit darker in colour, as if we might be approaching a combustion area. The archaeological layer was not particularly rich (compared to what we found later), and it was only a few cm thick, but it was promising, especially since it seemed to be located on top of a dalle, just like layer D in the Tranchée Moulin.

Figure 8.2 shows the tranchée L after a few weeks of excavation. The large rock on the left of the entrance to the trench is visible, as is the dalle poking through the deposits at the front. The trench itself takes up only the metre-width of band L, and some breccia is visible within it (the paler and greyer layer). The large rock above the rear of the trench was going to become a problem. In 1988 the square behind that rock, L6, was excavated into layer 2, but not much was found, and it was difficult to dig back there: dark, confined and chilly.

By 1989 we had dug far enough back that we were starting to get underneath the rock (Figure 8.3), which was an untenable situation. The rock was thick and deeply buried in the sediments, but it was supported only by loose sediments (layer 2), so we could not be sure that it would continue to stay up as we removed the sediments underneath it. Tunnelling underneath it was not an option!

Figure 8.2. The tranchée L after a few weeks of excavation. (My apologies for the blurriness of the photo.) ***La tranchée L après quelques semaines de fouille. (Mes excuses pour la qualité inférieure de la photo.)***

Figure 8.3. The tranchée L in 1989. ***La tranchée L en 1989.***

Figure 8.3 also shows that we were contending with breccia: the "step" in the middle part of the trench, for instance, and the whitish part of the right-hand (western) wall are thick, hard breccia layers. Not shown here, because it is protected by a tarp in front of the "step", is layer 4, which was richer and darker the further back we went. We knew we wanted to continue to follow it, so at first we tried broadening the trench into band M (Figure 8.4).

Going into band M exposed more of the very dark, rich layer 4, and exposed some swaths of darker colour in the sediments above layer 4, as if the essence of 4 had been squeezed into the overlying layers. We tried jackhammering the rock, but only got off a corner of it: it was very thick, so dismantling it in that way would have been a huge job. As Serge described it in the 1993 site report, ". . .il couvrait plus de 7 m^2 dans les axes 7-8-9 et était épais de plus de 1.5 m, enclavé dans la couche 2, et était sans fissure" (Lebel, 1993). Explosives were also not an option, as they would carry with them the risk of disturbing the archaeological layers – and perhaps dislodging a few more of those giant rocks from the rock shelter's roof. Finally, in 1993 Serge found the solution: expanding cement.

At the beginning of our field season in 1993, Serge arrived with about 60 kg of a product called Nitro-Bickford (celtamex) (Lebel, 1993). It is a powder which is mixed with water to form a slurry which is then poured into holes.

Figure 8.4. The tranchée L expanded into band M, in 1990.
La tranchée L, élargie jusqu'à la bande M, en 1990.

Over the next 8 hours it dries out and expands, generating 6,000 tonnes per m^2 of pressure, which is enough to gently crack limestone boulders into pieces. To use it, we had to drill about 30 holes, 40 mm wide, into the rock, using a pneumatic jackhammer with a rotary drill attachment instead of the usual hammer (Figure 8.5). The holes had to go 90% of the way into the rock (Lebel, 1993), so that the pressure would be exerted within the rock, rather than escaping out the top or through the bottom.

After a couple of days drilling holes, we were ready to mix the celtamex (Figure 8.6) and pour it into the holes, to leave it to do its work overnight. Serge and I were up very early the next morning, eager to see what had happened, and we went down by ourselves to look while the rest of the team was still sleeping. We were delighted by what we saw (Figure 8.7): the rock had cracked into fragments. Carrying those chunks away took a team of 8 to 10 people another two days to accomplish (Figure 8.8), sometimes using a jackhammer to break them into smaller pieces. This was complicated by the need to avoid the cables of our "overhead" grid system, which were only at about knee- to waist-height to anyone working on top of the deposits, and were spaced one metre apart. Still, in the end the rock was gone, and the way was opened for us to dig all the way to the back of the rock shelter (Figure 8.9).

Figure 8.9 also shows a few other significant features. At the very front of the trench, in 1988 we had cut through the fine-grained breccia in band L (but not in M): that is the extension of the dalle upon which layer 4 and layer D both, seemingly, lie. Also, on the back wall of the excavated area, on the side in band L, the dome-like nature of the breccia is visible. The breccia is much thicker at the K/L wall (right-hand side of the picture), and has an approximately horizontal base, but a steeply sloping upper surface, so that it pinches out before reaching the L/M side. The left-hand wall of the trench is composed of much softer, unconsolidated, sediments (with a tendency to collapse).

As Figure 8.10 shows, we later excavated the tranchée L all the way to the back wall of the rock shelter, primarily following layer 4. Layers 3 and 2 turned out the be interesting, too, especially layer 2 at the back of the site, once we were able to get at it better. We also expanded into bands N, on one side, and K, J and even I, on the other side. Nothing is simple, however, and the rock shelter does not just end smoothly at the back of the tranchée L – as we found out when we expanded into bands O and P (see chapter 9).

That is the basic story of how the tranchée L came to be excavated, from front to back and from top down. Since this book is describing the site in stratigraphic order, from bottom to top, the next job is to deal with the lowest layer in the tranchée L, layer 5, even though that was the last layer we found. After almost all of the rest of the work was done (except the O-P expansion), and layer 4 had been excavated down to the dalle and back to the wall, Serge decided to see what was underneath the dalle, so we excavated a test pit through it in the zones K4-5, L4-5, and M2-5, and found layer 5.

8.5. Sediments and the evolution of the site to layer 5

Layer 5 is located at z values between about 320 and 340 cm under zØ, similar to layer D in the Tranchée Moulin, which just goes to show that z values have to be used in concert with stratigraphic observations: 5 is nothing like D, in position or characteristics. The dalle that layer D is on top of is not horizontal across the site, and it is higher in the tranchée L than in the Tranchée Moulin, so layer 5 is underneath it. The only information available about the sediments that might be under layer 5 comes from extrapolations from the Tranchée Moulin (of dubious value, given the potential for significant lateral variations), and observations at the very front of the tranchée L. There we have the dalle, and under that a series of layers of medium-grained breccia, and "cailloutis lessivé": a highly porous and permeable layer of small (1-2 cm), angular rocks with no interstitial matrix or cement. In places the "cailloutis" was as much as 30 or 40 cm thick. However, the front of the trench is unlike the back of the trench: there was no sign of layer 4 in front, for instance, so neither the front of the tranchée L nor the Tranchée Moulin can really tell us what might be under layer 5.

Figure 8.5. Jean-Pierre Rey drilling holes into the big rock in 1993. ***Jean-Pierre Rey perçant des trous dans le gros bloc, en 1993.***

Figure 8.6. Pouring the Nitro-Bickford (celtamex) mixture into the holes drilled into the big rock. ***Versement du Nitro-Bickford (celtamex), mélangé avec de l'eau, dans les trous percés dans le gros bloc.***

Figure 8.7. The big rock the next morning, cracked by the celtamex expanding as it dried. ***Le gros bloc le lendemain matin, fissuré par l'action expansive du celtamex.***

Figure 8.8. Taking the chunks of the big rock out of the excavation area, complicated by the cables of the grid system. ***Enlèvement des morceaux du gros bloc de la zone de fouille, rendu difficile par les câbles du carroyage.***

Figure 8.9. The tranchée L cleaned up after removal of the big rock, showing the big hole that it left, and the dark layer 4 at the bottom of the excavated area. ***La tranchée L, nettoyée après l'enlèvement du gros bloc. On voit le gros trou que le bloc a laissé, ainsi que la couche 4, noire, au fond de la tranchée.***

Figure 8.10. The tranchée L, excavated all the way to the back wall of the rock shelter. ***La tranchée L, fouillée jusqu'au fond de l'abri.***

Layer 5 itself is distinctly different from what we found in the rest of the tranchée L, and in the Tranchée Moulin. It consisted of mainly loose (occasionally slightly cemented or indurated), coarse rocks (up to 10 cm), stone tools and bones, with a yellowish fine-grained matrix (Lebel 2000). The layer and its components seem to be mainly unburned. The layer is about 30 to 40 cm thick, but the archaeological material gets smaller towards the west, and the layer disappears near the K4/K5 boundary line. Lebel (2000) describes this as "au contact de la dalle", which implies that layer 5 pinches out against the overlying dalle. In places there seemed to be a black breccia showing up underneath the layer, but it was not consistent, and was not excavated.

8.6. Pollen and fauna

No pollen data are available for layer 5, but the faunal picture is better. Large, unburned, bones and teeth give archaeozoologists a good chance to identify the species they came from, so despite the small area excavated, the finds were numerous enough and well-preserved enough that Fernandez (2001) was able to determine at least 6 species of herbivores, and later three carnivores were added (A. Argant, 2006; Fernandez, 2006; Crégut-Bonnoure 2011; Crégut-Bonnoure et al., 2011) (see Table 8.1). Based on the teeth, the aurochs individual was an adult, around 20 years old, while one of the horses was an adult around 10 years old, and the other was a juvenile, about 3 years old or younger (Fernandez, 2001).

These animals range from relatively small to large, and from fairly gentle to downright dangerous. The cave lion was represented by a damaged fragment of the crown of a lower left canine (piece number M3 5 487) (A. Argant, 2006; Fernandez, 2006; Crégut-Bonnoure, 2011), and is the only trace of cave lion that we found in our excavations, but the "*Felis leo spelaea*" reported by Moulin (1904) and de Lumley (1971) from the Tranchée Moulin is an older name for the same thing. The lynx is one of three remains from our excavations of the Bau. One other comes from layer 2 but was determined as *Lynx lynx* by Fernandez (2006); the third is reported by A. Argant (2006) as coming from layer 7, but we did not have a layer 7, so that must be a marking error. In any case, given the scarcity of the remains, we cannot know whether the carnivores were present on their own, or were hunted by the Neandertals. Given the large size of the cave lion and the bear, though,

Table 8.1. Faunal results for layer 5, from Fernandez (2001, 2006); A. Argant (2006); Crégut-Bonnoure (2011); Crégut-Bonnoure et al. (2011): Minimum Number of Individuals (MNI) for each species. ***Liste faunique pour la couche 5, d'après Fernandez (2001, 2006); A. Argant (2006); Crégut-Bonnoure (2011); Crégut-Bonnoure et al. (2011): Nombre Minimal d'Individus (MNI) de chaque espèce***

Taxon	common name	MNI
Capra sp.	ibex	1
Bos primigenius	aurochs	1
Equus sp.	horse	2
Cervus elaphus	red deer	1
Dama dama	fallow deer	1
Megaloceros giganteus	Irish elk	1
Ursus arctos	brown bear	1
Panthera spelaea	cave lion	1
Lynx spelaea	cave lynx	1

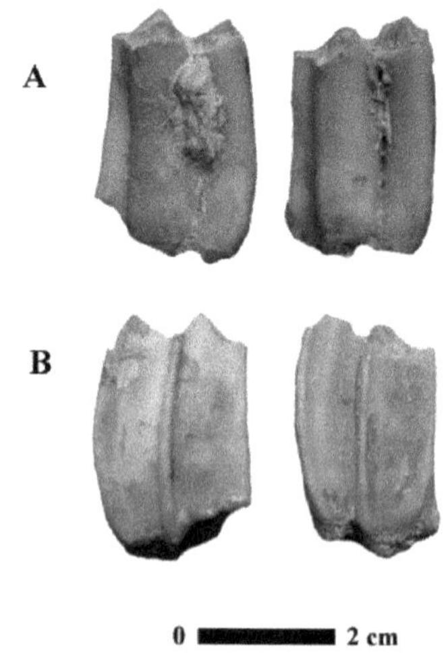

Figure 8.11. ***Capra caucasica praepyrenaica.*** **Upper M3: A, from left to right, M4 4 1044, right, and L4 5 878, right, lingual face. B, same series, same references, vestibular face. Photos P. Fernandez.** ***Capra caucasica praepyrenaica. M3 supérieures: A, de gauche à droite, M4 4 1044, droite, et L4 5 878, droite, face linguale. B, même série, mêmes références, face vestibulaire. Clichés P. Fernandez.***

the former explanation seems more likely, at least for these species.

The herbivores must have been brought in by the hominins, and are dominated by the larger fauna: two horses, an aurochs, an Irish elk, and a red deer outweigh an ibex (Figure 8.11) and a fallow deer. The two horses, one adult and one juvenile, may have been from a family group. Such a variety of prey implies versatility in hunting practices, and in hunting grounds: open plains, forests, and craggy mountainsides. This combination also suggests that the climate must have been cool, but not too cold (still some forest), similar to what we saw in the Tranchée Moulin.

8.7. Lithics

Layer 5 provided a respectably-sized lithic assemblage; my raw materials list includes 546 pieces, and when we studied the typology we found 335 pieces, which should be a representative sample. Table 8.2 gives the numbers and percentages of pieces in each broad typological category, and Table 8.3 describes the individual pieces (other than tamis/collectif) in a bit more detail. The total number of pieces in Table 8.3 is higher than the "non-tamis/collectif" total in Table 8.2 because some of the flakes and blades serve as supports for other tools, and were only counted under "Liste Bordes" in Table 8.2. The number of Levallois flakes and points is high, compared to the other layers we have seen so far, and this relatively small assemblage has more Mousterian points (e.g., Figure 8.12) than any layer examined so far in this book: even more than in the very large assemblage from layer H (chapter 5), which had 288 pieces just in the Liste Bordes alone. In fact, this combination of points and scrapers, with plenty of blades, reminds me of what I have seen in the old collections of material from the Tranchée Moulin.

Table 8.2. Lithic assemblage of layer 5, summarised by typological category. ***Assemblage lithique sommaire de la couche 5, par catégorie typologique***

Category	Typology (n)	Typology (%)
Liste Bordes	45	13.4
Debris	0	0
Eclats	19	5.7
Lames	15	4.5
Tamis/Collectif	249	74.3
Nucléus	7	2.1
Total	**335**	**100**

Perhaps those collections are more representative than I give them credit for. In any case, I think we can call this a progressive Mousterian industry.

Layer 5 was dug fairly quickly, as a test pit, so we excavated it using the same method as we had been using for several years in layer 4 above it. Layer 4 is extraordinarily rich (see chapter 9), and the usual way of digging, where objects are uncovered, left in place, and coordinated individually (forming a décapage), while the sediment in the bucket is screened to become the "tamis", would have left us bogged down for years. We therefore modified it. We still dug by décapages, but only the largest objects were left in place to be coordinated. Each décapage had four associated collectives, one for each quadrant of the square metre. A single catalogue number was attributed to all lithics in a collective, with another number for bones, and another for teeth. This speeds up excavation, as it leaves most of the "sorting out" of objects for the lab, after the collective has been screened and the sediments have been washed off.

Table 8.3. Typologically identified pieces in layer 5. ***Pièces identifiées typologiquement de la couche 5***

Typology	n	%
Eclats levallois	18	20.0
Lames et lamelles	15	16.7
Pointes levallois	7	7.8
Pointes moustériennes	8	8.9
Racloirs simples	11	12.2
Autres racloirs	7	7.8
Grattoirs	0	0.0
Autres types Paléo. sup.	3	3.3
Denticulés	5	5.5
Encoches	3	3.3
Bulbe enlevé	6	6.7
Nucléus levallois	3	3.3
Nucléus non-levallois	4	4.4
Total	**90**	**100**

Figure 8.12. Piece M3 5 481, a Mousterian point made on a Levallois flake, which has also had its bulb removed and its proximal end retouched bifacially. It is patinated and unburned. ***Pièce M3 5 481, une pointe moustérienne sur éclat levallois, dont le bulbe a été enlevé et le talon retouché bifacialement. Elle est patinée et non-brûlée.***

Since we did not have a lot of time to devote to layer 5, we used the collectives technique there, too. As a result, most of the lithics were not given individual numbers: they are part of collectives, and share numbers with 10 or 20 or 30 other pieces. (In layer 4, the collectives can have hundreds of pieces each.)

Breaking the typological categories down into raw material groups is therefore very difficult, because there are very few cases where numbered pieces from the typology list match identical, individual, numbered pieces on the raw materials list. Instead, I have the raw material details for, say, 17 pieces in a collective, and typological information that one of them is a side-scraper: but which one? Occasionally, over the years, while going through the raw material identifications, I would write a brief note about a particularly interesting or unusual tool type – but I generally did not have the time to do that. Occasionally, more recently, while going through the typology identifications, I would note down that a piece was an unusual or particularly good example of a raw material type – but I generally did not have the time to do that, either. However, that does allow me to add a couple of pieces to the typology list in this layer. Tables 8.4 and 8.5 give the rather frustrating results. Only 13 pieces from the Liste Bordes category have firm raw material identifications, along with 10 flakes and 3 blades. Seventeen of those 26 pieces are in unidentifiable raw material (patinated or burned). Incidentally, although the remains and the sediments in layer 5 did not appear particularly burned while we were digging, 57 pieces (10.4%) of the lithics were classified as burned raw material types, so there was fire around somewhere.

What is perhaps surprising about Tables 8.4 and 8.5 is that, even allowing for the very small numbers of tools and flakes, the raw material identifications in those categories are not wildly out of synch with the provenances of the entire assemblage. Pieces from Murs and Méthamis are present, but not abundant; other source areas are scarce or absent. Layer 5's assemblage is not very diverse, in terms of raw material types. Only the main source areas are represented, with nothing much that is unusual, nothing unknown, and nothing at all that is new.

One other note in my lists is worth mentioning. One piece, a double straight side-scraper made on an atypical Levallois flake (piece number L5 5 1080) has a double patina: the original flake was made, abandoned, and patinated before being found and retouched to make the scraper. This is classified as a raw material type A, patinated and therefore of unidentifiable source. In this case, however, attributing that piece to its ultimate geologic source would likely have been misleading, anyway. There is every chance that the hominin who made the scraper did not pick up the flake at the original source; he or she may have found it at the Bau, for all we know. Recycling of stone tools is known from other sites, too, even older ones such as Qesem Cave (a Lower Palaeolithic site in Israel, see for instance Parush et al., 2015), and I did make note of it from time to time in various assemblages at the Bau. It would probably be interesting for someone to undertake a more detailed study of the Bau assemblages to see how common such behaviour was.

When we eliminate the unidentifiable pieces, and look at percentages of pieces from geographic regions, we get the

Table 8.4. Number of pieces by typology and source area in layer 5. ***Nombre de pièces par catégories typologiques et groupes de sources des matières premières dans la couche 5***

5 n	Liste Bordes	Debris	Eclats	Lames	TamisColl	Nucléus	non-id	Total
Unidentifiable	9	0	8	0	96	0	223	336
Unknown	0	0	0	0	0	0	0	0
Faraud etc.	1	0	0	1	6	0	14	22
Méthamis etc.	1	0	1	2	11	0	33	48
Les Sautarels etc.	0	0	0	0	0	0	2	2
Sault	0	0	0	0	1	0	11	12
Murs	2	0	0	0	29	0	89	120
Local	0	0	0	0	1	0	2	3
St. Jean de Sault	0	0	0	0	0	0	0	0
Roussillon	0	0	0	0	0	0	0	0
Murs-Bezaure Olig.	0	0	1	0	1	0	0	2
Ravin de la Treille	0	0	0	0	0	0	1	1
St. Trinit	0	0	0	0	0	0	0	0
Durance	0	0	0	0	0	0	0	0
Tertiary Calavon	0	0	0	0	0	0	0	0
Tertiary Murs VdeV	0	0	0	0	0	0	0	0
Mormoiron	0	0	0	0	0	0	0	0
R de Guérin	0	0	0	0	0	0	0	0
N.Aurel	0	0	0	0	0	0	0	0
Total	**13**	**0**	**10**	**3**	**145**	**0**	**375**	**546**

results in Table 8.6. Layer 5 has a very low proportion of material from the NE, compared to other layers, and is dominated to an unusual extent by material from the SW, followed by the WSW, with a relatively small proportion from the Near SW. This pattern (illustrated in Figure 8.13) does not resemble what we have seen in any of the other layers so far.

Layer 5's assemblage was big enough and diverse enough to give a reasonable result in the GLM analysis (Wilson & Browne, 2014), with Large rocks and AOSISA (Area Of the Sources in the Source Area) showing up as the most important individual factors (Table 8.7). When the Wald Chi-Square values are converted to percentages (Table 8.8), the combination of Terrain variables accounts for just over 33.2%, and the Raw Material variables make up 55.3%. As was discussed in chapter 7 for layers E and C, in the lower layers of the site, up to and including E, the Terrain variables always dominate (recall that layer E's values were 56.1% for Terrain and 33.5% for Raw Material). Layer C, which we considered out of stratigraphic sequence because it was in the Tranchée Moulin, was the first we had seen with that pattern inverted, and Raw Material variables making up more than the majority. Layer 5 is the layer that most likely should follow E in stratigraphic sequence, and the switch in importance of the two subsets shows up here. Every layer in the site above E has the Raw Material variables dominating the Terrain variables, suggesting rock characteristics are now more influential than ease of access in choices of raw material procurement. Even though layer 5's assemblage includes the usual main suspects for types and sources of raw materials, the different proportions that we saw in Table 8.6 translate into different procurement strategies when seen through the lens of the GLM.

8.8. Conclusion

Good planning and educated evaluations of where and how to dig are essential when one undertakes the excavation of an archaeological site. Good luck helps, too. The fact that Robert Brandi had a winch and could move a large boulder out of the way meant that we exposed what turned out to be the least brecciated part of the upper deposits, and discovered a series of rich and interesting archaeological layers. The lowest of these, layer 5, is both typical of the Bau, and unlike any other layer. The faunal list includes many of the main species found in other layers, especially among the herbivores, but only one or two individuals of

Table 8.5. Percentages of pieces by typology and source area in layer 5. ***Pourcentages de pièces par catégories typologiques et groupes de sources des matières premières dans la couche 5***

5 %	Liste Bordes	Debris	Eclats	Lames	TamisColl	Nucléus	non-id	Total
Unidentifiable	69.2	0	80.0	0	66.2	0	59.5	61.5
Unknown	0	0	0	0	0	0	0	0
Faraud etc.	7.7	0	0	33.3	4.1	0	3.7	4.0
Méthamis etc.	7.7	0	10.0	66.7	7.6	0	8.8	8.8
Les Sautarels etc.	0	0	0	0	0	0	0.5	0.4
Sault	0	0	0	0	0.7	0	2.9	2.2
Murs	15.4	0	0	0	20.0	0	23.7	22.0
Local	0	0	0	0	0.7	0	0.5	0.5
St. Jean de Sault	0	0	0	0	0	0	0	0
Roussillon	0	0	0	0	0	0	0	0
Murs-Bezaure Olig.	0	0	10.0	0	0.7	0	0	0.4
Ravin de la Treille	0	0	0	0	0	0	0.3	0.2
St. Trinit	0	0	0	0	0	0	0	0
Durance	0	0	0	0	0	0	0	0
Tertiary Calavon	0	0	0	0	0	0	0	0
Tertiary Murs VdeV	0	0	0	0	0	0	0	0
Mormoiron	0	0	0	0	0	0	0	0
R de Guérin	0	0	0	0	0	0	0	0
N.Aurel	0	0	0	0	0	0	0	0
Total	**100**	**0**	**100**	**100**	**100**	**0**	**100**	**100**

Table 8.6. Percentages of identified pieces (excluding unidentifiable and unknown source pieces) from each geographic region for layers K2, K1-J, I, H, G, E, C and 5. ***Pourcentages des pièces identifiées (excluant pièces indéterminables ou de sources inconnues) de chaque région géographique pour les couches K2, K1-J, I, H, G, E, C et 5***

Geographic area	K2 %	K1-J %	I-I4 %	H-H2 %	G %	E %	C %	5 %
Local	0.2	3.4	1.3	0.7	0	0	0.5	1.4
NE	19.4	20.4	12.3	15.7	41.5	31.0	39.5	5.7
E	0	0.3	0.1	0	0	0	0	0
Far SE	0	0	0	<0.1	0	0	0	0
Far S	0	0	0.1	0	0	0	0	0
S	0	0	0	0	0	0	0.5	0
SSW	0	0	0.1	0	0	0	0	0
SW	44.3	54.1	35.0	30.0	48.4	25.4	17.6	59.0
Near SW	23.0	15.0	29.1	34.8	9.6	29.9	30.7	10.5
WSW	12.4	6.6	21.9	18.6	0.5	12.7	11.2	22.9
NW	0.7	0.1	0.1	0.2	0	1.0	0	0.5
Total	**100**	**100**	**100**	**100**	**100**	**100**	**100**	**100**

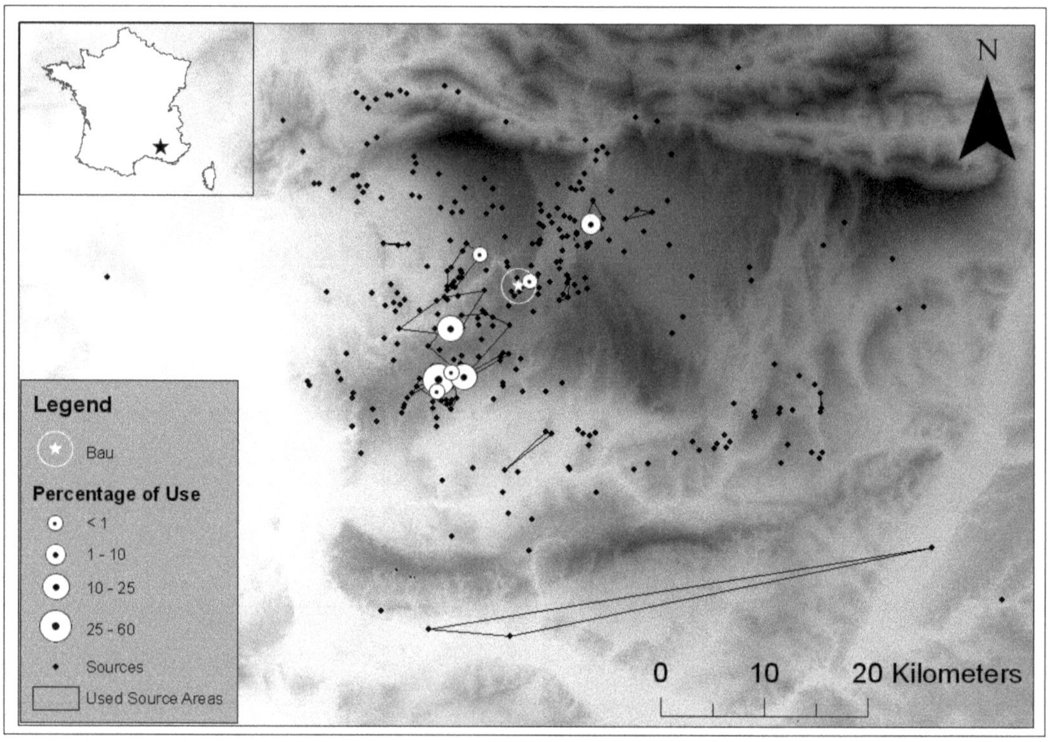

Figure 8.13. Map of the raw material source areas used for layer 5. ***Carte des aires de sources de matière première utilisées pour la couche 5.***

each, giving a very diverse set of animals. The stone tools are visually reminiscent (at least to me) of the collections from the early excavations in the Tranchée Moulin, but their typology differs from what we ourselves found in the Tranchée Moulin, and from what we found in lower layers. Their raw materials are completely unsurprising in terms of types and sources, until you look at the proportions of the types, where a very different pattern emerges. What little we can interpret of the climate at the time seems similar to other layers in the middle part of the stratigraphy of the site, but the hominins' behaviours seem to have been different. And unlike most of the other layers to which it might first be compared (E, D and 4), it is unburned.

The easiest way of interpreting the layers in the upper part of the site would be to draw equivalences between those in the Tranchée Moulin and those in the tranchée L, and assume that each layer extends across the site from one trench to the other. Easy ways so rarely work. Layer 5 is the first chance we have at that, and it is not going to cooperate. It is directly under the dalle. There is some justification for treating the dalle as a stratigraphic marker, because it can be traced across the front of the deposits, and it has been uncovered and followed along the rear of the deposits. Assuming, therefore, that layer 5 being under the dalle means we should be able to line it up with an equivalent layer in the Tranchée Moulin, what do we get? It should be equivalent to layer E, which is directly under the dalle in the Tranchée Moulin. Layer E (see chapter 7) is highly burned, very black, and very fine-grained. Layer 5 is none of those things, and their lithic assemblages differ in many ways.

Perhaps layer 5 should be matched with layer F, which is coarser and not burned. We do not know much about layer F (see chapter 6), but at least the fauna can be made to match: the species present in F are also present in 5. The lithics do not match as well, however, either typologically or in raw materials. Layer 5's industry contains a much higher proportion of Levallois technology and of blades than F has. The raw materials in F are sadly unidentifiable in better than 90% of cases, but in the few remaining pieces the same sources are used as in 5, but the proportions are different. Barring further excavations, therefore, layer 5 will have to remain distinct: a typical Bau layer, unique in its own way.

8.9. Sommaire du chapitre

Ce chapitre permet d'introduire la tranchée L, dans la partie centrale du site, et de décrire la couche 5, qui est la plus profonde des couches identifiées dans cette tranchée. Nous avons commencé ce travail dans la partie centrale du Bau parce que la Tranchée Moulin était compliquée à fouiller (voir chapitre 7). Avec l'aide de Robert Brandi qui avait l'équipement nécessaire, nous avons pu enlever de gros rochers devant le sommet du remplissage. Ceci a permis de laisser un passage ouvert, une "entrée", vers l'intérieur des dépôts.

Le sondage initial dans la partie la plus au nord de la tranchée L n'était pas très intéressant, mais plus on progressait vers le fond de l'abri, plus on voyait apparaître des sédiments de couleur foncée, comme si nous nous approchions d'une aire de combustion. Cette couche archéologique

Table 8.7. Generalised linear model for raw material source choices in 5. P-values in bold font are statistically significant. Sample: 208 pieces from 8 source areas

Omnibus test: likelihood ratio chi-square = 355.587, df = 10, p = <0.001

Variable	Beta	S.E.	Wald Chi-Square	P-value	AIC
Quality	2.988	1.7050	3.071	0.080	76.556
Extent	1.647	1.2272	1.802	0.179	
Difficulty (Cal/km)	0.012	0.0487	0.056	0.812	
Small rocks	-11.252	8.5261	1.742	0.187	
Medium rocks	-6.786	5.3217	1.626	0.202	
Large rocks	5.650	2.2338	6.397	**0.011**	
Very large rocks	2.591	2.6185	0.979	0.322	
Calories	0.000	0.0019	0.014	0.905	
AOSA	-4.2 e-7	3.1 e-7	1.894	0.169	
AOSISA	0.000	6.2 e-5	4.554	**0.033**	
(Intercept)	-16.455	9.7435	2.852	0.091	

Table 8.8. Wald Chi-Square values as percentages, assemblage from layer 5

Variable	Wald Chi-Square	Wald Chi-Square as %
Quality	3.071	12.3
Extent	1.802	7.2
Difficulty (Cal/km)	0.056	0.2
Small rocks	1.742	7.0
Medium rocks	1.626	6.5
Large rocks	6.397	25.6
Very large rocks	0.979	3.9
Calories	0.014	<0.1
AOSA	1.894	7.6
AOSISA	4.554	18.2
(Intercept)	2.852	11.4
Total	**24.897**	**100**

était sus-jacente à une dalle, comme la couche D de la Tranchée Moulin. Alors pourquoi ne pas continuer? Pour fouiller cette couche potentiellement intéressante, il fallait traverser une épaisseur d'au moins un mètre de brèche. De plus, la tranchée menait directement vers un gros rocher posé sur la surface du remplissage alors que la couche devenait de plus en plus riche et noire près du gros bloc.

Au lieu de fouiller sous le rocher, nous avons commencé la fouille de la bande M, en élargissant notre tranchée d'un mètre. Ceci nous a appris deux choses: d'abord, la couche que nous avons nommée couche 4, devenait de plus en plus riche et de plus en plus noire vers le fond de l'abri, même dans la bande M: il fallait donc poursuivre. Par ailleurs, la brèche était plus mince dans la bande L que dans la bande avoisinante. Par chance, nous avions orienté notre tranchée entre deux dômes de brèche: celle-ci devenait plus épaisse des deux côtés de la bande L. Ainsi, il a fallu enlever le gros bloc, pour ouvrir le passage.

Le rocher était trop épais pour être cassé au marteau-piqueur, et le dynamiter était évidemment hors de question. En 1993, Serge a obtenu un produit appelé Nitro-Bickford (celtamex) qui était un ciment expansif. Nous avons mis quelques jours à percer le rocher d'une trentaine de trous qui ont été remplis par le mélange de poudre celtamex avec de l'eau. Pendant la nuit le mélange a sèché et développé 6 000 tonnes de pression au mètre carré, ce qui a lentement fracturé le rocher. Le lendemain matin nous l'avons trouvé réduit en plusieurs grands morceaux, pour ensuite passer deux jours entre 8 et 10 personnes à enlever ces blocs!. La voie était maintenant libre pour atteindre le fond de l'abri. Plus tard, quand nous avons fouillé dans le fond de l'abri, nous avons trouvé que ses contours étaient très irréguliers (voir chapitre 9).

Dans cet ouvrage, la description des couches archéologiques est donnée en partant du bas, à l'exception des couches de la Tranchée Moulin, mais ce n'est pas l'ordre dans lequel nous les avons découvertes. Dans ce chapitre, j'ai commencé par la tranchée L et son histoire générale. Je m'attacherai ensuite à la couche 5 qui est la plus profonde de la tranchée et qui a été trouvée après plusieurs années de fouille de la couche 4. Cette couche 5 se trouve à une profondeur d'environ 320 à 340 cm sous zØ et est située sous la dalle qui semble se poursuivre depuis la Tranchée Moulin. Etonnamment, la couche ne ressemble pas du tout aux couches de la Tranchée Moulin: elle comprend des sédiments assez grossiers (cailloux jusqu'à 10 cm) contenus dans une matrice fine jaunâtre. Ils sont normalement meubles mais parfois un peu indurés ou cimentés, apparemment non-brûlés. Ces sédiments n'ont pas été échantillonnés pour leurs pollens, mais nous avons de bons résultats pour la faune (Tableau 8.1). Les différentes espèces ont dû être chassées selon des stratégies variées et plutôt sélectives. Par ailleurs, le climat devait être froid, mais pas extrêmement rigoureux.

La couche 5 a livré un bon nombre de pièces lithiques relativement riche en pointes moustériennes, racloirs et lames: une industrie moustérienne progressive par conséquent. Les pièces sont souvent patinées, et pour le reste on retrouve les types de matières premières habituelles que nous avons vues dans les autres couches, mais avec beaucoup plus de matériel du SW et du WSW, et très peu du NE.

En résumé, la couche 5 livre des espèces fauniques que nous avons déjà vues dans d'autres couches. Chaque espèce est représentée par un ou deux individus et l'assemblage est ici plus diversifié; on y trouve par exemple plus de carnivores que dans d'autres couches. Les détails typologiques des pièces lithiques indiquent des ressemblances avec celles issues des collections anciennes de la Tranchée Moulin. Elles diffèrent cependant de ce que nous avons trouvé nous-mêmes dans la Tranchée Moulin ainsi que dans les couches inférieures. Les mêmes types de matières premières que dans les autres couches ont été utilisées (voir Tableau 8.6) mais avec des proportions que nous n'avons pas vu ailleurs. A la différence des autres couches auxquelles on pourrait la comparer (les couches D, E et 4) et qui sont en positions stratigraphiques proches, la couche 5 n'est pas brûlée. La couche 5 est donc typique du Bau, et en même temps unique.

9

La couche 4

9.1. Introduction

If the Bau de l'Aubesier is known for anything, it is known for this layer: the couche 4. Rich in lithics and bones, it is 15 or 20 cm thick in most places (sometimes thicker), black and highly burned. Our excavation did not find the full extent of it, despite excavating it over more than 40 square metres from the back of the site forward in bands I through P (see Figure 1.11). It seems to continue on both sides of our excavated area, especially to the east (i.e., beyond band P), but towards the front it does become thinner, paler and less rich at about band 10, and disappears by band 13 or 14. How rich is it? The complete raw materials identification list that I compiled over the years includes 40770 pieces; 20167 (49.5%) of them are from the couche 4. The proportion would be higher, but every summer I did not have enough time to do all of what we had found in the couche 4. The typological list we put together in the summers of 2013 and 2014 includes 62953 pieces; 40048 of them (63.6%) were from the couche 4.

The couche 4 was thus the dominant contributor to the lithics we found at the Bau. Its identifiable faunal component is not as strong, but still dominates the upper part of the site, and it also yielded 6 Neandertal teeth. Furthermore, it has a stratigraphical peculiarity, in that it rises up the back wall of the rock shelter: the material within it is horizontal on the main layer, but becomes oriented vertically at the wall, as if it was a too-large carpet spread over the rock shelter floor that had to roll up the wall, too. The main part of the couche 4 is located at z values between about 280 and 305 cm below zØ, but at the back of the rock shelter it rises up by 90 to 100 cm above that level. Such a clearly important layer of course received extra attention, with palynological, micromorphological, sedimentological and dating studies focussed upon it. This chapter will discuss and synthesise the results of all of this work, and try to explain how this layer formed, as well as what it can tell us about the hominins involved in its formation.

To start with, here is an obvious question: how old is it? I have a handwritten note in my documents from 1995 showing the couche 4 dated to 90,000 years by ESR and 110,000 by U/Th dating, but I have not been able to find the original source(s) of that information. Hélène Valladas and her team (who provided dates for layer H1) (Lebel et al. 2001) also attempted to get dates for the couche 4, using thermoluminescence dating (TL). The couche 4 seems like an excellent candidate for TL: full of highly burned flint. However, the first attempt came back with dates that were clearly wrong, showing 4 as older than H1 (which is more than 3 metres lower in the stratigraphy). We tried again, with a new series of dosimetres inserted into the deposits, but I do not have any final result for that. In publications concerning the Neandertal teeth (e.g. Lebel & Trinkaus 2001), the couche 4 is described as dating to the "early Late Pleistocene", on the basis of the faunal assemblages described by Fernandez et al. (1998) and Fernandez (2001). That matches fairly well with those ESR and U/Th dates. The other relevant result is the U/Th date on the speleothem in zone I5, mentioned in previous chapters: it was dated to 141,854 (+5257, -4956) years (Lebel 2006). The couche 4 was above that speleothem, so would be younger.

9.2. Sediments, evolution of the site, and excavation of the couche 4

This section will start with the dalle upon which the couche 4 sits, before discussing the couche 4 itself and then the raised part, which we called the couche 4P (P is for "paroi", the rock wall), followed by the extension eastwards into bands O and P, and finally the contacts between 4 and layer 3 above it.

The dalle has already featured in two other chapters of this book: it is a fine-grained, highly cemented layer, 20 to 30 cm thick in general, that apparently extends across the upper part of the deposits from the Tranchée Moulin to the tranchée L, and perhaps beyond. It has an eroded upper surface, and the cone-shaped speleothem at the back of the site, in zone I5, was built up by water flowing out onto this dalle. It seems thus to have been a stable and longstanding surface of the deposits, a floor, upon which later hominin activity occurred, resulting in the build up of the couche 4 and of layer D.

At the front of the tranchée L, where we started digging, my notes from 1990 say that this surface "présente une structure feuilletée, compacte et dure, brune à grisâtre." Either the dalle, or perhaps the sedimentary layer on top of it at the front of the trench, was labelled layer 4A. That layer is described in my notes as "sédiment très meuble et grumeleux, gris-brun à particules blanchâtre[s]. Traces d'occupation et ossements bien conservés." Elsewhere the dalle was consistently lacking in archaeological material, but if 4A was the dalle, then it did contain material. I am more inclined to believe that 4A was the greyish sediments on top of the dalle, however. The only lithics attributed to layer 4A came from zones K12, L11, M9 to M12, and N9, all towards the front of the trench. They consisted of a total of 273 pieces, all tamis. Identifiable bones from 4A were also found and are mentioned in Fernandez (2001), but no zones are given for them in that work. The material from 4A is included in the overall layer 4 assemblage below.

Midway along the trench, in zone M8, in May 2006 M.-A. Courty took some samples for micromorphological analysis (Courty 2006). These included sample IV-4, described as coming from the 4/5 transition, at z= 295 cm, and consisting of a "brèche granulaire calibré cohérent sans cryoclastes ni artéfacts." One sample in M. Wilson's (1997) Master's thesis, sample MM22, is also from M8, at -290 cm, and described as coming from the base of the couche 4. She says "Son contenu anthropique (10%) est beaucoup moins riche mais la cendre micritique est abondante . . . Les fragments calcaires grossiers . . . sont fortement altérés par l'action des micro-organismes (injections, lichens, bactéries). Plusieurs phases (5-6) de cristallisation successive (jusqu'à 2,4 mm) sont observés . . ." (pp. 185-186).

At the rear of the trench, the test pit that reached layer 5 (chapter 8) went through the dalle. In my notes from July 25th 1995 I talk about the test pit being dug in the southeastern quadrant of zone L4. As the dig progressed, the digger told me that the sediments became less black, then became indurated and almost brecciated. Lower still, pieces of breccia composed of large clasts without much fine sediment appeared: these were "assez facile à casser." The breccia was not evenly distributed across the square, and seemed to persist longer in the northeastern part of the quadrant. By a depth of a little more than 15 cm below the previous level (i.e., top of the dalle/base of 4), the digger reported that "il arrive sur une couche de plus grosses roches dans la partie ouest du quadrant. Il trouve plus de matériel au sud, dans les sédiments friables, qu'au nord dans la brèche." The "couche de plus grosses roches" was layer 5. Other than in the northeastern corner, the base of the breccia or dalle had been reached by a z value of 298 cm below zØ. This leaves open the possibility that there was archaeological material in the dalle, but if so it was labelled as layer 4 (not 4A).

The test pit showed that despite the rather complicated series of layers along the wall of the rock shelter, which we will deal with below, the dalle itself does extend to the back wall, and the couche 4 is on top of it, with a clear transition between the black layer 4 and the grey, hard, dalle beneath it. The brecciated dalle contains, if anything at all, less archaeological material than the layers above and beneath it, and it is also broken up into fragments in that area near the wall. At times it seemed as if it might have served as a barrier between areas that were archaeologically more or less rich. That impression was never proven, but the idea kept cropping up, so is worth mentioning here.

Opening up the zones to the west, between the Tranchée Moulin and the tranchée L, soon showed that there was an obstacle in the way: a stalagmitic accumulation referred to here as the speleothem. Figure 9.1 shows the situation in 1994, when we were digging in J4 and K4. The top of the speleothem is visible in the middle-left side of the picture, attached to the rock shelter wall (a more recent view, from the opposite direction, is in Figure 7.5 in chapter 7). In 2006 Serge Lebel fully exposed it in order to take samples. He describes it (Lebel 2006) as being 75 cm high and 15 cm wide at its summit. It is oriented north-south (as we see in Figure 9.1), and disappears into the sediments of the section wall between bands 5 and 6, making it at least 75 cm long. It covers a large part of the zone I5. "Ce spéléothème possède une structure laminée formée de minces couches de calcite superposées. Il est composé de nombreuses lamines de croissance verticales variant de quelques millimètres d'épaisseur (sur le pourtour) à ½ centimètre (au centre) et les lamines entourent progressivement les premiers stades de sa formation. . . Aucun vestige archéologique ou particules de charbon pouvant être liées aux activités humaines en surface n'ont été piégés dans la stalagmite ou les concrétions" (Lebel 2006).

The deposits thus seem to have had a stable upper surface for some length of time, a surface with its own architecture: the speleothem would have been an obstacle, a natural divider, or perhaps something to lean against, for the hominins whose activities created the couche 4. The rest of the surface, the dalle, would as far as we can tell have formed a gently sloping, gently undulating, floor. Just as we used to think that the site consisted only of the upper deposits, the hominins of that earlier time may well have thought of the rock shelter as consisting only of that floor, the walls around it and the roof above it. The rest of the deposits were just the slope you climb to get to the site.

A lot of hominins must have climbed that slope, over an unknown period of time. The composition of the couche 4 is dominated by archaeological remnants: there is very

Figure 9.1. Looking from the tranchée L westwards towards the Tranchée Moulin in 1994. The top of the speleothem in zone I5 is visible on the left, spreading down and out from the rock shelter wall. ***Vue vers l'ouest, de la tranchée L jusqu'à la Tranchée Moulin en 1994. Le sommet du spéléothème de la zone I5 est visible à gauche, descendant et s'élargissant à partir de la paroi rocheuse.***

little sediment. The layer is not as rich at the front of the trench, but from about band 9 back to the wall of the rock shelter, it is thick and black and crammed with material. There were some areas of brecciation of the deposits, and sometimes we would think that maybe we were finding an internal structure to them. . . but that never seemed to materialise. Until we got to the back wall, the couche 4 proper was an undifferentiated, unstructured, continuous, horizontal layer of soft, humid, plastic black material surrounding a whole lot of bones and flint.

We dug by décapages and collectives. Each décapage involved every available zone, so that we would see a complete "carpet" of material. Objects were left in place if they were larger than about 3 cm, and digging would proceed until there was no open space left to dig. Figure 9.2 shows a close-up of part of such a décapage. Meanwhile, each zone was divided into four 50x50 cm quadrants, and smaller material from each quadrant was collected in bags (one each for lithics, bones and teeth), which were kept associated with the bucket which received the sediments from that quadrant. These were going to be part of a collective, and the material from the bags and the sediments would be assigned to one of three catalogue numbers, for lithics, bones and teeth.

In each décapage, one square metre zone could contain several hundred coordinated objects – and we had at least five décapages. The thickness of a décapage depended on the abundance of objects over 3 cm in size, but usually a décapage would be two or three cm thick, and as soon as we coordinated the objects of that décapage, as many more would start turning up underneath them. They were almost always lying flat, although they could be oriented in any direction, as Figure 9.2 shows.

The smaller material, in its separate bags, would be kept associated with its respective bucket of sediments as they went through the sieving and washing process. First, the sediments were sieved through two stacked sieves, giving a coarser and a finer fraction. Then each fraction was washed, still in the sieve, in a tub of water at the site, and put into a plastic bag. Its identifying label was put into a smaller plastic bag and placed inside the larger bag. All this was brought up from the site to our lodgings at the top of the hill, where the coarser fraction was usually washed again in running water. Everything was placed on plates, with the labels clipped to them, to dry in the sun. Figure 9.3 shows a typical serving of couche 4 coarse fraction, on its plate. Once it was dry, the material on the plate would be sorted into lithics, bones and teeth, and bagged with the rest of the collective. Most plates had almost no actual sediment left over on them once that was done. All this took so much time that we did not always finish sorting material before the season was over, so we would start the next summer's excavation with bags of material that the new diggers would sort, which had the benefit of familiarising them with what couche 4 material looks like.

In amongst all this archaeological material, though, there was still some sediment, so in order to get a better idea of the composition of the couche 4, I took a sample from M8 in 1992 and had it analysed geochemically. First, I extracted every visible archaeological piece; that reduced the weight of the sample by half. The remnant was analysed, and Table 9.1 gives a summary of the results, converted to approximate percentages of the complete sample. The "others" listed in Table 9.1 include elements such as potassium, sodium, iron, aluminum and various other metals.

There are only very small amounts of iron and manganese in the sample, indicating that its black colour is due to burning, not due to the colouring effect of those elements. The state of the flint and bones had already told us that, anyway: the bones cover the full range of burned colours (brown, black, white, even blue), and the flint pieces are often rubefied, cracked or have indentations where potlids (thermoclastic fragments) have popped out. We often found the potlids themselves, too. The next question is: what was burned? The deposits were full of tiny carbonaceous particles, but we rarely found anything identifiable as

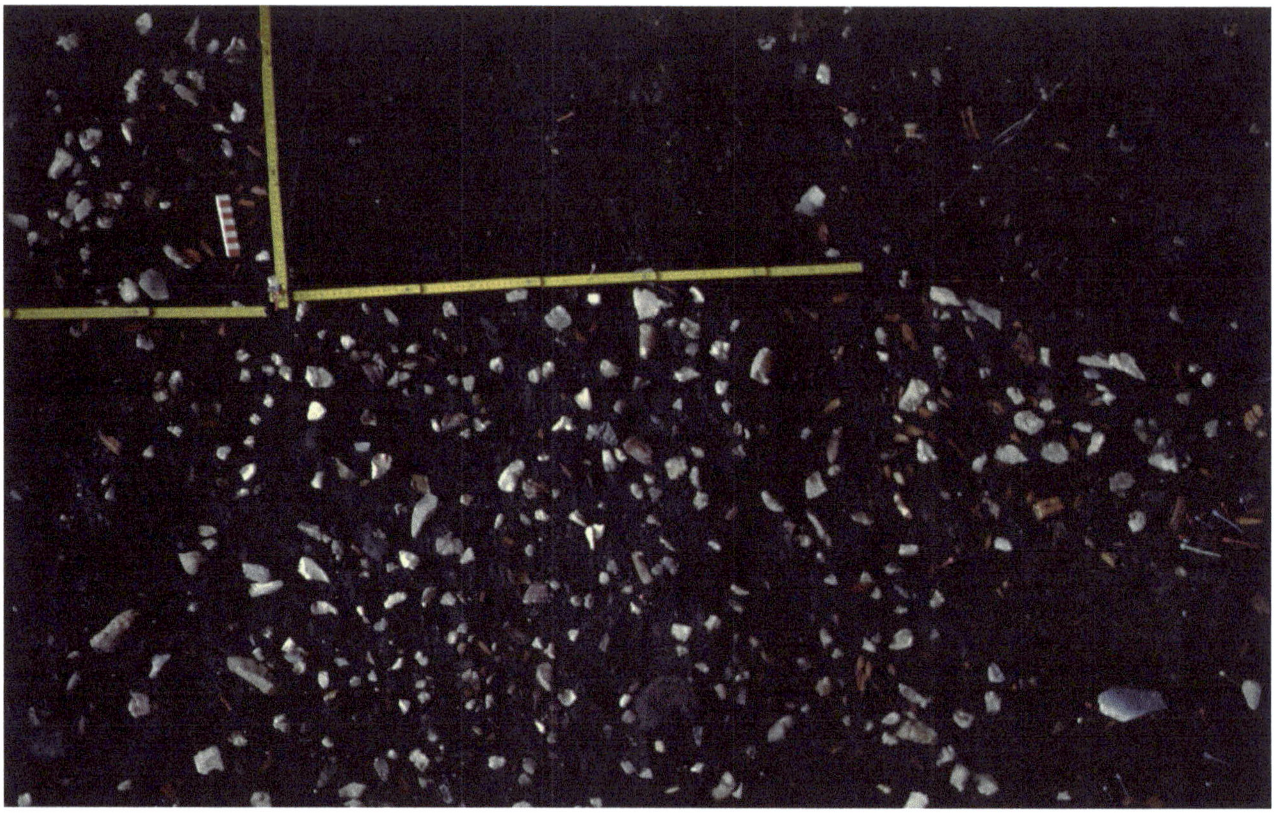

Figure 9.2. One décapage of the couche 4 during coordination, with one quadrant already finished. *Un décapage de la couche 4 en train d'être coordonné, avec un quadrant déjà completé.*

Figure 9.3. Sieved but unsorted coarse fraction tamis components from the couche 4, washed and drying on a plate. They consist almost entirely of burnt bone (black and brown colours) and flint (paler colours). *La fraction grossière d'un tamis de la couche 4, lavée et en train de sécher sur une assiette avant d'être triée. Elle se compose presqu'entièrement de morceaux d'os brûlés (noir ou brun) et silex (plus pâle).*

Table 9.1. Summary geochemical analysis of layer 4 sediments from zone M8, sample 208, collected in 1992. ***Résultats géochimiques sommaires de l'analyse de sédiments de la zone M8, couche 4, échantillon 208, pris en 1992***

Component	Approximate % by weight	Probable source
visible archaeological material	50	
humidity	20	
$CaCO_3$	15	rock shelter limestone
$Ca_3(PO_4)_2$	10	bones and teeth
organic	3	plant and animal material
SiO_2	1.5	flint; possibly some from plants
others	0.5	from clays, etc.

charcoal, and almost nothing more than 1 cm in size. The geochemistry can help us to some extent with this question, even though the couche 4 is not ideal for such studies: most of the potential markers of fire and fuels have other origins in this case. For instance, calcium, which could indicate plant ash, is problematic when the sample comes from a limestone rock shelter. Silica can come from plants, especially grasses, but the abundance of flint in this layer obscures any such contribution. Aluminum can be a marker of the use of plant material, rather than animal, but there is very little here and it is more likely to come from the insoluble clay component of the limestone making up the rock shelter. On the other hand, the P/K ratio may be significant. The relatively high amount of phosphorous (P) present is due to all the bones. Potassium (K) can indicate the use of plant material, especially wood, but there is very little here. The dominance of P over K therefore implies that animal remains (bones, meat and fat), rather than plant material, served as the fuel for the fires.

The fires seem to have been structureless: that is, not confined within a circle of rocks, not placed in holes dug down into the sediments, not even limited to particular areas. We had more than 40 square metres of burnt deposits, and try as we might, we never found any visible anthropogenic differences within that extent, other than the decline in abundance of material north of about band 9. Some variations in degree of burning of flint pieces will be described below, in the lithics section, but these variations were not visible at the dig. We got a bit excited about a lighter-coloured circle in M5 in 1998: it was 10-15 cm in diametre, and was located underneath a rock of about 15 cm size, but it did not persist below a couple of cm depth, and had no significant differences in its contents. It might have been due to water being channeled under the rock and leaching out some of the smaller burnt material. The only other features of note were due to brecciation. In some places the burnt deposits themselves were cemented together, which made them very difficult to dig. On one memorable occasion in 1993, one digger trying to deal with brecciated black sediments (in a deep trench in a part of the rock shelter that never gets sunlight) got fed up with not being able to see what he was doing, took a block of the breccia and brought it out to the front part of the site – which of course is *not* proper technique! Then he flipped his block over, and discovered that he had the entire side of a horse's jaw, teeth and all, but broken into thousands of millimetric fragments. If it had not been cemented, it would have just seemed to be thousands of tiny bone and teeth fragments. . . like the rest of the couche 4 contained.

The lack of discernable structure in the deposits led us to conclude that they are a palimpsest, an intermingled accumulation of material from numerous separate episodes of occupation of the site, and that during those occupations fires were lit directly on the floor of the site. The deposits may be a mixture of residues from fires of different sizes and intensities, and of the "utilisation multiple et successive de foyers" (Lebel 1996). M. Wilson (1997) examined the idea of the couche 4 being a palimpsest, and determined that it must be, based on several lines of evidence. There are, for instance, "tapis de sol" in her micromorphology samples from some areas (she cites N8 and L7), showing that the deposits had been walked on, which means the fires were not burning at the time. The various samples contained mixed combustion residues, limestone fragments and both burned and non-burned flint pieces; the sediments had been rubefied, and the larger grains had "coiffes litées" around them, composed of a build-up of ash, possibly linked to phases of freezing and thawing in a periglacial environment. The bones in some cases showed orange colours that imply that they were burnt in the presence of oxygen, while at the same time the charcoal present (even if it is very small fragments) implies a lack of oxygen. The samples also contain white ash, which results from intense combustion with a lot of oxygen, and grey ash, which results when the intensity of the fire dies down. The variety of colours in the bones translates into them having been burned at temperatures between 100 and approximately 650 degrees C. Some show signs of having been burned more than once (Wilson 1997, pp. 133-136). Thus these deposits must have built up over a long period of time, through repeated but separate occupation events.

One "outcrop" of breccia was located along the eastern wall of the trench, in the N5/M5 area. We found it convenient for stepping on, and placing objects on, as can be seen in Figure 9.4. It is composed of yellowish sediments, and postdates the couche 4. Figure 9.4 also shows two other very interesting features: the black layer goes up the rock wall at the back of the trench (and I will come back to that topic), and there is a large swath of dark colour rising up the eastern wall of the trench, in zone M4. That swath consisted of more than dark colour: it also contained archaeological material, which was recorded as coming from layer 4M (M for "mur"). The material from 4M included bones (Fernandez [2001] lists an MNI of 1 for

Figure 9.4. The couche 4, second décapage, 1995, showing also some breccia along the eastern side of the trench (under the sign), the dark swath of layer 4M rising up from square M4 behind the breccia, and the couche 4P rising up the wall at the back of the trench. ***La couche 4, deuxième décapage, 1995, avec de la brèche visible le long du mur est de la tranchée (sous le panneau), la montée noire de la couche 4M dans la zone M4, derrière la brèche, et la couche 4P qui monte la paroi rocheuse en arrière.***

red deer in 4M), and lithics (300 pieces in the typological list, of which 285 are tamis/collective). That material is included in the overall couche 4 assemblage below, but the existence of such a swath deserves some attention.

The presence of swaths of darker colour in the sediments above the couche 4 had already been described before we got to 4M, but this is by far the most important of these. Its components seem no different from those of the main couche 4, so its origin is very likely to be due to some sort of squeezing action or injection of couche 4 material into the overlying sediments, after they had been deposited. It is located just to the north of a series of broken-up blocks of breccia, which may have constrained that squeezing action. M. Wilson (1997) describes a micromorphological sample of it (MM16, taken at z=-250 cm) as having fewer crystalline structures than the main couche 4 samples, as being very deformed, and notes that despite this it contains undisrupted "coiffes litées" around the grains within the sediment. She attributes it to a slow-acting "phénomène d'injection" in a periglacial environment (p. 148), without specifying what that means.

Chapter 8 described how we removed a large rock from the top of the deposits, opening up the rear of the tranchée L to excavation. As we dug down at the back wall of the rock shelter, while we should still have been in layer 2, we started to find a thin black line parallel to the wall, but separated from it by a couple of cm thickness of yellowish sediments. (Incidentally, there was no breccia near the back wall: we were beyond the dome.) This persisted and thickened as we dug down another few cm, so we decided to leave the sediments near the wall in place, until we reached the couche 4 in the rest of the area. Figure 9.5 shows this situation. As we dug in the zones near the wall, we found that as we got lower, dark sediments kept turning up a little farther out from the wall. Finally, as Figure 9.6 shows, we had the newly available area cleared everywhere down to the contact with the couche 4. It was continuous: the couche 4 covered the entire area, lying flat until it got within a short distance of the back wall, and then rising up on a steep angle, until it was vertical. The top of these raised sediments was between 90 and 100 cm higher than the main level of the couche 4. We found that the back wall curved southwards into zone M2, so we stopped digging at the southern limit of that zone, but the black line carried on into those sediments, too. There was even a pocket (8 or 9 by 5 cm in size) of porous sediments in the centre of that south wall which seemed to have a faint draft of air coming out of it, which gave promise of the rock shelter continuing for some distance.

We named this tilted part the couche 4P, and excavated it from the outside inwards, towards the rock wall, rather

Figure 9.5. The upper part of the couche 4P (black) showing through the grey-yellow sediments at the back wall of the rock shelter (with white chisel marks). ***La partie supérieure de la couche 4P (noire) visible dans les sédiments gris-jaunes près de la paroi de l'abri (avec des marques blanches de ciseau).***

than from the top surface downwards, since it would have been ridiculous to treat it as if it was a normal horizontal layer. This got complicated. It was just as rich in material as the normal couche 4 was, as Figure 9.7 and the closer view in Figure 9.8 show. The tools and bones were horizontal on the normal couche 4, and followed the change in orientation of the surface of the bed as it rose to become 4P, where the tools and bones were vertical (while still lying "flat", so to speak, against the surface of the layer; that is, they maintained the same attitude as the layer itself). We used coloured plastic picks to try to get the objects to stay where they were while we completed each décapage. Figure 9.7 shows that we also adopted that technique for the couche 4M.

The couche 4P produced a prodigious amount of material, since each square metre contained far more of the archaeological layer than was present in a square where the layer was lying flat. The material is, as far as I have been able to tell, indistinguishable from the material of the couche 4, or 4M. I have some incomplete notes as to which numbers of pieces correspond to which décapage, but for this, too, I have been able to find no real differences. The material is essentially the same all the way through, in terms of typology, raw materials, and faunal composition.

The couche 4P itself, however, is not simple. The easiest way to describe it is as the couche 4 rising up to follow the rock wall, but that is not really accurate. Our first discovery of it was as a black layer separated from the rock wall by a yellow layer, but as we proceeded to excavate it, we found that in part of zone L3, for instance, the black layer was directly in contact with the rock wall in places where the rock wall bumped outward, while elsewhere there could be four layers in sequence: black, yellow, black, and then yellow again, before reaching the rock wall. In M2 the black sometimes gave way to more reddish colours, and in M3 the layers were black and then dark grey. In some places there was a white layer between two dark layers. Figure 9.9 shows a section across 4P on the western side of L3: the horizontal black layer takes a sharp turn upwards. . . but what is that other black layer behind it? From left to right, we see the rock wall, some yellowish sediments, a black layer, a grey layer, and another black layer forming an L-shape, with more yellowish sediments filling in the L, and the dalle underneath. The front black layer was the richest in archaeological material, but there was some material in the other layers, although according to my notes of July 22nd and 24th 1995, the material from those layers was unburned and highly patinated, and there was very little material in the second black layer. Was the front black layer the couche 4, with the others being something else? Or did the couche 4 somehow bifurcate into two black layers?

In terms of sediments, my notes from July 1995 mention that their texture varied along the rock wall, from compact

Figure 9.6. The upper surface of the couche 4 in the zones newly exposed in 1994. The layer is flat in the front part of the trench, but rises steeply up the back wall, and a dark line shows that it continues into the unexcavated sediments south of zone M2 at the rear of the trench. ***La surface supérieure de la couche 4 dans les zones de fouille exposées en 1994. La couche est à plat en avant de la tranchée, mais monte abruptement près de la paroi de l'abri, et une ligne noire sert de témoin qu'elle se poursuit dans les sédiments non-fouillées au sud de la zone M2, en arrière de la tranchée.***

in some places to more porous elsewhere. There seemed to be larger rocks (4 to 8 cm rather than 1 to 3 cm) at the base of the slope, where the black layer was farther from the rock wall. The grey layer was more sandy or granular than the black layer. In the eastern half of zone L3 there was a clayey, compact, layer near the rock wall, but farther from the wall the sediments were a mix of clay, sand, pebbles and sometimes larger rocks, with greater porosity. In the southeastern area (zone L2), there was breccia containing archaeological material, which took up a larger area as the excavation dug deeper and expanded into M3.

The micromorphological samples analysed by M. Wilson (1997) consisted of three slides taken from the zone K4, all at a depth of about z= -215 cm. She describes the area as having had two black layers, with a yellowish layer in between them, and another yellow layer between the second black layer and the wall. Then she describes both black layers as "la couche anthropique", and finds it to be very similar to the couche 4 (Wilson 1997, pp. 143-144). The other two layers seem to be grouped together, too, as "la couche sableuse", and this is described as being highly porous, composed of carbonate particles, and without secondary cementation (Wilson 1997, p. 144). Contacts between the two layers are described as "progressive" (p. 144), with an interpenetration of components, but with no indication of the distance over which this interpenetration occurs, nor any clear statement of whether this is true of both yellow layers. She also says (p. 146) that the "niveau sableux" would have formed in a cold, dry climate, while "les niveaux anthropiques" formed in a more humid, less cold, environment.

In 1999 we expanded the excavated area at the rear of the trench towards the east, into the zones N2, N3, O2 and P2, and discovered the very irregular contours of the rock shelter wall (see Figure 9.10). The couche 4P dwindled away and did not rise far up the rock wall in zone N2, and was practically gone by the time we reached O2 and P2, beneath the shelf-like concavity. On the other hand, the couche 4, on the floor of this area, did continue and seems likely to continue beyond P2: but it is brecciated in these zones, which made it harder to dig.

Finally, the upper surface of the couche 4 is in contact with layer 3. Although M. Wilson (1997) describes the contact as gradual (based on one sample, from N12), she also says that the surface of the couche 4 had been colonised by lichens and mosses, meaning that it had been abandoned long enough for that to happen, before being buried under more sediments (p. 123). In my personal observation, that would take a few years at most – not a significant length of time. However, it is true that as we dug, we were always able to observe an abrupt transition to the couche 4, so that the order to "dig *down* to but not *in*to" the couche 4 was never very hard to follow. Figure 9.6 (above), for instance, shows the result of such an approach.

Now that we have seen the extent of the couche 4, and its contacts with surrounding layers, it is necessary to make some attempt at explaining how this happened. How the couche 4 itself formed will be left for later, after presentation of the pollen, faunal, hominin and lithic results, but at this point we need to address the question of the couche 4P. Why does it go up the wall?

One possibility to consider is that 4P did not go anywhere: 4 did. Could the couche 4 itself have started out at the level of the top of 4P, but subsided, leaving this sloping layer near the wall? This seems to me to be very unlikely. It would require a metre's worth of subsidence, and there are no signs of any of the underlying layers being anything but roughly horizontal (as is the couche 4 itself, away from the wall): so where would all that other material have gone? Also, the couche 4 is located on the dalle, and that seems to have been a stable platform. Therefore it seems more probable that the couche 4 stayed more or less where it was, but 4P rose up the wall.

Figure 9.7. A décapage of the couche 4P. ***Un décapage de la couche 4P.***

Figure 9.8. Closer view of a décapage of the couche 4P. The coloured picks are meant to keep objects from falling down the slope during the excavation. ***Un décapage de la couche 4P, vue détaillée. Les pics colorés ont été utilisés pour garder les objets en place sur la pente de la couche pendant la fouille.***

Figure 9.9. Cross-section through the couche 4P, eastern side of zone L3, with the rock wall on the left, and a series of layers from left to right: yellow, black, grey, black again, and more yellow. The outer black layer shows a clear bend, from nearly horizontal on the right, to overturned (beyond vertical) above. The dalle forms the base for all of this. For scale, both black layers are about 3 to 5 cm thick. ***Coupe dans la couche 4P, section Est de la zone L3, avec la paroi rocheuse à gauche, et une série de couches de gauche à droite: jaune, noire, grise, une deuxième noire, et jaune encore. La couche noire extérieure est pliée; presqu'horizontale à droite, elle se renverse (dépassant la verticale) plus haut. A la base on voit la dalle. Chaque couche noire mesure environ 3 à 5 cm d'épaisseur.***

Figure 9.10. The rear of the tranchée L in 1999, showing the concavity of the rock shelter wall. ***Le fond de la tranchée L en 1999, montrant la concavité de la paroi rocheuse.***

Based at least partly on the micromorphology, M. Wilson (1997, pp. 145-146) sees the action of differential swelling due to freezing and thawing as having caused the couche 4P to rise up. She postulates that it occurred "suite à un gonflement cryogénique différentiel dans un secteur hors contrainte ou de gélivité plus faible, combiné au phénomène de paroi. . . Les injections sont des déformations plastiques résultant d'une contrainte exercée par le gonflement cryogénique des couches les plus gélives (pressions cryogéniques), lesquelles s'injectent progressivement dans les secteurs hors contraintes (plan de faiblesse) et dans les couches moins gélives . . ." But M.A. Courty (2006) disagrees. She agrees that there must have been frost action, but does not agree that it could have resulted in so much displacement: "La présence d'une micro-organisation en agrégats subarrondis à revêtements micro-lités grossiers enrobant les agrégats de sol témoigne de transformation liées au gel/dégel. L'aspect jointif de nombreux agrégats et fragments osseux indique cependant que la cryo-reptation au cours du dégel ne s'est faite que tout au plus sur quelques centimètres. Elle ne peut avoir générer une remontée de la couche IV, voire une injection le long de la paroi." She suggests that the yellow layer may have originated as a loess deposit, and that there may have been "une dissolution marquée de la couche IV, ou tout au moins une compaction, cependant plus atténuée sur le fond de l'abri resté bien drainé. La bréchification en masse dans le secteur central a considérablement contribué à masquer toute évidence de microstratification, dont seule quelque reliques peuvent être reconnues en lames minces." All of this may be partly true, but I propose an additional factor.

In the Tranchée Moulin we saw that layer D was on the dalle, and that breccia was on top of that. There was evidence that the breccia had become detached from the rock shelter wall, leaving scraps of itself behind, and apparently fallen onto the underlying layers with enough force to create some faults in the dalle. Layer D, being soft and plastic, did not provide enough structural support to hold up the breccia once it got thick enough and heavy enough to overcome its connection to the rock wall. The space left between the fallen breccia and the rock wall was subsequently filled in with loose sediments in a periglacial environment.

In the tranchée L, we have a similar situation, with some important differences. The couche 4 is also (other than a few brecciated areas) soft and plastic. It is also located on the hard surface of the dalle. It was also covered with a thick sedimentary layer, much of which was cemented into a very hard breccia. However, the tranchée L was between what I have described as two domes of breccia: the eastern one starting in band M and going up the alphabet from there, and the western one, which gets thicker from band K over towards the western part of the site. In the band L, the breccia is at its thinnest, only about a metre thick. The breccia that collapsed in the Tranchée Moulin scenario is the western dome. That dome also disappears towards the back of the site in the tranchée L, so that the southernmost zones had no breccia above the couche 4.

When that western dome collapsed onto the dalle, what happened to the couche 4? The movement of the collapsing dome would have come from the southwest. I suspect that it pushed the soft layer 4 outwards in all directions. The front part of the layer would have been squeezed outwards and is gone now. The material that became layer 4M was squeezed up against the breccia of the eastern dome and rose up as a swath into the overlying sediments, while that thinnest part of the breccia reacted to the torsion inherent in one side collapsing by breaking into large chunks. It is noteworthy that the faults in that area are roughly horizontal, as if due to stress from the side, not from above. At the back of the site, where there was no breccia, the couche 4 material was squeezed up into the overlying sediments, or between them and the rock wall. Where the rock wall protruded farther, the black layer stuck right up against it. Farther away, into the bands N, O and P, the impetus of the collapse had diminished, and in any case the black layer had itself been cemented, so it could not spread out, and we found that the couche 4P disappeared.

That scenario does not explain everything. I am not claiming that periglacial injection of freezing and thawing material, as proposed by others (e.g., Wilson 1997), did not play a role: I am sure it did. I just do not think it is sufficient to explain 4P, since it would also have been active in other areas, where there is no equivalent to 4P. The collapsing dome scenario also does not explain the multiple layers in the area of 4P. Is the second black layer part of the couche 4, or a different layer that also got squeezed up? How did the yellow layer get between the black layer and the rock wall? At the back of the Tranchée Moulin it was clearly post-event deposition that filled in the gap between the breccia and the rock wall; the yellowish layer behind the second black layer in Figure 9.9 is the same type of sediment. But how can there have been a gap between the rock wall and a soft, plastic layer? Did that material all stick together and tilt away from the rock wall as the mass of sediments settled down later? I do not have all the answers. I suspect it is a complex situation, due to the thumping impact of the collapse, combined with processes including squeezing, sticking, injecting, percolating, leaching, settling, freezing and thawing over and over, and a lot of time going by. The couche 4 got smeared in the process, obscuring much of its internal organisation, crushing bones into tiny fragments, and leaving the archaeological material lying horizontally in the sediments.

9.3. Pollen and fauna

J. Argant (2000) took at least 4 samples of the couche 4 for palynological analysis, two from the zone K6 and two from N2, but analysed only one of them: sample 19, from K6 (z=-264 cm). She reported that herbaceous plants were the most abundant, especially the *Cichorioideae* (at 30% of the entire sample), and the *Poaceae* (14%). The former include various flowering plants, such as dandelions, and the latter are grasses. Among the trees, pine was the most common (17.3%), but *Abies* (fir), *Fagus* (beech), *Quercus* (oak), and *Tilia* (linden) were also present. She points out that the presence of these trees, which prefer a mild climate, in a sample of pollen that otherwise suggests steppic conditions, is surprising. She suggests that there may have been mixing of sediments, which is supported by the presence of numerous undetermined and apparently older conifer pollen (not included in the percentages). Altogether, she feels that the sample represents formation during a time of a cold, dry climate.

The couche 4 provided a seemingly endless supply of small bone and teeth fragments, attesting to intense use of animals, and burning of their remains. A relatively small proportion of these were identifiable as to their species. Nonetheless, Fernandez (2001) was able to identify numerous individuals from several species, as Table 9.2

Table 9.2. Faunal results for layer 4, from Fernandez (2001) for the ungulates, and Crégut-Bonnoure et al. (2011) for *Ursus arctos*: Minimum Number of Individuals (MNI) for each species. ***Liste faunique pour la couche 4, d'après Fernandez (2001) pour les ongulés, et Crégut-Bonnoure et al. (2011) pour*** **Ursus arctos*****: Nombre Minimal des Individus (MNI) de chaque espèce***

Taxon	common name	MNI
Capra cf. caucasica	ibex	6
Rupicapra rupicapra	chamois	1
Bos primigenius	aurochs	7
Equus sp.	horse	7
Capreolus capreolus	roe deer	2
Cervus elaphus	red deer	10
Dama dama	fallow deer	3
Megaloceros giganteus	Irish elk	1
Sus scrofa	boar	1
Ursus arctos	brown bear	4

shows, and in some cases could determine their age at death. In his thesis he combines layers, so 4 (and 4A, 4M and 4P) are combined with layers 5, E, F, F2 and G to make up his "niveau 4". As a result, I cannot use the percentage representation of each animal in the total assemblage. However, he also gives details of Minimum Numbers of Individuals (MNI) by layer for all species except the ibex, so I have reported these in Table 9.2, combining all variants of 4 but excluding other layers. For the ibex, I worked through the list of teeth to see which could match up with each other, and came up with an MNI of 6, which seems reasonable given that Fernandez (2001) says his "niveau 4" could have an MNI of 8 (based on the lower teeth) to as many as 16 (with the addition of the upper teeth). I did not "double up" the lower and upper teeth, since the most parsimonious measure assumes that both halves of each animal's mouth were brought to the site. Post-doctoral work by Fernandez (2006) identified the ibex to sub-species level as *Capra caucasica praepyrenaica* (Fernandez, 2006, Table 50), and that work also confirms the MNI of 6, using results for individual teeth in Table 50.

Fernandez (2001) reports that the presence of ibex rather than tahr (which disappears from the Bau in these upper layers) shows that these layers date to the Würm glaciation, rather than the earlier Riss glaciation. None of the ibex specimens were old when they died, and in fact they were mainly juveniles. The life pattern of the ibex suggests therefore that they were most likely hunted during the late spring to early fall period (April to October), when the juveniles are found in groups, perhaps accompanied by an adult female (Fernandez, 2001, p. 85). This is their period of highest abundance and vulnerability, since they are weak from lack of food in the winter, and are expending energy as they move to new areas in search of springtime vegetation. The age profile of the assemblage suggests a selective hunt of such groups, rather than scavenging of naturally dead carcasses, since there are no old individuals present.

There is not much to say about the chamois, which is represented by one bone. The aurochs were once again an important prey animal to the hominins of this layer, with an MNI of 7 including at least one very young animal, one fairly old, and the rest prime-aged adults. These clearly were selectively hunted, as they are the least likely to have just died and been scavenged. Their presence suggests that the climate was fairly cold, so that the area was not heavily forested, which matches the pollen data. The presence of one boar tooth does not contradict that, and instead reaffirms that the climate was likely to have been relatively dry: boars like forests, but they can live in more open areas as long as the snow is not too deep (Fernandez, 2001).

The remains of the horse also provided an MNI of 7, as this animal was also abundantly used. However, the material was too fragmentary for Fernandez (2001) to be able to draw firm conclusions about which species they may represent. The individuals are spread across all age classes, and the only seasonal information is that one horse died in July (Fernandez, 2001).

The deer species (see Figure 9.11 for fallow deer teeth) present are able to live in a variety of environments, although the Irish elk probably prefers more open areas, rather than dense woods. Of the two roe deer, one was a juvenile and one was an adult. Finally, the four bear individuals are represented primarily by teeth and tooth fragments, along with one fragment of a humerus (Crégut-Bonnoure et al., 2011).

9.4. From The Chest: The Notched Bone

One small piece of bone found in the couche 4 deserves to be mentioned, despite the lack of detail that I can provide. I remember seeing it, probably in the late 1980s. It was 2 or 3 cm long, lightly burned (to a medium brown colour), and had 3 or 4 distinct v-shaped notches carved into it. Serge took it to Canada that year, and brought it back the next year, telling me that studies had shown the notches were carved by a flint tool. I can not find any written reference to it, nor photographs, and I have no idea what might have happened to it. It is worth mentioning because the notches seemed to be unrelated to any incidental marks caused by butchery or hide-skinning. They were regularly spaced, and (for the size of the bone) fairly deep. Were they intentional?

9.5. Human remains

The couche 4 provided six Neandertal teeth, numbers 5, 6, 7, 8, 9 and 12 on the human remains list (Table 9.3). The first five are described in Trinkaus et al. (2000a), with #5 additionally described in Trinkaus et al. (2000b). The last of them, #12, is described in Lebel & Trinkaus (2001, 2002b). Four of these are deciduous teeth: an upper first and second molar, an upper canine, and a lower first incisor (numbers 5, 7, 6 and 8 respectively). Those four are not

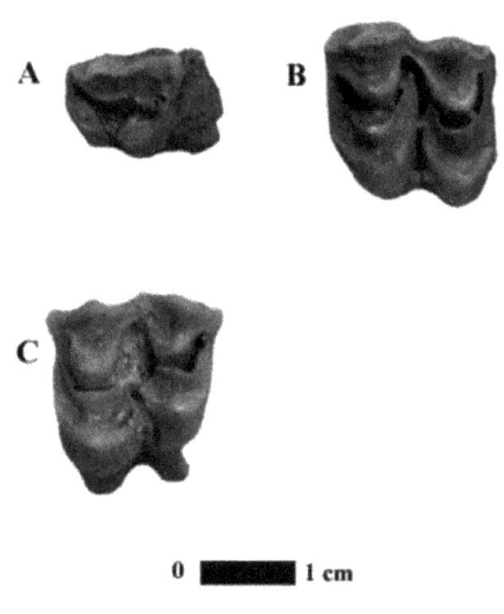

Figure 9.11. ***Dama dama.*** **A, left P2/ K7 4 185. B, right M1/-M2/ L3 4P 734. C, left M3/ L10 4 316; occlusal face. Photos P. Fernandez.** ***Dama dama. A, P2/ gauche K7 4 185. B, M1/-M2/ droite L3 4P 734. C, M3/ gauche L10 4 316; face occlusale. Clichés P. Fernandez.***

Table 9.3. Human teeth from the couche 4, after Trinkaus et al. (2000a) and Lebel & Trinkaus (2002b). ***Dents humaines de la couche 4, d'après Trinkaus et al. (2000a) et Lebel & Trinkaus (2002b)***

Hominin remains number	Year discovered	Name of discoverer	Site catalogue number	Tooth	Side
5	1994	J.A. Labelle	K4 4P 470	dm^1	right
6	1994	B. Murphy	M2 4P 100	dc^1	left
7	1995	T. Bourris	L4 4 770	dm^2	left
8	1998	J. Tousignant	K5 4 127	di_1	right
9	1998	J. Tousignant	K4 4P 864	I^2	right
12	2000	J.P. Rey	O2 4 83	M^1 or M^2	right

necessarily from the same individual, especially since two of them (the molars) were shed due to normal development during life, and the other two became separated from their tooth sockets after death.

Specimen #5 (Figure 9.12) is a well-worn deciduous upper first molar, similar in size and shape to other known Neandertal dm^1s. It is noteworthy because it is one of the very rare cases of a Neandertal tooth with dental caries, in this case a small carious lesion in the inner (tongue) side of the crown (Trinkaus et al., 2000a, 2000b), which penetrates through the enamel and into the dentine. It also has a crack in the enamel that appears to have occurred during life, and to have stopped the expansion of the cavity on that side. The presence of this cavity is evidence that this Neandertal child was eating some sort of carbohydrate, which created the conditions in the mouth necessary for the bacteria that cause caries to thrive. Since the tooth was lost in the normal way as the child developed, however, it did not cause any long-term harm.

The deciduous left canine was also well worn: its apex had been worn off enough to expose dentine, and the inside (lingual) surface was smoothed by wear, but had no dentine showing (Trinkaus et al., 2000a). It is large, but within the normal size range and morphology of later Middle Pleistocene and Neandertal maxillary deciduous canines (Trinkaus et al., 2000a).

Tooth #7, the second molar, was also shed in the normal way during life and was fairly well worn, with dentine showing in the cusp areas of the crown. It was big (but not unusually so) for a Neandertal dm^2. We will never know, unfortunately, if this left-side dm^2 is a match for the now-lost right-side dm^2 found by Moulin, which he describes as also being lost during life in the course of normal development (Moulin, 1904). The final deciduous tooth, #8, is a relatively large (but again, not significantly so, for a Neandertal), slightly shovel-shaped, incisor, with the incisal edge worn down enough to expose dentine (Trinkaus et al., 2000a).

Tooth #9 consists of a largely complete crown and root, meaning it was lost after death, and it was burned, which

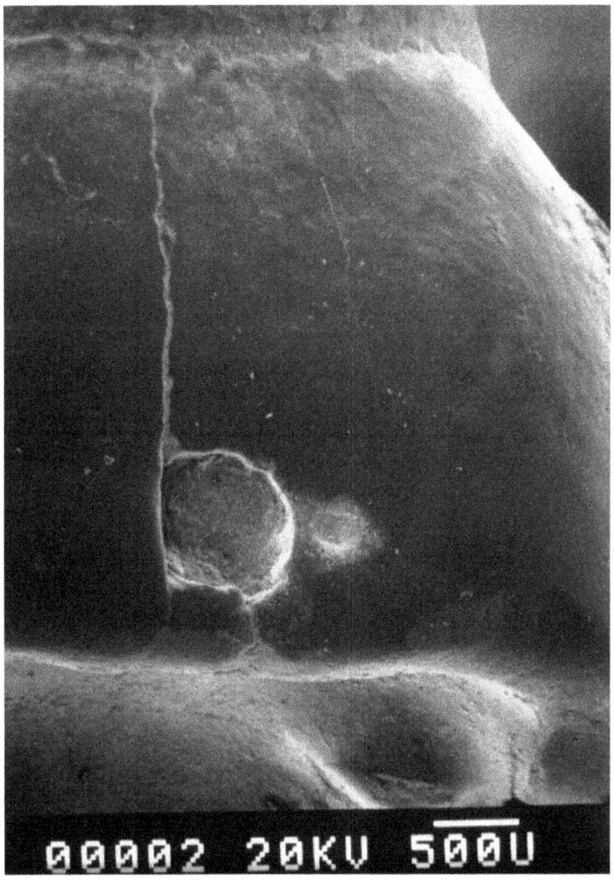

Figure 9.12. Scanning Electron Microscope (SEM) image of the dental caries and crack in Aubesier 5 right dm^1, K4 4P 470, buccal view. Photo E. Trinkaus. ***Image par microscope électronique à balayage (MEB) de la carie dentaire et fissure dans la dm^1 droite Aubesier 5, K4 4P 470, vue buccale. Cliché E. Trinkaus.***

is evidence that the person died before at least one fire occurred. There is no reason to think the tooth was burned deliberately. It has also been partly covered by a thin deposit of calcite, and chipped and fissured post-mortem. It is a moderately worn shovel-shaped incisor, and it is relatively small (Trinkaus et al., 2000a). Based on the amount of wear, it probably came from someone who died in early adolescence (Lebel & Trinkaus, 2000b). Interestingly, the premolar found by B. Mary in the Tranchée Moulin and studied by M.-A. de Lumley is also described as being relatively small (de Lumley, 1973; Trinkaus et al., 2000a).

Finally, tooth #12 (Figure 9.13) is an upper permanent first or second molar, of normal size for a Neandertal and minimally worn (Lebel & Trinkaus, 2002b). If it is a first molar, the amount of wear would indicate that it came from someone who died in early adolescence – conceivably, the same individual(s) who had tooth numbers 6 (the deciduous canine) and/or 9 (the adult incisor) (Lebel & Trinkaus, 2002b), but if it is a second molar (which grows in to the mouth later), the amount of wear indicates death in later adolescence or early adulthood, which does not match any of the other teeth. It is also possible that it came from the mouth of the individual who had earlier shed the two deciduous molars. This possibility is especially intriguing because #12 provides a second example of a Neandertal tooth with dental caries – even worse than #5. In this case, "the distolingual corner of the cervical half of the root has been eroded by a caries, which extends around the corner of the root, up into the crown, and into the pulp chamber" (Lebel & Trinkaus, 2001, p. 20). Thus, if this is from the same individual as #5, that individual had "multiple carious lesions" (Lebel & Trinkaus, 2002b, p. 557); if not, more than one Neandertal at this site had caries. Either scenario is surprising, given the rarity of dental caries in Neandertals (Lebel & Trinkaus, 2001, 2002b; Trinkaus et al., 2000a, 2002b).

These six teeth can be assorted in various ways, as we have seen, to give at least two or three individuals. One or two of those individuals died either at the site (which seems more likely), or elsewhere, and in that case, for some reason at least these few teeth were brought to the site. One of those teeth ended up in the fire. We have no evidence of what happened to the rest of the bodies. It is interesting that all of these teeth were found near the back wall of the site, not anywhere else in this extensive layer. Interesting, but for the time being, inexplicable.

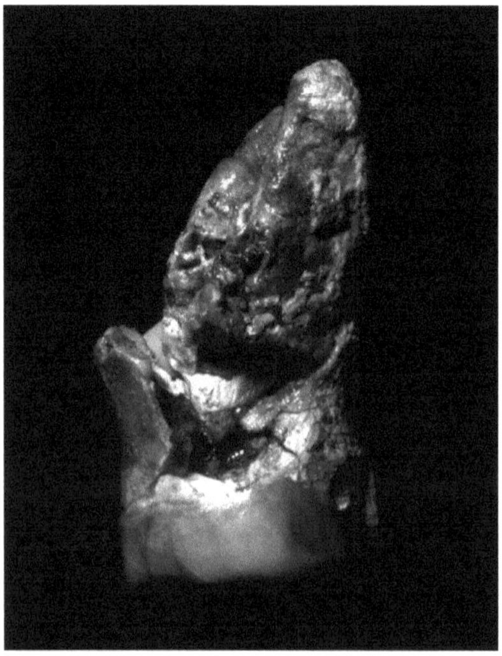

Figure 9.13. Dental caries in the Aubesier 12 right M^1 or M^2, O2 4 83, distal lingual view. Photo E. Trinkaus. ***Carie dentaire dans la M^1 ou M^2 droite, O2 4 83, vue distale linguale. Cliché E. Trinkaus.***

9.6. Lithics

The lithics of the couche 4 were as abundant as the bone fragments, perhaps more so, and in better condition: they were not the fuel for the fire, although a lot of them were in it. Still, the lithic assemblage is greatly dominated by tamis/collectif (Table 9.4), which is both because of the high quantities of tiny pieces, and because of the manner in which we excavated the layer, as has been described. Coordinating only lithics larger than about 3 cm in size did, however, create thousands of individually numbered pieces, and subsequent study of the assemblage sorted other tools out from the collectives, giving the results in Table 9.4. All segments of the *chaîne opératoire* are present and abundant, demonstrating that toolkits were made and maintained at the site.

The assemblage can also be broken down into the categories of the Bordes list (Bordes, 1961), although this gets complicated because many pieces are two or more things at once. For instance, if a piece includes two kinds of sidescraper and a notch, with a Levallois flake – with its bulb removed – as their support, it would be counted in "other sidescrapers", "notches", "bulb removed" and "Levallois flake". This inflates the numbers, shown in Table 9.5, but at least everything gets counted. The results show that there is a high proportion of sidescrapers and of use of the Levallois technique, along with a large number of points (Figure 9.14), some endscrapers and other tools. Once again, we can easily call this Mousterian, and perhaps assign some "progressive" tendencies to it, given the endscrapers, blades, and assorted others including a couple of burins.

The typological sample includes more than twice as many pieces as the raw material sample, which totals 20,167 pieces. This is mainly because I did not have time to do

Table 9.4. Lithic assemblage of layer 4, summarised by typological category, excluding 2 galets. ***Assemblage lithique sommaire de la couche 4, par catégorie typologique, 2 galets non-compris***

Category	Typology (n)	Typology (%)
Liste Bordes	1045	2.6
Debris	168	0.4
Eclats	5275	13.2
Lames	486	1.2
Tamis/Collectif	32667	81.6
Nucléus	405	1.0
Total	**40046**	**100**

Table 9.5. Typologically identified pieces in layer 4. *Pièces identifiées typologiquement de la couche 4*

Typology	English name	n	%
Eclats levallois	Levallois flakes	298	11.0
Lames et lamelles	Blades and bladelets	627	23.2
Pointes levallois	Levallois points	248	9.2
Pointes moustériennes	Mousterian points	71	2.6
Racloirs simples	Single sidescrapers	387	14.3
Autres racloirs	Other sidescrapers	215	7.9
Grattoirs	Endscrapers	77	2.8
Couteaux à dos	Backed knives	32	1.2
Denticulés	Denticulates	63	2.3
Encoches	Notches	94	3.5
Bulbe enlevé	Bulb removed	141	5.2
Autres	Others	47	1.7
Nucléus levallois	Levallois cores	109	4.0
Nucléus non-levallois	Non-Levallois cores	296	10.9
Total		**2705**	**100**

Figure 9.14. A selection of points from the couche 4. *Un assemblage de pointes de la couche 4.*

the raw materials for all of the tamis and collective bags while we were excavating, and many of them contained hundreds of pieces. The biggest difference is in L9, whose typological sample includes 7072 (mainly tiny) pieces, while I saw 919 when counting the raw materials. At least 919 pieces is likely to be a representative sample. A worse situation affects band 7, where I somehow managed to miss most of the pieces. My raw material sample includes 7 pieces from K7, versus 869 in the typological sample, 284 for L7 (versus 681 in the typological sample), and only 3 pieces from M7, versus a whopping 1900 in the typological sample. Most zones do not show such flagrant discrepancies, however, and in fourteen zones the raw material sample is larger, so overall the raw materials sample should be representative of the whole assemblage.

In order to attribute raw material sources to the pieces on the typological list, it is necessary to match them up by their catalogue numbers, where we run up against the problem of many typologically identified pieces having been extracted from collectives, so that multiple pieces have the same number as each other. Unfortunately, the only way of dealing with that is to lump them all back together: put them back in their collectives. Tables 9.6 and 9.7 give the result, and Figure 9.15 shows the overall distribution of source areas used. The tables include 19332 pieces (rather than the 20,167 of the complete raw material list), which is a good proportion, but we must bear in mind that the Tamis/Collectif numbers include many pieces which figure on the typological list in more specific categories, from Mousterian points to cores. Also note that the category of Eclats/Lames Levallois includes Levallois flakes and blades which were not included in the Liste Bordes categories (to save time, they were not given the full descriptive treatment), so they are not being double-counted here. They were included under Eclats levallois in Table 9.5, however.

Table 9.6. Number of pieces by typology and source area in the couche 4. ***Nombre de pièces par catégories typologiques et groupes de sources des matières premières dans la couche 4***

4 n	Liste Bordes	Debris	Eclats	Eclats / Lames levallois	Galets	Lames	TamisColl	Nucléus	Total
Unidentifiable	293	43	2046	41	0	137	9541	98	12199
Unknown	3	0	4	0	0	0	29	0	36
Faraud etc.	54	22	617	5	0	44	952	46	1740
Méthamis etc.	58	12	281	8	0	34	469	29	891
Les Sautarels etc.	10	0	19	1	0	0	72	1	103
Sault	66	18	413	6	0	23	1048	24	1598
Murs	108	17	603	11	0	55	1277	61	2132
Local	11	0	34	0	1	3	458	0	507
St. Jean de Sault	0	0	1	0	1	0	5	0	7
Roussillon	1	0	0	0	0	0	1	0	2
Murs-Bezaure Olig.	11	0	10	0	0	3	30	1	55
Ravin de la Treille	11	0	7	0	0	0	9	1	28
St. Trinit	10	0	4	1	0	1	8	0	24
Durance	0	0	0	0	0	0	0	0	0
Tertiary Calavon	0	0	0	0	0	0	1	0	1
Tertiary Murs VdeV	0	0	2	0	0	0	1	0	3
Mormoiron	0	0	0	0	0	0	1	0	1
R de Guérin	0	0	1	0	0	0	3	0	4
N.Aurel	0	0	0	0	0	0	1	0	1
Total	**636**	**112**	**4042**	**73**	**2**	**300**	**13906**	**261**	**19332**

Table 9.7. Percentages of pieces by typology and source area in the couche 4. ***Pourcentages de pièces par catégories typologiques et groupes de sources des matières premières dans la couche 4***

4 %	Liste Bordes	Debris	Eclats	Eclats / Lames levallois	Galets	Lames	TamisColl	Nucléus	Total
Unidentifiable	46.07	38.39	50.62	56.16	0.00	45.67	68.61	37.55	63.10
Unknown	0.47	0.00	0.10	0.00	0.00	0.00	0.21	0.00	0.19
Faraud etc.	8.49	19.64	15.26	6.85	0.00	14.67	6.85	17.62	9.00
Méthamis etc.	9.12	10.71	6.95	10.96	0.00	11.33	3.37	11.11	4.61
Les Sautarels etc.	1.57	0.00	0.47	1.37	0.00	0.00	0.52	0.38	0.53
Sault	10.38	16.07	10.22	8.22	0.00	7.67	7.54	9.20	8.27
Murs	16.98	15.18	14.92	15.07	0.00	18.33	9.18	23.37	11.03
Local	1.73	0.00	0.84	0.00	50.00	1.00	3.29	0.00	2.62
St. Jean de Sault	0.00	0.00	0.02	0.00	50.00	0.00	0.04	0.00	0.04
Roussillon	0.16	0.00	0.00	0.00	0.00	0.00	0.01	0.00	0.01
Murs-Bezaure Olig.	1.73	0.00	0.25	0.00	0.00	1.00	0.22	0.38	0.28
Ravin de la Treille	1.73	0.00	0.17	0.00	0.00	0.00	0.06	0.38	0.14
St. Trinit	1.57	0.00	0.10	1.37	0.00	0.33	0.06	0.00	0.12
Durance	0.00	0.00	0.00	0.00	0.00	0.00	0.00	0.00	0.00
Tertiary Calavon	0.00	0.00	0.00	0.00	0.00	0.00	0.01	0.00	0.01
Tertiary Murs VdeV	0.00	0.00	0.05	0.00	0.00	0.00	0.01	0.00	0.02
Mormoiron	0.00	0.00	0.00	0.00	0.00	0.00	0.01	0.00	0.01
R de Guérin	0.00	0.00	0.02	0.00	0.00	0.00	0.02	0.00	0.02
N.Aurel	0.00	0.00	0.00	0.00	0.00	0.00	0.01	0.00	0.01
Total	**100**	**100**	**100**	**100**	**100**	**100**	**100**	**100**	**100**

Less than two-thirds of the assemblage is so patinated or burnt that its rock types are unidentifiable, so the remainder includes a comfortably large number of pieces. Among these, we see that the couche 4 has the largest number of source areas represented of any layer, but that the major source areas are the same as in the other layers. Depending on the typological category, the Murs area usually provided the largest proportion of the material, but it is not even close to dominant in any category, reaching a high of 23.37% in the cores, and making up 11.03% overall. Several source areas are represented by all parts of the *chaîne opératoire*, from cores to tools and tamis, although Méthamis has relatively low tamis/collectif numbers, compared to its values in other categories. Méthamis also seems to have higher than expected numbers in the Levallois category (see Figure 9.16 for an example). Faraud has high numbers in most categories but not in the Liste Bordes column, which suggests that it is "higher up the chain": more early-stage pieces, not as many end-of-the-line tools. At least a few pieces also give hints about their history in the form of remnant patinas, although no systematic study of this has been done. Figure 9.17, for instance, shows a piece which has a patina on a flat surface, while the rest of the piece is not patinated. It was knapped either from a pre-existing artifact, or from a naturally exposed and broken nodule.

The minor source areas are generally represented by incidental pieces in the tamis/collectifs or as a flake or a tool. They are evidence of toolkits made in other areas, and of how diverse those other areas were, located to the south, the east and the west of the Bau. Table 9.8 recalculates the percentages of pieces from each geographic area, excluding pieces of unidentifiable or unknown source, and places them in comparison to the values in other layers. This is based on the full raw material sample, not the sample matched to the typology, and includes a total of 7454 pieces.

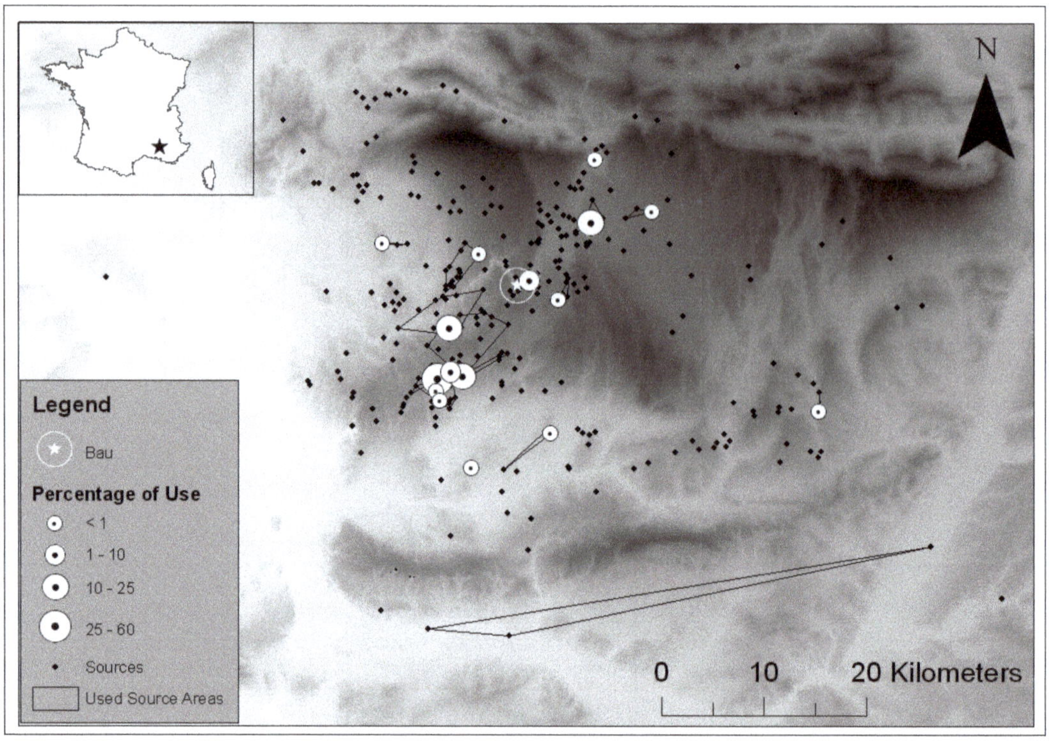

Figure 9.15. Map of the raw material source areas used for the couche 4. ***Carte des aires de sources de matières premières utilisées pour la couche 4.***

Figure 9.16. Tool number M9 4 241 and 261 (two pieces of the same tool coordinated separately), a sidescraper made on a Levallois point, in raw material type H from Méthamis. ***Outil numéro M9 4 241 et 261 (deux morceaux de la même pièce coordonnés séparément), un racloir sur pointe levallois en type de matière première H, de Méthamis.***

Figure 9.17. Piece number K4 4 797, a sidescraper on a thick flake, with a remnant of an older patina, in raw material type R from Sault. ***Pièce numéro K4 4 797, un racloir sur éclat épais, avec une surface de patine plus ancienne, dans le type de matière première R de Sault.***

The first thing that strikes me as I look at Table 9.8 is that, at last!, there is a respectable percentage of pieces from the local source. Since the couche 4 is just as patinated as most of the other layers, this confirms that the explanation for the "missing" local pieces in the other layers is not that they were preferentially patinated. Also, since I was looking at pieces from various layers every year, it is unlikely that I was unable to recognise local material in some layers but could recognise it when it came from the couche 4. That leaves two possibilities for its lack in other layers, one geologic, the other human. The first is that in earlier time periods, the flint from the local source had not

Table 9.8. Percentages of identified pieces (excluding unidentifiable and unknown source pieces) from each geographic region for layers K2, K1-J, I, H, G, E, C, 5 and 4. ***Pourcentages des pièces identifiées (excluant les pièces indéterminables ou de sources inconnues) de chaque région géographique pour les couches K2, K1-J, I, H, G, E, C, 5 et 4***

Geographic area	K2 %	K1-J %	I-I4 %	H-H2 %	G %	E %	C %	5 %	4 %
Local	0.2	3.4	1.3	0.7	0	0	0.5	1.4	6.94
NE	19.4	20.4	12.3	15.7	41.5	31.0	39.5	5.7	22.95
E	0	0.3	0.1	0	0	0	0	0	0.09
Far SE	0	0	0	<0.1	0	0	0	0	0.05
Far S	0	0	0.1	0	0	0	0	0	0
S	0	0	0	0	0	0	0.5	0	0.01
SSW	0	0	0.1	0	0	0	0	0	0.04
SW	44.3	54.1	35.0	30.0	48.4	25.4	17.6	59.0	32.18
Near SW	23.0	15.0	29.1	34.8	9.6	29.9	30.7	10.5	24.68
WSW	12.4	6.6	21.9	18.6	0.5	12.7	11.2	22.9	12.61
W	0	0	0	0	0	0	0	0	0.01
NW	0.7	0.1	0.1	0.2	0	1.0	0	0.5	0.42
Total	**100**	**100**	**100**	**100**	**100**	**100**	**100**	**100**	**100**

yet been exposed (or not much), so it was not available to be used. By the time of formation of the couche 4, that flint was exposed, and was used. The problem with that is that the other upper layers that we have already looked at, and in particular layer C, there is not much use of the local source.

The other explanation is the human one: the hominins did not choose to use it in some layers, but in the couche 4 they did. Why? Perhaps the use of the site is the answer. The couche 4 is the largest and richest layer in the site, and must represent a significant period of occupation, even if that in fact means repeated visits not just one long one. Longer-term use of a site is likely to have two results: a larger amount of lithic remains, due to more tool-making going on; and a larger amount of local material, because of increased exploitation of the easily available nearby raw material (Wilson, 1988). If you are arriving at a site from somewhere else, carrying your tools, some blanks, a spare core, and half of a horse, why bother picking up a local flint nodule? And if you are passing through, but do collect some local flint, chances are you will take it away with you rather than leaving it at this site. But if you are hanging around at a site for a few days or a season, there is a chance you will explore the area and pick up some local flint to work with.

The only other difference between the couche 4 and the other layers is its diversity. It has more source areas represented than any other layer, but none of the values are surprising. The main source areas have values that are in the middle range compared to other layers, with the SW having the highest value – which is the most common situation. The other source areas are represented by very small quantities of material, and all of them but the Western area (Mormoiron) have already been found in at least one other layer, but with no pattern as to which other layer that is. This is the first time we have seen a piece from the Western area, so that extends the territory of use in that direction (Figure 9.15), but we cannot say much more than that about one piece. It is definitely a distinctive rock, however: type AO from my list (see Table 1.1 in chapter 1), a translucent, pale to white siliceous rock from the upper layers of the ochre deposits around Mormoiron (Table 1.2, Figure 1.3).

In our Generalised Linear Model analysis of the couche 4 (Wilson & Browne, 2014), we found that the raw material variables accounted for more than half of the weight of the model (Tables 9.9 and 9.10), led by Quality. Small rocks and Medium rocks had a negative Beta value, meaning they were selected against (sources with more abundant small and medium rocks were more likely to be avoided), as did Calories, while AOSISA (Area Of the Sources in the Source Area) was a positive, and all four of these had about the same weight. Overall, it seems that the hominins were emphasising obtaining good quality raw materials of adequate size, more than the ease of obtaining them, but the latter aspect was still important.

Finally, the lithic sample can give us some insight into the spatial arrangement of the couche 4. We have already seen that it is an expanse of burned material, thick and black, but getting thinner and lighter-coloured towards the front of the tranchée L, north of about band 9. We have also seen that the deposits have been subject to a great deal of post-depositional perturbation, which has smeared them out and forced them up into parts of the overlying sediments. The six

Table 9.9. Generalised linear model for raw material source choices in 4. P-values in bold font are statistically significant. Sample: 7439 pieces from 16 source areas

Overall:
Omnibus test: likelihood ratio chi-square = 882.238, df = 10, p = <0.001

Variable	Beta	S.E.	Wald Chi-Square	P-value	AIC
Quality	1.990	0.2937	45.907	**<0.001**	288.463
Extent	0.997	0.3086	10.445	**0.001**	
Difficulty (Cal/km)	-0.032	0.0169	3.687	**0.005**	
Small rocks	-10.395	2.0230	26.404	**<0.001**	
Medium rocks	-5.532	1.0739	26.539	**<0.001**	
Large rocks	1.284	0.3781	11.538	**0.001**	
Very large rocks	1.212	0.7370	2.703	0.100	
Calories	-0.002	0.0004	25.649	**<0.001**	
AOSA	-2.2 e-7	6.2 e-8	12.089	**0.001**	
AOSISA	6.6 e-5	1.3 e-5	27.036	**<0.001**	
(Intercept)	8.974	2.8426	9.966	**0.002**	

Table 9.10. Wald Chi-Square values as percentages, assemblage from the couche 4

Variable	Wald Chi-Square	Wald Chi-Square as %
Quality	45.907	22.7
Extent	10.445	5.2
Difficulty (Cal/km)	3.687	1.8
Small rocks	26.404	13.1
Medium rocks	26.539	13.1
Large rocks	11.538	5.7
Very large rocks	2.703	1.3
Calories	25.649	12.7
AOSA	12.089	6.0
AOSISA	27.036	13.4
(Intercept)	9.966	4.9
Total	**201.963**	**100**

human teeth were all found towards the back of the layer, near the rock shelter wall, but I have no complete information on the distribution of faunal remains. I do, however, have information on the distribution of the lithic material by zone. Figure 9.18 shows the distribution of lithics, by number of pieces per zone, coded to make the richer zones stand out. The numbers are the totals in either the raw materials list or the typology list, whichever is higher, for each zone. The figure is oriented with North at the bottom instead of the top, because this is the way I "see" the tranchée L: as if I am standing in front of it looking back towards the rear wall.

The smaller number of pieces in some zones on the edges is due to their size (cut off by the rock wall) or the fact that they were not completely excavated (N3 had not been excavated from the top down, and only a small corner of it fell out of the section wall and had to be recovered as if excavated). Nonetheless, the fall-off in quantity of material north of the band 9 is clear, although M11 does have a respectable concentration of lithics. Band 9 itself is the richest of all, due to the very large Tamis/Collectif numbers from those squares. It should be noted that the contents of these are mainly small flakes, such as would result from flint knapping, not broken fragments which could be due to post-depositional effects. There are also other concentrations near the back wall (remember that it cuts diagonally across, from M2 to I4), and in band M from zones 6 to 9. There are surprisingly low numbers (compared to neighbouring squares) in M3, K5 and L8.

The lithics list can also be used to attempt to see how burned the deposits are, in two ways. When I recorded the raw material types, some pieces were attributed to types which are, by definition, burned. This is not a complete count of burned pieces, because burned pieces that were still recognizably some other type were counted as that type, and their burned state almost never separately noted. In zones where everything was more heavily burned, however, it seems plausible that the burned types would be more common. The second way of judging burning involves the typological list. As we went through the pieces to describe their typology, for some categories we noted down whether the piece was burned, based on criteria such as reddening of their colour, cracking, and the dimple-like holes from which potlids had been ejected (Figure 9.19). Again, this is not a complete list: we did not do it for Tamis/Collectifs, for instance. Furthermore, even if we had noted whether every single piece showed

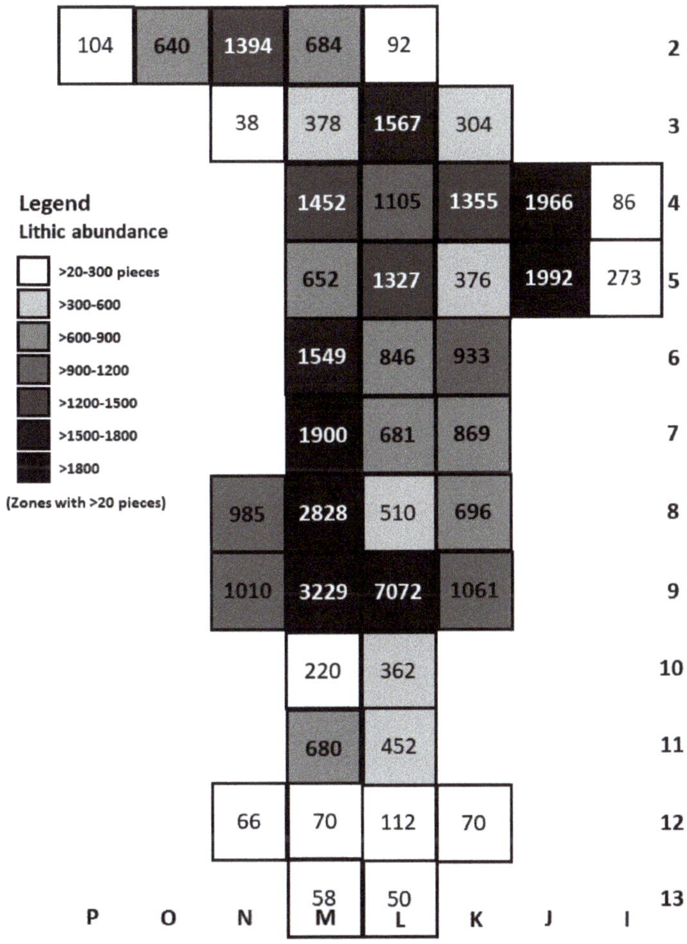

Figure 9.18. Distribution of numbers of lithic pieces by zone in the tranchée L. Zones L2 and K3 make up less than one m² each, due to the location of the rock shelter wall, and N3 was only partially excavated. ***Distribution du nombre de pièces lithiques par zone dans la tranchée L. Les zones L2 et K3 font moins d'un m² pour chacune d'entre elles, en raison de la position de la paroi rocheuse, et la zone N3 n'était fouillée que partiellement.***

Figure 9.19. A pile of flakes from the couche 4, many of which (based on colour changes, and the dimple-like holes left where potlids have formed) were burned. ***Un assemblage d'éclats de la couche 4, dont un bon nombre ont été brûlés, d'après leurs couleurs altérés et la présence de cupules thermiques.***

signs of burning, we would not necessarily have reached "the" correct answer. Recent work (Agam et al., 2020) has shown that different types of flint react differently to heating, and show – or do not show – different changes when exposed to high temperatures. Still, either of these measures should be indicative of relative proportions of burning within the layer, and it is especially reassuring that the two sets of numbers, derived so differently, are strongly correlated with each other (p=<0.001).

Figure 9.20 shows the results from the raw material list, including only zones where more than 10 pieces had been counted. Figure 9.21 shows the results of the typological evaluation, again only including zones where there was data on more than 10 pieces as being burned or not. In both cases the zones are colour-coded to make the more burned zones stand out more. As there was with the numbers of pieces by zone (Figure 9.18), there are concentrations of more-burned material towards the back of the layer near the wall, and again farther forward, towards bands 7, 8 and 9. The most intense burning, judging by the typology results shown in Figure 9.19, was in zones L7 and L8, while the highest numbers of pieces (shown in Figure 9.18) were in zones L9, M8 and M9: in other words, right next to the most burned zones. Despite the post-depositional perturbations of this layer, it does still seem to be showing different zones of activity across the excavated area, and intense burning seems to have been correlated somehow either with flint knapping, or with disposal of knapping remains.

9.7. Conclusion

The couche 4 is the biggest, richest and most burnt layer in the Bau: more than 40,000 lithics; many thousands of pieces of bone and teeth, from at least 42 individual animals (but probably many more), of a variety of species; 6 Neandertal teeth; all mixed in with highly burned sediments to form a layer around 20 or more cm thick and covering about 40 square metres – that we know of. It gives every sign of extending farther, especially to the east. What explains all of this?

We will never know how many hominins were involved in creating the couche 4, but it must have been a group of them, not just a couple. We have teeth from two or three individuals, after all, at least one of whom (probably two) died. We will never know how much time they spent at the Bau, but it must have been more than a one-night campsite to accumulate that amount of material. We will

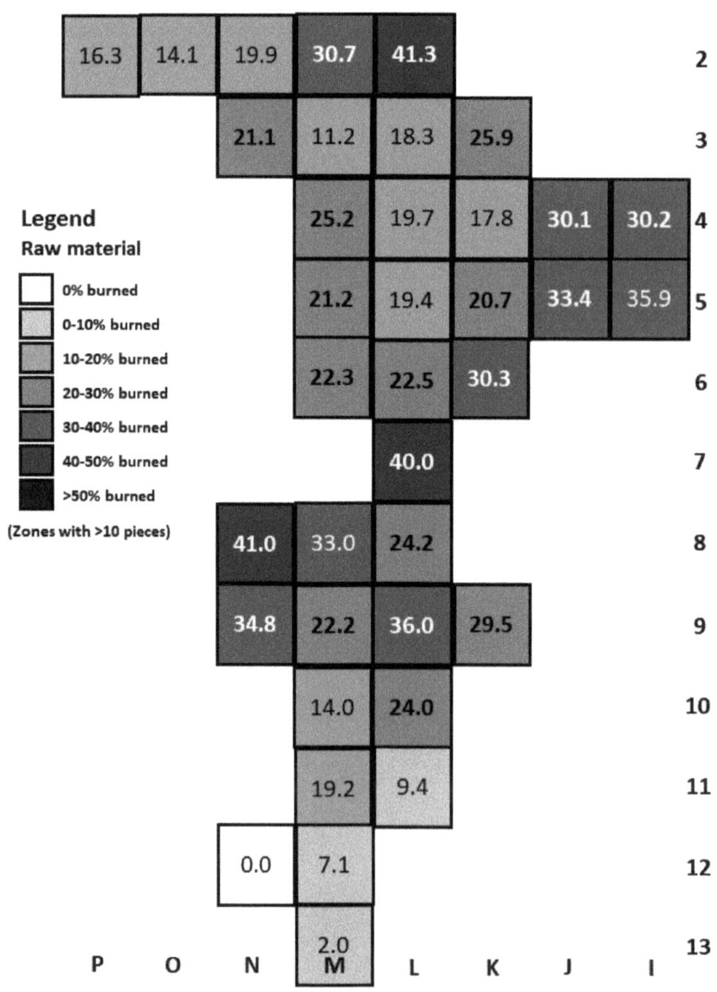

Figure 9.20. Percentages of lithic pieces categorised in the "burned" types, including only zones with >10 pieces on the raw materials list. See text for explanation. ***Pourcentages des pièces lithiques des types "brûlés", dans les zones avec plus que 10 pièces comptabilisées sur la liste des matières premières. Voir texte pour plus de détails.***

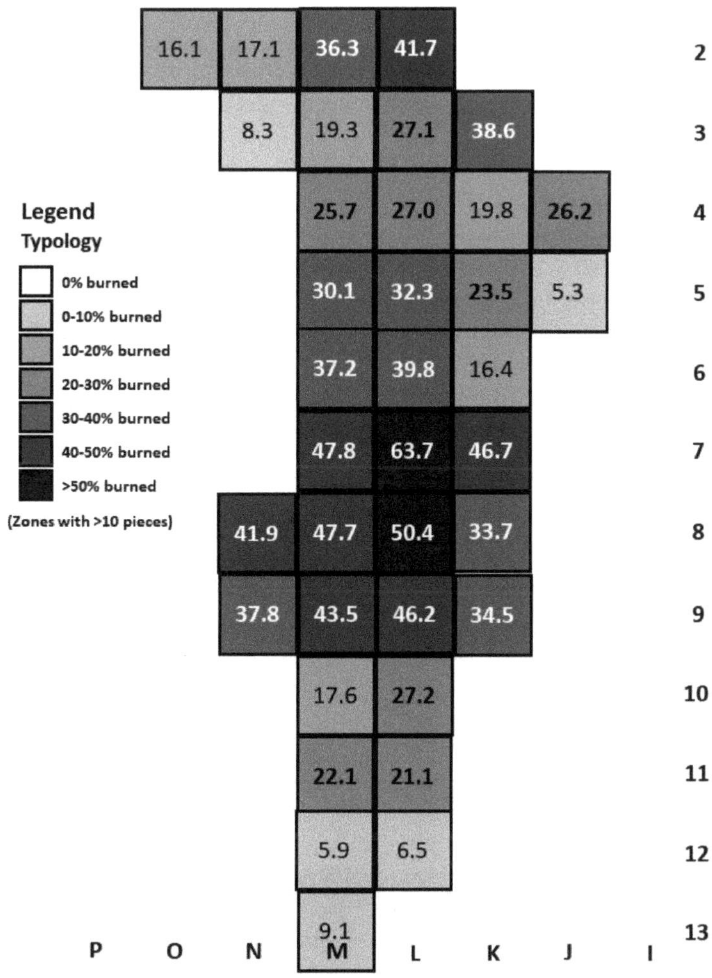

Figure 9.21. Percentages of lithic pieces described as "burned" vs. "not burned" in the typological list, including only zones with >10 pieces so evaluated. See text for explanation. ***Pourcentages des pièces lithiques notées "brûlés" vs. "non-brûlées" dans les résultats typologiques, incluant seulement les zones avec plus que 10 pièces comptabilisées de cette manière. Voir texte pour plus de détails.***

never know for sure how many times they came and went, but it is likely that the layer is a palimpsest of multiple occupations. And we will never know how big a role this site played in their lives – but it was probably smaller than in mine! Recent modelling of land use by Cornel Pop (C. Pop, pers. comm.) has shown that the Bau must have been just one of many (probably hundreds) of sites scattered across the Vaucluse, to accommodate even small groups of Neandertals living in the region over a period of 100,000 years or so. There just is not enough sign of them at the Bau for it to be otherwise. They did not use this site all the time: in fact, they must have used it only occasionally, at intervals punctuating decades to centuries of abandonment. Almost all of the time, throughout prehistory, hominins were *not* at the Bau.

However, they were certainly there when the couche 4 was formed. I suspect that the couche 4 represents use of the site not only by a family group, but by a larger, extended, family group, which came together for a few weeks. It is likely that this was repeated more than once, one or more years apart, but we are not be able to tell this, because there would not be enough natural sedimentation over a year, or even a few years, to cover the remains from the previous occupations. The only seasonal indicators we have suggest that they were there in the warmer part of the year, from spring through fall (based on the ibex) and in summer (one horse died in July) (Fernandez, 2001). At that time, the weather is warm, the days are long, and some animals are at their most abundant and most vulnerable. Plant food is also more available, and one or two of these Neandertals had to be eating something other than meat, because the conditions for dental caries require dental plaque, and carbohydrates to feed the cariogenous bacteria under the plaque (Trinkaus et al., 2000b).

What were these hominins doing, while at the Bau? Butchering and presumably eating animals, making and using stone tools, and using fire. Wherever this layer falls on the spectrum of "number of people" vs. "length of occupation", and whether it represents one or several occupations, it is clear that intense activity went on here, and that it involved fire. There are several common reasons why a group of people might have a fire: for warmth, for light, for cooking, to scare away predators, to burn wastes, and for other, let's call them "industrial", purposes. While all of these are possible in this case, some are more likely than others.

If it is true that the couche 4 formed mainly during spring-summer-fall occupations, there would have been less need of fire for warmth than there would be in the winter, although even now, in a warm climate, a bit of heat of an evening is often welcome. Also, if they were there in the summer months, the days would have been long, reducing (but not necessarily eliminating) the need for supplemental light. Cooking is a year-round activity, so that is a possible reason for the fire – but why so much? Was this one huge communal barbecue? As for scaring away predators, that does not seem like an issue for a relatively large group of competent hunters, so it could have been a side-effect of these fires, but really does not seem to justify so much burning on its own.

The fires were certainly used for burning wastes, in the sense that the bones seem to have constituted a major part of the fuel for them. But again – why so much? It is sometimes suggested that hominins may have brought grasses into their campsites to serve as bedding, and then burned the grass to keep down insects and other pests, but there is no sign of that here. The geochemistry suggested a preponderance of bones as fuel, rather than plants, and neither the palynological nor the micromorphological studies make any mention of phytoliths, which one would expect to find if large quantities of grass had been present.

So we are left with "industrial" purposes, which matches the industrial quantities of lithics and bones we found. I suggest that the hominins were taking advantage of the strategic location and large size of the Bau to get together in a large group in order to hunt and preserve the products of the hunt: smoking and drying meat, and preparing hides. This accounts for the large number of stone tools, the variety and number of animals, the burning of bones as the carcasses were processed, and the extent of the fires. The fires were not just the result of normal daily life, they were due to preparation for the winter to come.

The size and internal "architecture" of the site allowed a large group to live and work there. The space they were in was not a trench as we see it now, but likely a wide-open area, with a floor (the dalle), the rock wall at the back, and within that the natural cavity behind O2 and P2, with its shelf-like hollow. The speleothem in I5 formed another variation in the topography. The activities were carried out both near the back of the cave, and about 5 metres farther forward, although not necessarily at the same time. At one time, or one year, a group might have used the back of the cave, while another year the next group worked farther forward. On the other hand, both areas might have been used more or less at once, but for different purposes. A living area at the back, with a cosy fire, would give them some light and warmth as they cooked their food. It is interesting that all of the hominin teeth were found near the back (although we all wonder where the rest of the bodies might be). Meanwhile the fires farther forward were the "industrial" area, for smoking the meat and for disposal of detritus. The post-depositional changes that smeared the couche 4 around have obscured its details, but some division of space into activity areas seems justified: and there was a whole lot of activity going on.

9.8. Sommaire du chapitre

Si le Bau est connu pour une chose, c'est sans aucun doute pour la couche 4. Très riche en matériel archéologique, épaisse de 15 à 20 cm et parfois plus, très noire, très brûlée, elle s'étend sur plus de 40 mètres carrés, et nous n'avons pas trouvé ses limites vers l'est et l'ouest. Plusieurs dizaines de milliers de pièces lithiques ont été découvertes (elle domine en cela largement l'assemblage du site au complet), et la couche a livré le plus grand nombre de restes fauniques des couches supérieures de même que 6 dents néandertaliennes. De plus, elle se singularise par sa position horizontale sur plusieurs mètres en remontant le long de la paroi rocheuse en arrière de l'abri à la manière d'un tapis de sol trop grand pour la pièce dans laquelle on l'aurait posé. Nous avons appelé la partie horizontale la couche 4, et la partie qui remonte, la couche 4P.

Nos tentatives de datations de la couche 4 ont été parfois infructueuses, mais nous pouvons dire qu'elle date du tout début du Pléistocène supérieur, grâce à un spéléothème situé dans la zone I5, sur lequel elle est posée, et donc plus jeune que 141,854 (+5257, -4956) ans.

La couche 4 se trouve sur la surface de la dalle décrite dans le chapitre 7, qui a pu servir de plancher. Un peu de matériel archéologique a été trouvé dans la partie la plus au nord de la tranchée L, soit à l'intérieur de cette dalle, soit dans des sédiments gris sur la surface de cette dalle, que nous avons appelé 4A (je crois que cette dernière hypothèse est la plus probable). Ce matériel est inclut dans ce que nous appelons la couche 4 dans ce chapitre.

La couche 4 proprement dite était très noire et riche, sans beaucoup de sédiment non-archéologique, et sans structure stratigraphique interne évidente. Nous l'avons fouillée par décapages et collectifs, nous donnant ainsi une vision étendue de toutes les zones ouvertes à la fois. Chaque mètre carré a été fouillé par quart de carré. Nous avons fouillé en laissant en place les objets de plus de 3 cm environ pour les coordonner tous ensemble. Les vestiges plus petits ainsi que les sédiments enlevés à la fouille constituaient les collectifs, qui allaient être triés plus tard après tamisage à l'eau. Ces collectifs étaient identifiés comme provenant d'un quart de mètre carré, avec un numéro pour tous les artefacts lithiques, un autre pour les os, etc. Même après avoir enlevé les pièces de plus de 3 cm de long, le "sédiment" restant représentait environ 50% du poids de matériel visiblement archéologique (Tableau 9.1). Le contenu chimique de la matrice qui restait après l'enlèvement de ces petites pièces suggère qu'il y avait aussi une bonne quantité de contenu archéologique invisible à l'oeil nu.

La couche 4P a livré tout autant de matériel que la couche 4 horizontale, et même plus, puisque chaque mètre carré était augmenté en proportion de son volume du fait du pendage beaucoup plus important vers la paroi. La couche 4P était en fait la continuation de la couche horizontale, mais elle était également très compliquée. La couche se caractérisait parfois par des alternances de couleur parallèles à la paroi (i.e., noir-gris-noir-jaunâtre) avec des variantes de limites et de couleurs. Ces variations semblaient être directement affectées par les contours de la paroi: moins si la paroi était bosselée et plus s'il y avait un renfoncement.

Plusieurs hypothèses concernant cette remontée de la couche 4P vers la paroi ont été émises: phénomènes d'injection, gonflement cryogénique, soutirage des sédiments inférieurs, etc. A mon avis, bien que l'injection et le gonflement aient pu jouer un rôle, c'est le déplacement de la masse brêcheuse, que nous avons vu dans la Tranchée Moulin, qui est vraiment à l'origine de cette remontée. La brèche s'est formée par-dessus la couche 4, et elle est devenue très épaisse (1 à 2 m), dense, lourde et imperméable. En revanche, la couche 4 était dans son ensemble plus meuble et plastique, bien que bréchifiée en quelques endroits, et par ailleurs coincée entre la brèche et la dalle sous-jacente. Le poids de la brèche l'a détaché de la paroi rocheuse, et la couche 4 n'a pas pu la soutenir: la couche 4 a été comprimée. La partie la plus en avant a pu sortir d'entre la dalle et la brèche, comme du dentifrice de son tube; la partie arrière a remonté dans l'espace laissé libre entre la brèche et la paroi. D'autres vagues de la couche 4 ont pu également remonter dans les sédiments supérieurs, coincées entre des blocs de brèche: c'est le cas pour la couche 4M.

Les pollens suggèrent un climat froid, et la présence du bouquetin au lieu du tahr pourrait indiquer un épisode froid du Würm plutôt que le Riss plus ancien. Les bouquetins étaient pour la plupart de jeunes individus chassés pendant la période du printemps jusqu'à l'automne, lorsqu'ils se retrouvent en groupes et qu'ils sont à la fois plus abondants et plus vulnérables. Les autres espèces présentes (Tableau 9.2) comprennent le cerf, le cheval, l'aurochs, etc.

Les six dents de Néandertaliens (Tableau 9.3) comprennent quatre dents lactéales, dont deux ont été perdues après la mort de l'individu. Elles ne proviennent probablement pas d'un seul enfant. Les deux autres dents sont des dents définitives. La dent Aubesier 5, une molaire lactéale perdue très probablement pendant la vie de l'enfant, est cariée: c'est l'une des seules connues à notre connaissance pour une dent néandertalienne. Sa présence indique que l'enfant mangeait des glucides qui sont nécessaires au développement des bactéries qui créent les caries. Malgré leur rareté chez les Néandertaliens, deux dents cariées ont été découvertes dans la couche 4 au Bau. La molaire définitive Aubesier 12 en est l'autre exemple, avec un état plus avancé que celui d'Aubesier 5. Aubesier 12 appartient à un adolescent ou un jeune adulte puisqu'elle n'est pas très usée. Il n'est pas impossible que ces deux dents (Aubesier 5 et 12) aient pu appartenir à la même personne laquelle aurait eu des caries multiples. L'autre possibilité, peut-être encore plus surprenante, serait la présence au Bau de deux personnes avec des caries. En tout cas, avec ces six dents nous avons des restes d'au moins deux ou trois individus.

Le matériel lithique est très brûlé et riche en débitage ainsi qu'en nucléus, outils, éclats, lames. . . bref, la chaîne opératoire complète. Il y a eu beaucoup de taille sur place, donnant une industrie moustérienne assez progressive. Les sources des matières premières sont plus diversifiées que dans les autres couches du site, mais ceci tient sans doute au fait que l'assemblage de la couche 4 est de loin le plus important. Les sources les plus éloignées ne sont représentées que par une ou deux pièces alors que la majorité de l'assemblage provient des sources habituelles (Murs, Sault, etc.), bien qu'il n'y ait pas une seule région dominante. Pour la première fois on note que le silex local est assez bien représenté dans cet assemblage (Tableau 9.8). Je ne peux pas expliquer pourquoi le silex local est si peu utilisée dans d'autres couches (choix des hominidés? manque de disponibilité géologique auparavant? problème d'identification de ma part? – je ne sais pas). Nos études suggèrent que les préhistoriques de la couche 4 cherchaient surtout des matières premières de bonne qualité de grand format, se concernant moins du terrain et de l'accès aux gisements.

La distribution des pièces lithiques laisse apparaître quelques traces possibles d'une organisation de l'espace, qui n'était pas visible à la fouille. En effet, la bande 9 contient la plus grande densité de pièces, peut-être en raison des activités de taille du silex dans ces zones. Il existe également des concentrations de matériel plus brûlé près de la paroi, et dans les bandes 7 et 8. Il est possible que les feux étaient concentrés dans ces deux endroits, peut-être pas en même temps (bien que ceci reste possible). Remarquons que le feu dans les bandes 7 et 8 serait tout près de la zone de taille des silex dans la bande 9: une activité pour la soirée, autour du feu?

Comment expliquer cette couche et sa richesse? Elle représente sans aucun doute un palimpseste: à la lecture finale, une accumulation mélangée de vestiges de plusieurs occupations bien séparées dans le temps. Ces occupations semblent avoir eu lieu le plus souvent pendant la partie la plus chaude de l'année, et je suggère qu'elles impliquaient des groupes non seulement familiaux, mais de familles étendues. Il est clair que pendant ces visites, ils s'engageaient dans des activités intenses utilisant le feu, en relation avec la pratique intensive de la chasse et de la taille du silex dont les restes nous sont parvenus en grandes quantités. Le feu peut servir comme source de chaleur, pour la cuisson, pour la lumière, pour éloigner des prédateurs, et dans ce cas je crois que l'on peut ajouter des activités "industrielles" aussi. La position stratégique du Bau et la grandeur de l'abri en fait un lieu idéal pour les groupes d'hominidés qui veulent se réunir, chasser et préserver les produits de la chasse pour l'hiver à venir: sécher et fumer la viande, ou encore préparer les peaux. Les processus post-dépositionnels ne permettent pas une lecture de leurs activités dans le détail mais on en perçoit, selon l'endroit, les différences. La richesse nous montre qu'il y avait beaucoup d'activités.

10

Layer 3

10.1. Introduction

Layer 3 is stratigraphically complex, but can be summed up as "the upper breccia layer", with variations in cementation, matrix presence or absence, and grain size, in various parts of the tranchée L. It was found at z values ranging from about 150 to 280 cm below zØ, making it more than a metre thick in some areas, and yet it disappeared completely at the back of the trench, because the breccia dome also disappeared. It was divided into 6 sub-layers (described below), two of which contained a few hundred lithic pieces each, while the others had a few dozen each. There were very few identifiable faunal remains. It has no direct dating results: all we can say is it must be younger than the couche 4, and must involve both a glacial climate phase and a warmer one, possibly within MIS 5, but perhaps MIS 4 and 3. Either way, these deposits fall somewhere within the time period from about 100,000 to 50,000 years ago (broadly speaking), which is fully within the classic period of existence of the Neandertals.

Although layer 3 is found in the tranchée L, it does not appear in the artifact lists for every zone. Combining all sublayers, lithic materials from layer 3 were found in zones K7, K9, K12, L6, L9-14, M7-14, N9, and N12 (see Figure 1.11), with the bulk of the material in band M (zones M8 through M14), followed distantly by L9 through L12. Of the 1660 lithic pieces in the typological sample, M8 alone had 834: nearly half. The band M had a total of 1276 pieces, which means that its neighbouring band, L, which was fully excavated, really had very little material in layer 3 (234 pieces). This not due to material having been lost somehow. I once spent the best part of two weeks excavating those layers in the front part of the tranchée L, sieving bucket after bucket of sterile sediments. There was nothing there. Furthermore, of the pieces that we did find, the great majority were tiny. The typological sample includes 1573 pieces in the tamis, with a total weight of 1513 g: on average, these pieces weigh less than one gram each. The raw material data has notes on some of the tamis, such as "etc.", and "et d'autres petits", because some of the pieces were so small I could not determine what raw material type they might be. We did *try* to find any archaeological material there was in layer 3 – it was just slim pickings. Nonetheless, this layer can tell us an interesting story, as we will see below.

10.2. Sediments, evolution of the site, and excavation of layer 3

In the broadest possible terms, the upper deposits of the Bau consist of a very thin layer of grey, modern soil (layer 1), on top of loose yellow cryoclastic sediments (layer 2, see chapter 11), on top of breccia, which is on top of the dark, rich, layer, the couche 4 (chapter 9). As we began excavating the tranchée L, therefore, the breccia was named layer 3. We soon found that layer 3 was not just breccia, and therefore we described sub-layers 3A through 3E. These do not simply line up in stratigraphical order, however, and the breccia (which eventually was named 3D) can be under some of them in one place and on top in other places. These therefore are better thought of as variations in the degree of cementation versus washing, due to differences in water flow through the sediments. The initial mass of sediments is mainly of cryoclastic origin, due to frost-cracking in a cold climate, and probably also includes wind-blown silt and sand. Post-depositional effects caused matrix to be washed away in some areas, and in other areas cement was precipitated, forming the breccia. During warmer times, percolating water likely carried organic materials into the sediments, which caused some sub-layers to be a bit darker in colour and finer-grained in matrix. These sub-layers are described below based mainly on my notes from 1988 and the site report for that year (Lebel, 1988).

Starting at the bottom, the contact between the couche 4 and layer 3 constitutes, as was noted in chapter 9, a clear transition. M. Wilson (1997, p. 123) described it as gradual, but also said that the top of the couche 4 had been colonised by lichens and moss, meaning that the top of the couche 4 was easily identifiable. That was our experience during the excavation, too, due to the abrupt colour transformation and the change from either the dense, hard breccia, or loose cryoclastic rocks above, to humid, plastic, fine-grained sediments below.

Within this overall layer 3, layer 3A was found only in zone L12. It was between layer 2 and the breccia, which limited its extension to the south and to the west. It consisted of loose sandy-clayey sediments with some layers of small (1 to 2 cm) angular clasts within it, which differentiated it from layer 2. There was no apparent discontinuity between 3A and 2, however. Layer 3A had areas of different colours (different shades of brown), without any apparent differences in structure.

Layer 3B was only found in band L from about L9 forward, underneath the breccia. It consisted of small angular clasts or granules (described by that valuable French word, *cailloutis*), with no interstitial matrix. It cut across other layers and tended to collapse easily when exposed, since it was not cemented, which left holes underneath the breccia. This suggested that it had been thoroughly washed by water flowing under the breccia, which carried away any finer particles. It disappeared in L13 ("*Lit. . .s'efface sous la brèche en L13*" according to my notes), and contained a few flint flakes.

Layer 3C was both more extensive and richer in archaeological material than 3B; it is one of the two relatively rich sub-layers of 3, the other being 3E. It was also a cailloutis, found underneath 3B where that was present, or under the breccia elsewhere. It was dominated by its coarser fraction, of which the (very angular) clasts were larger (2 to 4 cm) than in 3B. These were surrounded by a brown sandy-clayey matrix, making the layer hold together better than 3B. It is described in my notes as having a clear transition with the couche 4.

The breccia itself was at first called layer 3, but later became known as 3D. It is a pale-coloured layer, very hard and seemingly homogenous. It is composed of angular clasts a few cm in size, with a few larger blocks, and also contained some lithics and bone fragments. All of this was strongly cemented together. It was one metre thick in L11, but got thinner towards the east and became more discontinuous, breaking up into separate blocks of breccia. It also became thinner towards the north. According to Lebel (1988) "La concrétion à ce niveau est différente et forme un bourrelet d'agrégats sableux, très fins et de granules qui sont très compactes et liées par un ciment induré, avec quelques témoins archéologiques." Thus we were, to some small extent, able to exploit lateral variations in the hardness and consolidation of the breccia as we excavated it: but it was mainly just very difficult to get through, especially before we got jackhammers to work with. Figure 8.2 (see chapter 8) shows excavation of the tranchée L in the early days, where most of the work had taken place in layer 3. Unfortunately, I do not have any more detailed pictures than that available.

The other relatively rich sub-layer was 3E, and more than 90% of its archaeological material came from one square metre, M8. Layer 3E was not described in the first couple of years of the excavation, and probably not until 1992, when we excavated M8. I have not been able to find any description of 3E in my notes or in the reports that I have. Based solely on my recollections, therefore, we decide to "create" a layer 3E when we encountered a relatively dark layer which was above and coarser-grained than the couche 4. It had a salt-and-pepper appearance, due to its combination of small limestone clasts and darker, fine-grained matrix. It was loose and easy to dig. My list of lithic pieces includes material attributed to layer 3E in zones K7, L9, L10, M8 and a single piece at the limit between N10 and N11.

The post-depositional effects that smeared out the couche 4 must have also affected layer 3, since the collapse of the breccia dome must obviously post-date the creation of the breccia, and the injection of parts of the couche 4 into overlying layers, to form 4M and 4P, also could only have happened after those overlying layers had formed. Those events seem to have happened at some time in the interval between the formation of layer 3 and the formation of layer 2 (which fills in gaps left near the rock wall). Given the localised nature of layer 3E, its dark colour, and the fact that 44.1% of the raw material sample from that layer consists of pieces in the "burned" rock types (versus 16.5% in layer 3C), it is possible that it, too, is an example of the couche 4 injecting itself into overlying sediments. For that reason, I will deal with 3E separately in some of the raw material analyses below, but I will combine it with the complete layer 3 assemblage for the typological analysis since all sub-layers provided mainly small pieces in the tamis.

It is interesting that layer 3 provided almost all of its archaeological material from band 8 (for 3E in M8) and band 9 northwards, while the couche 4 was richest from band 9 on south. Although we can envisage the collapse of the dome causing couche 4 material at the front of the site to be squeezed outwards, where it could be eroded away, it does not seem possible that the post-depositional effects could have eliminated archaeological material south of band 8 or 9 (i.e., towards the interior of the rock shelter), if there had been any there. The deposits of layer 3 therefore must represent occupations near the front of the rock shelter, rather than back towards the wall.

10.3. Pollen and fauna

In 1999 J. Argant took four samples from layer 3: two from zone N2, which were not analysed, and two from K6, one of which (#12) was analysed. It came from the breccia (layer 3D), at a z value of -194 cm. This sample did not produce enough pollen for statistical analysis, nor even (as she puts it) "pour apprécier le résultat" (Argant, 2000; Lebel, 2000: p. 71c). However, her Tableau 5 does list the pollen that were found: 4 of *Pinus*, 1 *Quercus*, 1 *Corylus*, 1 *Juniperus*, 1 undetermined conifer, 2 Poaceae and 23 Cichorioideae. This combination of trees including pine, oak, hazel and juniper along with grasses and flowering plants of the family that includes dandelions, chicory, etc., is, indeed, not particularly distinctive or evocative. Most of these can tolerate a wide range of environmental conditions. In addition there is every likelihood that pollen accumulated both during the time of deposition of the cryoclastic clasts in the breccia, and during the warmer, wetter time when they were cemented together, carried in by the percolating fluids. The pollen results therefore really do not tell us much at all.

The faunal results are equally sparse. We did find bone fragments, but the only identifiable remains listed in the tables in Fernandez (2001) are at least one aurochs incisor (giving an MNI of one), and one deciduous tooth, a lower D3 premolar, from a roe deer (again, giving an MNI of one, and indicating that the individual was young). The roe deer tooth was in layer 3C. In addition, two other aurochs teeth are listed in Fernandez (2001) as coming from layer B3. As was mentioned in chapter 7, there is no other information about anything coming from a layer B3, so it is possible that this was a marking error, and should have been 3B: without knowing the zone, we cannot tell. Two more aurochs teeth would not add much to the sample, however, and they are not listed in Table 10.1. No carnivore remains were found, either.

What do these animals tell us? Assuming they existed at the same time, the environment was partly forested and partly open – but layer 3 covers a long time span, and these animals may have lived thousands of years apart. In terms of hunting, killing a young roe deer is a different task than tackling an aurochs, so we can suspect some versatility in hunting techniques, or at least opportunism. According to the information in Fernandez (2001), the presence of this lower D3 premolar indicates that the roe deer was killed anywhere from June to March, which does not narrow that down much, either.

10.4. Lithics

The assemblage that has been characterised typologically consists of 1662 pieces, greatly dominated by the tiny pieces of the tamis, as Table 10.2 shows. The 16 pieces of the Liste Bordes group include 5 sidescrapers (4 simple, 1 double) overlapping with 6 multiple tools, which also include an endscraper. There are also 4 Levallois flakes, 2 Levallois points, a Levallois blade, an atypical endscraper and a denticulate. Including 2 Levallois cores, a total of 9 of the 89 non-tamis pieces are of Levallois technology (10.1%). The most unusual characteristic of this small assemblage is that 10 of these 16 Liste Bordes pieces have had their bulbs removed (62.5%) – exactly the same ratio as in layer C, where 5 of 8 pieces had their bulbs removed. The bulb removals are scattered through the sub-layers (3 in sub-layer 3, 1 in 3B, 4 in 3C and 2 in 3E). In contrast, no other layer at the Bau has more than 22.7% of the bulbs removed (layer J). Perhaps this really is a characteristic of the industry left at the Bau during the time of accumulation of the upper breccia. Bulb removal thins the butt end of the piece, which might make it more suitable for hafting. However, only one of the pieces is a point, so we must not imagine that these were necessarily spear heads; that piece (N9 3C 269) is a Levallois point that not only had its bulb removed but was retouched to form a double convex sidescraper, just missing becoming a Mousterian point.

The raw materials assemblage is much smaller than the typology one: 660 pieces versus 1662. Almost all of that difference is due to the tamis, however: 578 pieces in the raw material assemblage, versus 1573 in the typological assemblage. This is partly due to me not doing the raw materials for some of the tamis bags, and partly because I could not always determine raw materials for all of the very tiny pieces in some of the tamis bags. All in all, 648 pieces in the raw material sample match up by their catalogue numbers to pieces in the typological sample, and it is those pieces which were used to generate Tables 10.3 and 10.4, which show the distribution of raw material source areas within each broad typological category. The source areas for the entire raw material sample are illustrated in Figure 10.1.

There are cores from five separate source areas, and those areas are usually also well represented in the tools,

Table 10.1. Faunal results for layer 3, from Fernandez (2001): Minimum Number of Individuals (MNI) for each species. ***Liste faunique pour la couche 3, d'après Fernandez (2001): Nombre Minimal des Individus (MNI) de chaque espèce***

Taxon	common name	MNI
Bos primigenius	aurochs	1
Capreolus capreolus	roe deer	1

Table 10.2. Lithic assemblage of layer 3, summarised by typological category. ***Assemblage lithique sommaire de la couche 3, par catégorie typologique***

Category	Typology (n)	Typology (%)
Liste Bordes	16	1.0
Debris	7	0.4
Eclats	49	2.9
Lames	8	0.5
Tamis/Collectif	1573	94.6
Nucléus	9	0.5
Total	**1662**	**100**

Table 10.3. Number of pieces by typology and source area in layer 3. ***Nombre de pièces par catégories typologiques et groupes de sources des matières premières dans la couche 3***

3 n	Liste Bordes	Debris	Eclats	Lames	TamisColl	Nucléus	Total
Unidentifiable	7	2	17	4	409	0	439
Unknown	0	0	0	0	2	0	2
Faraud etc.	1	2	6	0	35	3	47
Méthamis etc.	1	2	0	0	4	2	9
Les Sautarels etc.	0	0	0	0	5	0	5
Sault	1	0	3	1	71	1	77
Murs	1	1	11	0	39	1	53
Local	0	0	0	0	0	0	0
St. Jean de Sault	0	0	0	0	0	0	0
Roussillon	0	0	0	0	0	0	0
Murs-Bezaure Olig.	0	0	0	0	12	1	13
Ravin de la Treille	0	0	0	0	1	0	1
St. Trinit	0	0	1	0	0	0	1
Durance	0	0	0	0	0	0	0
Tertiary Calavon	0	0	0	0	0	0	0
Tertiary Murs VdeV	0	0	0	1	0	0	1
Mormoiron	0	0	0	0	0	0	0
R de Guérin	0	0	0	0	0	0	0
N.Aurel	0	0	0	0	0	0	0
Total	**11**	**7**	**38**	**6**	**578**	**8**	**648**

flakes, and in the small flakes found in the tamis, with two exceptions: Méthamis, and Murs-Bezaure Oligocene. Méthamis is represented by only 9 pieces, two of which are cores. It has a particularly low number of pieces in the tamis (4), and no flakes or blades, but does include two pieces of debris and one piece on the Liste Bordes: a flake with its bulb removed. The cores in layer 3 are fairly small, averaging just over 30 g in weight, but the two from Méthamis are definitely the largest: one weighs 75 g and the other weighs 65 g. Despite their size, however, they are among the most used cores in the layer, with little to no cortex left and a minimum of 10 and 14 visible flake scars, respectively. They are both from layer 3C. This combination of used-up cores, one bulb-removed flake, and little sign of knapping at the site, suggests that the Méthamis material was part of a personal toolkit, which was discarded at the Bau, presumably to be replaced by other pieces that were carried away. The one core from Murs-Bezaure is accompanied by a dozen small flakes in the tamis. That core, also from 3C, is very small (11 g), burned, has little or no remaining cortex and has at least 12 flake scars visible on it. It is tempting to think of that one, too, as having been used up and discarded, perhaps after one last attempt to extract something useful from it.

Three source areas show the full *chaîne opératoire*: Faraud, Sault and Murs (the usual "big three"). Sault has the highest number of pieces, but most of them are in the tamis. Murs comes in second, but has the largest number of flakes. Faraud has more cores (3) than any other source area, and the one piece on the Liste Bordes is a large piece with a lot of retouch all over it, including some bifacial retouch, and of course it has had its bulb removed. The other source areas are represented by incidentals: a flake, a blade, a few chips in the tamis. The whole assemblage seems to result from personal toolkit maintenance undertaken during brief visits to the site.

As mentioned earlier, it is possible that layer 3E is another injection of the couche 4 into overlying sediments, so it is interesting to look at its characteristics versus 4 and 3C, to see where it fits. The numbers are too small (and too dominated by the tamis) to break down into tool-type categories, so Table 10.5 contrasts the total percentages for each category of raw material source, excluding those source areas that were unused in either 3C or 3E. All in all, 3E is probably less like the couche 4 than 3C is! Judged this way, if 3E is not part of an overall layer 3, then it deserves to be considered on its own, not joined in with the couche 4.

Table 10.4. Percentages of pieces by typology and source area in layer 3. ***Pourcentages de pièces par catégories typologiques et groupes de sources des matières premières dans la couche 3***

3 %	Liste Bordes	Debris	Eclats	Lames	TamisColl	Nucléus	Total
Unidentifiable	63.6	28.6	44.7	66.7	70.8	0.0	66.7
Unknown	0	0	0	0	0.3	0	0.3
Faraud etc.	9.1	28.6	15.8	0	6.1	37.5	7.3
Méthamis etc.	9.1	28.6	0	0	0.7	25.0	1.4
Les Sautarels etc.	0	0	0	0	0.9	0	0.8
Sault	9.1	0	7.9	16.7	12.3	12.5	11.9
Murs	9.1	14.3	28.9	0	6.7	12.5	8.2
Local	0	0	0	0	0	0	0
St. Jean de Sault	0	0	0	0	0	0	0
Roussillon	0	0	0	0	0	0	0
Murs-Bezaure Olig.	0	0	0	0	2.1	12.5	2.0
Ravin de la Treille	0	0	0	0	0.2	0	0.2
St. Trinit	0	0	2.6	0	0	0	0.2
Durance	0	0	0	0	0	0	0
Tertiary Calavon	0	0	0	0	0	0	0
Tertiary Murs VdeV	0	0	0	16.7	0	0	0.2
Mormoiron	0	0	0	0	0	0	0
R de Guérin	0	0	0	0	0	0	0
N.Aurel	0	0	0	0	0	0	0
Total	**100**	**100**	**100**	**100**	**100**	**100**	**100**

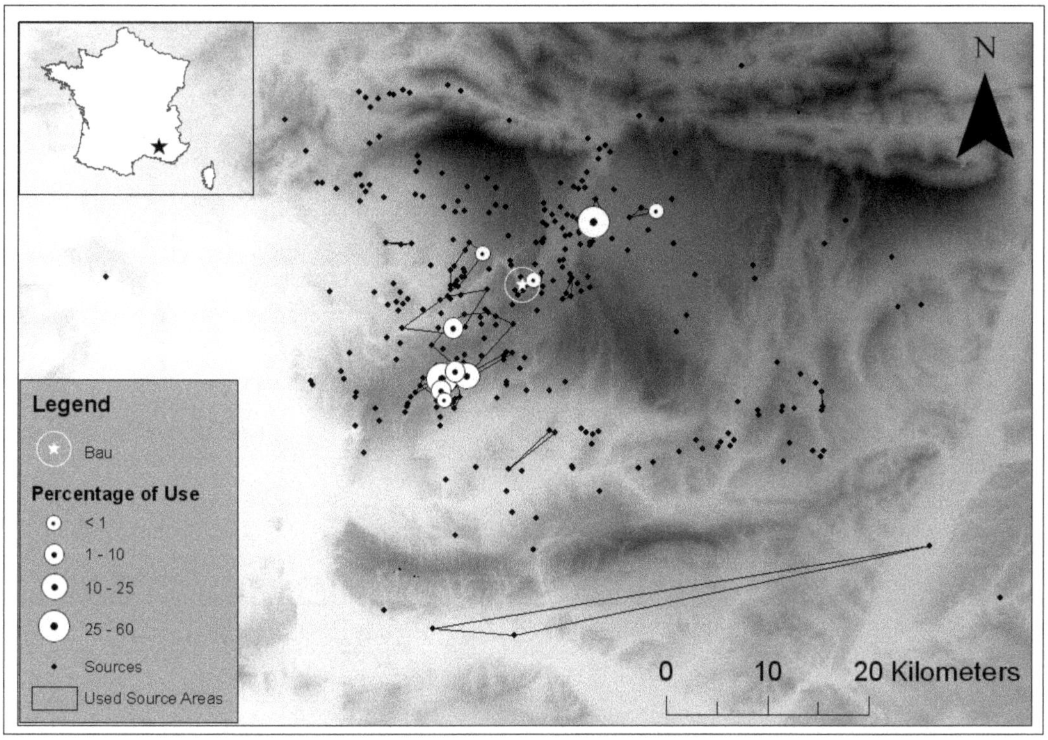

Figure 10.1. Map of the raw material source areas used for layer 3. ***Carte des aires de sources de matières premières utilisées pour la couche 3.***

Table 10.6 adds layer 3 to the table that has been growing chapter by chapter, showing use of geographic areas, recalculated excluding unidentifiable and unknown pieces. These numbers are based on the entire raw material sample, not just pieces that match the typological list. The main difference that causes is that a piece from the local source makes its appearance (number K9 3 3). Layer 3 shows a fairly high percentage of material from the NE, fairly low use of the WSW, and intermediate use of sources to the SW or Near SW, compared to the other layers. The assemblage does not closely match any of the other layers, but as Table 10.5, above, showed, sub-layers 3C and 3E have different values from each other, pulling the values in opposite directions. Table 10.7 therefore repeats the information for the layers from G up, and contrasts them with 3C and 3E, and Table 10.8 provides the results of a Kruskal-Wallis comparison analysis, where 1.00 would mean that two assemblages are identical. These layers all resemble each other to some extent, but with some variation. Layer 3E, although darker than the rest of layer 3, does not match layer 4 very well (a value of 0.74 in Table 10.8, its second-lowest similarity value). In fact, layer 3E is closest to layer G (0.90), and layer 3C is closest to layer E (0.97), neither of which makes any stratigraphic sense. And although, based on their stratigraphic positions, we might look for some resemblance between 3C and layer C of the Tranchée Moulin, it is not strong (0.78). What we do see, however, in Table 10.7, is that layers G, C, 3C and 3E share a tendency to have fairly high proportions of material from the NE, and fairly low amounts from the WSW. What these layers also share is that they have small lithic assemblages, resulting from short-term, spotty, occupations of the site.

Table 10.5. Percentages of pieces by source area in layers 3C, 3E and 4. ***Pourcentages de pièces par groupes de sources des matières premières dans les couches 3C, 3E et 4***

%	3C	3E	4
Unidentifiable	66.4	71.5	63.10
Unknown	0	0.4	0.19
Faraud etc.	8.0	5.3	9.00
Méthamis etc.	2.6	0	4.61
Les Sautarels etc.	0.6	1.1	0.53
Sault	10.3	14.1	8.27
Murs	8.8	6.5	11.03
Murs-Bezaure Olig.	2.8	0.8	0.28
Ravin de la Treille	0.3	0	0.14
St. Trinit	0	0.4	0.12
Tertiary Murs VdeV	0.3	0	0.02
other	0	0	2.72
Total	100	100	100

In our detailed models of factors influencing raw material choice (Wilson & Browne, 2014), layer 3 had very interesting results (Tables 10.9 and 10.10). Despite its small sample size, six of the factors were individually significant, and the overall model showed good results. According to the Wald Chi-Square values, the most heavily weighted factors were the availability of Large rocks, and the Quality of the raw material. These were

Table 10.6. Percentages of identified pieces (excluding unidentifiable and unknown source pieces) from each geographic region for layers K2, K1-J, I, H, G, E, C, 5, 4 and 3. ***Pourcentages des pièces identifiées (excluant les pièces indéterminables ou de sources inconnues) de chaque région géographique pour les couches K2, K1-J, I, H, G, E, C, 5, 4 et 3***

Geographic area	K2 %	K1-J %	I-I4 %	H-H2 %	G %	E %	C %	5 %	4 %	3 %
Local	0.2	3.4	1.3	0.7	0	0	0.5	1.4	6.94	0.5
NE	19.4	20.4	12.3	15.7	41.5	31.0	39.5	5.7	22.95	37.0
E	0	0.3	0.1	0	0	0	0	0	0.09	0
Far SE	0	0	0	<0.1	0	0	0	0	0.05	0
Far S	0	0	0.1	0	0	0	0	0	0	0
S	0	0	0	0	0	0	0.5	0	0.01	0
SSW	0	0	0.1	0	0	0	0	0	0.04	0
SW	44.3	54.1	35.0	30.0	48.4	25.4	17.6	59.0	32.18	34.7
Near SW	23.0	15.0	29.1	34.8	9.6	29.9	30.7	10.5	24.68	22.7
WSW	12.4	6.6	21.9	18.6	0.5	12.7	11.2	22.9	12.61	4.6
W	0	0	0	0	0	0	0	0	0.01	0
NW	0.7	0.1	0.1	0.2	0	1.0	0	0.5	0.42	0.5
Total	**100**	**100**	**100**	**100**	**100**	**100**	**100**	**100**	**100**	**100**

Table 10.7. Percentages of identified pieces (excluding unidentifiable and unknown source pieces) from each geographic region for layers G, E, C, 5, 4, 3C and 3E. ***Pourcentages des pièces identifiées (excluant les pièces indéterminables ou de sources inconnues) de chaque région géographique pour les couches G, E, C, 5, 4, 3C et 3E***

Geographic area	G %	E %	C %	5 %	4 %	3C %	3E %
Local	0	0	0.5	1.4	6.94	0	0
NE	41.5	31.0	39.5	5.7	22.95	30.5	51.4
E	0	0	0	0	0.09	0	0
Far SE	0	0	0	0	0.05	0	0
Far S	0	0	0	0	0	0	0
S	0	0	0.5	0	0.01	0	0
SSW	0	0	0	0	0.04	0	0
SW	48.4	25.4	17.6	59.0	32.18	37.3	29.7
Near SW	9.6	29.9	30.7	10.5	24.68	23.7	18.9
WSW	0.5	12.7	11.2	22.9	12.61	7.6	0
W	0	0	0	0	0.01	0	0
NW	0	1.0	0	0.5	0.42	0.8	0
Total	**100**	**100**	**100**	**100**	**100**	**100**	**100**

Table 10.8. Kruskal-Wallis analysis of similarity between the layers in Table 10.7, where 1.00 = identical. Kruskal-Wallis chi-squared = 4.8592, df = 6, p-value = 0.562. All individual p-values are significant to the 0.05 level. ***Analyse Kruskal-Wallis des ressemblances entre les couches données dans le Tableau 10.7; 1.00 = identique. Kruskal-Wallis chi-carré = 4.8592, df = 6, p-value = 0.562. Les valeurs-p individuelles sont significatives au niveau de 0.05***

r_s	G	E	C	5	4	3C	3E
G	1.00	0.91	0.87	0.86	0.83	0.94	0.90
E	0.91	1.00	0.81	0.85	0.84	0.97	0.84
C	0.87	0.81	1.00	0.81	0.77	0.78	0.81
5	0.86	0.85	0.81	1.00	0.92	0.89	0.67
4	0.83	0.84	0.77	0.92	1.00	0.86	0.74
3C	0.94	0.97	0.78	0.89	0.86	1.00	0.84
3E	0.90	0.84	0.81	0.67	0.74	0.84	1.00

followed by Medium rocks, which had a negative Beta value, so sources with abundant medium-sized rocks were less likely to be used. Three other significant factors, with very similar weights, were Extent of the source area, Area Of the Sources In the Source Area (both with positive Beta values), and the abundance of Small rocks (which had a negative Beta value). Converting the Wald Chi-Square values to percentages (Table 10.10) and adding up the factors in the two groups of Raw Material factors vs. Terrain factors, gives a total of 68.8% for Raw Material characteristics, and only 29.4% for Terrain values. These values continue the tendency we have seen in all of the upper layers, of having Raw Material factors outweigh the Terrain factors, but in the most extreme way yet: there is a larger difference between the two sets than there was in layer C, and a much larger difference than in the couche 4. In fact, of all of the layers tested at the Bau, layer 3 has the second-highest Raw Materials value (layer 2 beats it by a slim margin; see chapter 11), and the lowest Terrain values total of all (Wilson & Browne, 2014). The choice of raw materials for these artifacts was driven by the availability of the preferred large nodule size and by rock quality for knapping; the ease with which the rocks could be obtained did matter, too, but they had to be worthwhile rocks, first and foremost.

10.5. Conclusion

Layer 3 has small lithic assemblages scattered in a variety of sediment types, almost certainly representing multiple short-term occupations spread over a long period of time. We have very little indication of the flora and fauna

Table 10.9. Generalised linear model for raw material source choices in 3. P-values in bold font are statistically significant. Sample: 215 pieces from 10 source areas

Overall:
Omnibus test: likelihood ratio chi-square = 333.992, df = 10, p = <0.001

Variable	Beta	S.E.	Wald Chi-Square	P-value	AIC
Quality	1.940	0.5012	14.976	**<0.001**	104.014
Extent	1.829	0.7316	6.253	**0.012**	
Difficulty (Cal/km)	-0.048	0.0295	2.643	0.104	
Small rocks	-6.780	2.7480	6.087	**0.014**	
Medium rocks	-4.501	1.5671	8.250	**0.004**	
Large rocks	3.737	0.9450	15.638	**<0.001**	
Very large rocks	0.542	0.9908	0.299	0.585	
Calories	-0.001	0.0009	1.667	0.197	
AOSA	-2.1 e-7	1.3 e-7	2.701	0.100	
AOSISA	6.8 e-5	2.8 e-5	6.098	**0.014**	
(Intercept)	-5.639	5.1571	1.195	0.274	

Table 10.10. Wald Chi-Square values as percentages, assemblage from layer 3

Variable	Wald Chi-Square	Wald Chi-Square as %
Quality	14.976	22.8
Extent	6.253	9.5
Difficulty (Cal/km)	2.643	4.0
Small rocks	6.087	9.2
Medium rocks	8.250	12.5
Large rocks	15.638	23.8
Very large rocks	0.299	0.5
Calories	1.667	2.5
AOSA	2.701	4.1
AOSISA	6.098	9.3
(Intercept)	1.195	1.8
Total	**65.807**	**100**

of that stretch of time, but based on the sediments the climate changed from cold to warmer and wetter, possibly more than once. During this time period, Neandertals occasionally visited the site. They brought with them at least part of an aurochs and part of a young roe deer: not necessarily at the same time. They brought their toolkits with them, too, and although they did not leave us any pieces of rock from any unusual sources, the pattern of use of the "same old" sources is unusual: very much focused on obtaining good material. They brought in highly functional toolkits, discarded the pieces that were no good any more, probably replaced them with newly made pieces (leaving us only the knapping by-products), and continued on their way to other areas.

10.6. Sommaire du chapitre

La structure de la couche 3 est complèxe dans le détail, avec variations en granulométrie, en degré de cimentation, et présence ou absence de matrice, mais on peut aussi simplement la décrire comme étant la couche supérieure brèchique de la tranchée L. Elle se situe à environ 150 à 280 cm sous zØ, avec une épaisseur de parfois plus d'un mètre de brèche dans certaines parties de la tranchée, mais s'amincissant pour disparaître complètement en arrière, près de la paroi rocheuse. Nous l'avons divisée en 6 sous-couches, dont 2 ont livré quelques centaines de pièces lithiques, tandis que d'autres n'en avaient que quelques dizaines. Il n'y avait que très peu de restes fauniques identifiables. La plupart du matériel archéologique a été trouvé à partir de la bande 8 ou 9, et plus au nord, suggérant que les occupations qui sont à l'origine de la couche étaient concentrées vers l'avant du site. La couche s'est formée d'abord sous un climat glaciaire, suivi par une période plus chaude (e.g., cimentation des sédiments cryoclastiques). Nous n'avons pas de dates, mais on peut sans doute situer la couche 3 pendant la période d'existence des Néandertaliens dits "classiques".

Les sous-couches les plus riches en pièces lithiques sont 3C et 3E; il y en a assez dans chaque sous-couche que j'ai pu les analyser séparément. Ceci est d'autant plus intéressant que les deux diffèrent beaucoup l'une de l'autre par leur couleur et leur sédiment respectifs. La sous-couche 3E est nettement plus foncée et plus brûlée que la sous-couche 3C. Nous aurions pu envisager une remontée du matériel de la couche 4 jusque dans les sédiments de 3E autour de la zone M8, pendant la génèse de la couche 4P (voir le chapitre 9). Cependant, l'analyse des matières premières a montré des différences non seulement entre 3E et 3C, mais également entre 3E et 4, et nous devons accorder à 3E une origine au moins partiellement indépendante.

Le peu de pollens trouvés ne donne pas d'information significative, et la faune à peine plus. Le paysage était boisé et/ou ouvert, la chasse a pu être sélective ou bien opportuniste, et les stratégies de chasse devaient varier selon l'animal: un aurochs représente une proie beaucoup plus redoutable qu'un jeune chevreuil. Quant à l'assemblage lithique, il est dominé par le débitage, c'est-à-dire les petites pièces que nous avons trouvées dans les tamis, mais comprend également quelques éclats et outils (racloirs, outils multiples, etc.). Parmi les 16 pièces classifiées selon la liste Bordes, on note que 10 (62.5%) ont leurs bulbes de percussion enlevés, une proportion beaucoup plus grande que dans les autres couches, à l'exception de la couche C dans la Tranchée Moulin, où la proportion est identique.

Les matières premières proviennent de plusieurs origines géographiques, et nous avons identifié des nucléus issus de 5 sources. Les nucléus sont en moyenne assez petits et portent les traces de beaucoup d'enlèvements; ils ont probablement été abandonnés sur le site parce qu'ils n'étaient plus très utiles. Trois sources sont représentées par la chaîne opératoire complète. La plupart des pièces provenant de Sault ont été trouvées dans les tamis, probablement parce que des outils dans cette matière première ont été retouchés sur place, tandis que nous avons plus de nucléus et éclats provenant des sources en direction du sud-ouest. Nos analyses montrent que le choix des matières premières était orienté beaucoup plus vers des roches de grande taille et de bonne qualité que vers celles qui sont facilement accessibles.

Les vestiges trouvés dans la couche 3 étaient éparpillés dans une accumulation importante de sédiments, et résultent de plusieurs occupations de courte durée pendant une longue période de temps. Pendant cette période le climat et l'environnement ont changé, peut-être plus d'une fois. Les Néandertaliens qui passaient par là de temps en temps ont amené leurs trousses à outils: outils très fonctionnels et matières premières de bonne qualité. Ils les ont retouchés et remplacés, nous laissant le débitage et les pièces abandonnées. Ils ont mangé et laissé en place les restes osseux, puis ils sont partis.

11

Layer 2 and (a little bit) Layer 1

11.1. Introduction

This chapter is mainly about layer 2, which is the highest layer with any real quantity of archaeological material at the Bau. Layer 1 will be described separately at the end of this chapter, because it has to go somewhere, but there is not much to say about it.

Layer 2 is on top of the breccia dome and the other sedimentary variants making up layer 3 (chapter 10), so it is of very uneven thickness, and is thickest towards the back of the rock shelter, where sediments could accumulate between the downslope side of the breccia dome, and the rock wall. Its z values extend from approximately 170 cm below zØ at the lowest to about 80 cm below zØ at its highest point. In essence that means that layer 2 reaches nearly to the top of the sediments: layer 1 is very thin (a couple of cm in most places), and our zØ value was installed comfortably above all of the sediments, so that all measurements would be in negative values.

Layer 2 was excavated in the tranchée L, in zones I4-5, J3-5, K3-5, K7-10, L2-12, M2-14, N2, and N8-9 (see Figure 1.11), although I do not have lithic samples from M8, M10-11 and M13 (possibly because there were none). The archaeological material is far more common at the rear of the site, which may be a function of preservation and sediment thickness, more than of placement of occupations by hominins. Layer 2 has not been directly dated, but is clearly younger than layers 4 and 3 – a lot had to happen to form layer 3 after 4 was laid down, and still more time passed before and during the accumulation of layer 2. It also clearly originated during a cold time period, and the faunal material combines with the lithic material to suggest that it is Middle Palaeolithic, likely with an age in the MIS 4, Würm II, about 65-75,000 years old. It has always been tempting to wonder whether it might be younger, maybe even into the Upper Palaeolithic. . . but there is no strong evidence for that.

Layer 2 is not a clear occupation layer. It has no living floor, no hearths, and only scattered concentrations of material. At first, as we dug the front of the tranchée L, it yielded only sparse finds, but the area behind the big rock we removed in 1993 (see chapter 8) was much richer in lithics and bones: "au total 1 705 pièces lithiques, 2 112 vestiges osseux et 93 dents de mammifères" just from the zones K4, L3, L4 and M5 in the excavation season of 1994, for instance, found in "concentrations lenticulaires" (Lebel, 1996). Together, these provide an interesting record of both depositional and post-depositional events. One human tooth has also been attributed to Layer 2.

11.2. Sediments, evolution of the site, and excavation of the couche 4

After the formation of the breccia in layer 3, and probably after some period of erosion of the top surface of that breccia dome (which seems generally smooth), more cryoclastic plaquettes and granules broke off of the rock shelter walls and ceiling, and piled up. These combined with sand- and silt-sized grains, possibly wind-blown, to create an accumulation of loose, yellowish-coloured sediments: layer 2. Unlike nearly every other layer at the Bau, none of this ever got cemented together, although in some areas it was so densely packed that the electric jackhammer was used to loosen it up (Lebel, 1996). At the front of the tranchée L we found occasional thin layers of *cailloutis*, composed of small (1-2 cm) angular clasts with little or no matrix (Lebel, 1987). There was not much archaeological material in layer 2 towards the front of the trench, but in 1990 "une concentration archéologique lenticulaire a été individualisée à l'horizon IIb, 70 centimètres au-dessus de la couche IV" (Lebel, 1990). Layer 2B was not continuous, but it was slightly darker in colour than layer 2, which allowed us to identify it in different parts of the tranchée L. Ultimately the large majority of pieces in layer 2B came from just two squares, K5 and J4, at the back of the trench, but lithic pieces from K4, J5, I4, L8, M7 and M9 were also attributed to this sub-layer. The 2B material is greatly outnumbered by material from layer 2 proper, however, and is combined with that in the analyses below.

M. Wilson (1997) analysed four samples from layer 2: MM10 (L6, z=-170 cm), MM11 (M3, z=-105 cm), MM12 (M3, z=-90 cm), and MM29 (K6, z=-130 cm). All samples contained fragments of limestone and grains of quartz, as well as "auréoles sparitiques (physico-chimiques)" and usually "[aureoles] micritiques (biochimiques)". She also commonly found "injections d'acide organique, oxidation de fer, moisissure" and in one case (MM11), "cristallisation secondaire dans la porosité". Thus, the samples had been affected by both biological and physical alterations after deposition. Despite this, no samples showed signs of post-depositional deformation ("aucune déformation post-dépôt"), which is surprising because such deformation was visible to the naked eye in the stratigraphic sections, as signs of cryoturbation, for instance, which changed the orientation of the larger clasts into v-shaped formations (pointed out to me at the time by G. Prichonnet). The only signs of hominin presence in M. Wilson's samples were in MM11, where she reported seeing "ossements craquelés, silex hématisés". All in all, M. Wilson (1997, p. 120) attributes layer 2 to a periglacial environment, with signs both of seasonal deep freezing, and periodic thawing of ice lenses resulting in washing of the sediments ("phénomène de lavage"), and recrystallisation of dissolved carbonates within the deposits. That recrystallisation was on a small scale, however, as we have seen that the sediments were not cemented together. Figure 11.1 gives an idea of the thickness and character of these unconsolidated sediments. In Figure 11.2 (which is the top half of Figure 9.1, from chapter 9), layer 2 is clearly visible in the northern wall of zone J5, and we can at least imagine seeing groups of the small plaquettes oriented to form v-shapes, here and there.

Further signs of post-depositional activity are evident in the archaeological material: the small bones and some lithics of layer 2 were sometimes rounded (their edges were smoothed off due to transport by water), and some of the lithics showed small chips along their edges due to movement and mechanical pressure within the sediments (Lebel, 2000). The rounding implies considerable movement in free-flowing water, which must have occurred after deposition by hominins but before the material was covered in sediments. The likeliest explanation for that is that the pieces were on the inward side of the breccia dome, and affected by water from snowmelt and/or heavy rainfall, that dripped into the site by percolating through the rock overhang. The irregular contours of the back of the rock shelter (Figure 11.3), combined with the surface contours of the strongly cemented breccia, would conceivably have provided areas for water to flow around in as it pooled then drained through some outlet. Figure 11.3 shows not only the naturally hollowed-out shelf-like area at the back

Figure 11.1. Sediments of layer 2 (the lighter-coloured part) in the section walls of zones J5 (north side) and K6 (west side), seen looking NW from zone K5. The scale bar (added) represents approximately 10 cm. ***Sédiments de la couche 2 (la partie pâle) dans les coupes des zones J5 (côté nord) et K6 (côté ouest), vue vers le nord-ouest de la zone K6. L'échelle (ajoutée) représente environ 10 cm.***

Figure 11.2. Layer 2 in the northern section wall of zone J5 in 1994. Some groups of the small- to medium-sized rocks (plaquettes) seem to be oriented to form v-shapes a few cm high, which if true would be due to cryoturbation: disturbance by freezing and thawing activity, characteristic of a periglacial environment. ***La couche 2 dans la coupe nord de la zone J5 en 1994. Quelques groupes de roches et plaquettes de petite à moyenne taille et de quelques cm de haut semblent orientés en "v" et auraient possiblement été formés par de la cryoturbation: déformation due au gel et dégel, caractéristique d'un environnement périglaciaire.***

Figure 11.3. The back of the tranchée L in 2019, with the rock shelter wall on the right and centre, and sediments on the left, topped with a large rock slab. Photo M. Ughetto, used with permission. ***Le fond de la tranchée L en 2019, avec la paroi rocheuse à droite et au centre, et des sédiments à gauche, surmontés par un gros rocher. Cliché de M. Ughetto, utilisé avec son aimable permission.***

of the trench, but also the unexcavated sediments of zones N3 and O3, and it looks like there are traces of another dome of breccia, whose top surface is inclined northwards, towards the lower left side of the photo. Layer 2 would be on top of that breccia.

The lithics also had calcareous accretions on them, due to reprecipitation of calcium carbonate within the deposits, and although many were patinated, many more had only incipient patination, slightly obscuring their petrographic nature (Figure 11.4). This patination is due to reactions between the flint and the circulating waters causing alteration of the silica. In fact, the assemblage from layer 2 is less patinated than any other, perhaps simply because it is younger, but perhaps it was also exposed to different geochemical conditions, due to the shorter sequence of climatic phases that this assemblage has "lived through" in the site sediments.

11.3. Pollen and fauna

According to her report in Lebel (2000), in 1999 J. Argant collected 17 sediment samples from various parts of layers 2 and 2B, as well as two others that might have been related to layer 2, or might have been more modern, from inside of the "shelf" area at the back of the trench.

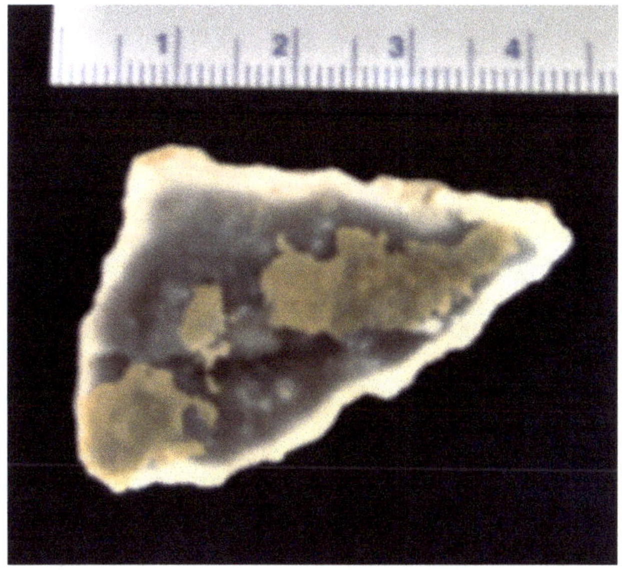

Figure 11.4. Piece number K4 2 200, a distal fragment of a Mousterian point, seen in ventral view. It has little or no cortex, was not burned, and was made from the local flint type. The raw material type is slightly obscured by a thin yellowish concretion that adheres to part of it, and by incipient patination, which is most obvious around the edges. (I apologise for the poor quality of the photo.) ***Pièce numéro K4 2 200, fragment distal d'une pointe moustérienne, en vue ventrale. La pièce a peu ou n'a pas de cortex, n'a pas été brûlée, et a été fabriquée dans un silex local. La nature de la matière première est légèrement cachée par une couche mince jaunâtre de concrétion, et par un début de patine, plus évidente sur ses bords. (Mes excuses pour la qualité inférieure de la photo.)***

She analysed one of those latter two, and one other from layer 2, but provides details only of the one from layer 2: sample number 4, from zone K6, at z=-114 cm. She says that it came from breccia, which does not seem to match layer 2, but the z value is plausible for layer 2. The sample provided 76 grains, dominated by pine, which she says is of *sylvestris* type (J. Argant, 2000). Her Tableau 5 shows that the grains consist of 56 *Pinus*, 1 *Betula* (birch), 1 *Juniperus*, 3 Poaceae (grasses), 2 *Artemisia* (mugworts, etc.), 1 *Ambrosia* type (a flowering plant), 1 *Plantago* (herbaceous plantain), 2 undetermined, 7 monolete spores and 2 trilete spores. She suggests that these could belong to a period of milder climate (J. Argant, 2000, p. 71c), but given the mixture of depositional effects in evidence in layer 2 (and the periglacial nature of the sediments), it seems unlikely that the bulk of the layer really dates to a time of mild climate. It is more likely that any mild-climate pollen infiltrated the older sediments, carried in by percolating water.

Despite its relatively poor preservation, the faunal assemblage from layer 2 is diverse, both in species present and in the environments they like to live in (Table 11.1). It is dominated by medium-sized animals such as ibex and roe deer, rather than the horse and aurochs of some lower layers, but those animals are also represented. Some of these species lived in the craggy, steep areas of the gorge and surrounding mountains; others, such as deer, prefer

Table 11.1. Faunal results for layer 2, from Fernandez (2001 and **pers. comm.) for the ungulates, A. Argant (2006) for carnivores, and Crégut-Bonnoure et al. (2011) for *Ursus arctos*: Minimum Number of Individuals (MNI) for each species. *Liste faunique pour la couche 2, d'après Fernandez (2001 et **comm. pers.) pour les ongulés, A. Argant (2006) pour les carnivores, et Crégut-Bonnoure et al. (2011) pour* Ursus arctos*: Nombre Minimal d'Individus (MNI) de chaque espèce*

Taxon	common name	MNI
*Capra caucasica praepyrenaica***	ibex	3
Rupicapra rupicapra	chamois	2
Bos primigenius	aurochs	1
Equus sp.	horse	1
Capreolus capreolus	roe deer	3
Cervus elaphus	red deer	1
Lynx spelaea	cave lynx	1
Ursus arctos	brown bear	2

partly wooded areas, and the horse and aurochs would tend to be found in more open plains areas, such as on the plateau above the site, or in the valley between Monieux and Sault. The aurochs seems to have been a young one (based on the presence of a deciduous lower incisor), and may have been acquired through opportunistic encounter hunting, but most of these animals were probably hunted deliberately, by lying in wait for them. The lynx is represented by a single piece (A. Argant, 2006; he does not say what it is), and the bears by two teeth: the worn point of a canine, and an unworn upper incisor (Crégut-Bonnoure et al., 2011). It is impossible to say whether these animals came into the site by themselves or not.

11.4. Human remains

One human tooth has been attributed to layer 2 "*remanié*", meaning from disturbed sediments; its stratigraphic context is unclear (see "From The Chest"). It is described in Trinkaus et al. (2000a, p. 212) as being "either terminal Late Pleistocene (Upper Palaeolithic) or Holocene in age; its fresh preservation and aspects of its morphology suggest that it may well be from now absent Holocene deposits." The tooth has been numbered as Aubesier 3 on the list of human remains, but its Bau catalogue number is J10 2 1: it is the only object that was ever found in zone J10 during our excavations. It consists of the complete crown and root of a right upper first incisor (Trinkaus et al., 2000a), with some dentine exposure along the incisal edge, but otherwise in good condition. It has a low amount of wear (Trinkaus et al., 2000a), and the beveling angle of the upper edge is also low, described as "close to those of recent human (and Neandertal) I^1s with similarly minimal incisal edge wear, indicating some procumbency of the tooth . . ." (Trinkaus et al., 2000a, p. 213). The other characteristics of the tooth are apparently not distinctive or particularly informative. Trinkaus et al.'s (2000a, p. 214) description concludes that it "is a small tooth, even compared to the recent European sample, although its crown dimensions (Table III) fall within the expected range of that sample." Although they do not say so, that is an extended range, allowing two standard deviations away from the average of the recent sample. By the same standard, its width could be made to fit within the extended range of both the Upper Palaeolithic and the Neandertal samples they cite, but it is much too thin (front to back) to fit within the ranges of those samples for that measurement. The conclusion has to be that this is the tooth of an anatomically modern human, Upper Palaeolithic at the oldest, more likely of Holocene age. That does not, however, mean that the rest of layer 2 is that young.

11.5. From The Chest: The Layer 2 Tooth

Aubesier 3 was found in 1993. I was surprised, several years later, to see my name listed as its discoverer (Lebel, 2000; Trinkaus et al., 2000a): I did not find it, but I was right on the spot when it was found. I am not going to give the name of its actual discoverer, however, because in my description below I might seem to be accusing him of some sort of skullduggery. I have always been suspicious of that tooth, but if there *is* something wrong about it, I do not know who might be responsible, so I do not want to seem to blame someone innocent. In case that impression is given, however, I think it is best to not name names. Here is the story:

In early July 1993 we used expanding cement on the large rock at the top of the tranchée L (see chapter 8), breaking it into pieces. Before we used the cement, we excavated around the rock, so that it would have space to expand and crack into, and not be held together by the pressure of the surrounding sediments. After it had been broken, we spent several days carrying away the pieces and cleaning up the space it had been in. Large pieces were manhandled out of there individually, but smaller pieces and loose sediments were loaded into buckets to be carried away. This means that a lot of bucket-loads of sediments were carried across the top of the site to the edge of the upper deposits, before being passed on to someone else and taken out to be dumped. One of the paths for carrying those buckets led across zone J10, which was 3 or 4 metres from the area of the big rock.

On July 7th, with clean-up underway, Serge took the day to go get groceries in Carpentras, a small city about 45 minutes drive away, and I was in charge of the dig. I happened to be near the clean-up area when a digger carrying a bucket passed me, then stopped and bent down, saying something like "What's this?" It was a human tooth. The real question is: where did it come from? Could it have come from the *in situ* deposits at the surface of J10? We cleaned the site every year when we arrived, removing vegetation and basically scraping the place down. I *think* we would have seen a tooth lying on the surface during that process. Perhaps the foot traffic loosened sediments there and allowed that tooth to wriggle its way to the surface? Maybe. Or it might have come from the sediments around and just under the big rock, and fallen out of a bucket on its way by J10, except that the area where it was found seemed clean – no little pile of spilled dirt. Could a single tooth jump out of the bucket by itself?

A third possibility is that someone planted it. It looked familiar to me, although I was not sure I had seen that particular tooth before. It reminded me, right away, of the little box of teeth that Serge had as a comparison set. I forget where he had gotten them, although I know a colleague of ours once gave him some bones and teeth from a Mediaeval cemetery, and I think his original set were of similar age. They were just a loose assortment of teeth from several individuals, for use in comparison to teeth we might find, to see if those might also be human. The colour and condition of this tooth from J10 reminded me of those teeth, which made me suspicious that someone had filched one from the box and brought it to the site. It certainly was not young: it had a small amount of sediment stuck to it, yellowish and fine-grained – but was that Bau sediment?

At lunchtime that day I went up to our lab and got out the box, but of course I had never memorised its contents, and I could not tell whether this tooth came from that set or not. When Serge came back from Carpentras I gave him the tooth and told him of my concerns. He pointed to the sediment and the wear on the incisal edge as evidence that it was old, and did not seem too concerned about potential mischief. . . but I have never been happy about that tooth. If it was a plant, it seems very risky to have left a tooth on the surface, in the hope that someone would notice it, rather than walk on it. But maybe that is what happened. Whether it was a joke that went wrong, someone being "helpful" in a misguided way, or a malicious act, I guess I will never know. And the possibility remains that it did legitimately come from the Bau, from the upper part of layer 2. It was not associated with any archaeological material at all, though, so its fairly young age does not tell us anything about the rest of layer 2.

11.6. Lithics

Layer 2 provided a sizable lithic assemblage, even though it did not have a single concentration of remains, or anything like a living floor. The typological sample that we analysed in 2013-14 included 2965 pieces, which makes it the third largest assemblage in the site, after layers 4 and H1. Table 11.2 gives the basic typological breakdown of the assemblage. It is dominated by pieces found in the tamis or recovered as collectives, but still contains a good number of other pieces, such as retouched tools, flakes, blades and cores. Table 11.3 gives a more detailed breakdown of the typology, concentrating on retouched pieces and blades, along with the percentages from the couche 4 assemblage for comparison. The numbers for layer 2 add up to a larger total than Table 11.2 would suggest, because Table 11.3 counts some pieces more than once. For instance, the blades category includes both unretouched blades and the blades that serve as supports for other tool types. Although neither table lists this as a category, multiple tools account for 24 of the 119 pieces from the Liste Bordes (20.2%), considerably less than in the couche 4 (33.0%).

Table 11.2. Lithic assemblage of layer 2, summarised by typological category. ***Assemblage lithique sommaire de la couche 2, par catégorie typologique***

Category	Typology (n)	Typology (%)
Liste Bordes	119	4.0
Debris	10	0.3
Eclats	142	4.8
Lames	50	1.7
Tamis/Collectif	2619	88.3
Nucléus	23	0.8
Galets	2	0.1
Total	**2965**	**100**

Since we wonder how old layer 2 really is, does its lithic assemblage give us any indicators? An Upper Palaeolithic assemblage might be expected to have plenty of blades, lots of endscrapers, and not much use of the Levallois technique, along with other special tools which are entirely absent from the Bau. Layer 2 does have a lot of blades. . . but then so does J4-K1, which includes 96 blades and retouched blades, in a slightly smaller assemblage. Adding up the Liste Bordes and Nucléus for J4-K1, as was done for layer 2 in Table 11.3, gives as a result that 43.4% of its total consists of blades and retouched blades. J4-K1 is definitely not an Upper Palaeolithic layer: it may be 200,000 years old (see chapter 3). The presence of blades can no longer be taken seriously as an indicator of an Upper Palaeolithic industry.

Layer 2 does have a slightly higher percentage of endscrapers than the couche 4 does, but far fewer in total numbers: the couche 4 has 77 endscrapers, so assuming that 13 of them in layer 2 indicates anything much seems like a stretch. Figure 11.5 shows one of these potentially Upper Palaeolithic tools, an atypical endscraper and notch combination. Layer 2 does not have as strong a presence of the Levallois technique as the couche 4 does, but there is some. What about the 7 "others" listed above? Two of them are truncated blades, two are truncated flakes, one is a *rabot* and two are miscellaneous retouched items. Nothing conclusive there, although truncated blades might suggest Upper Palaeolithic influences – but the *rabot* is more Middle Palaeolithic. Of course, it is likely that layer 2 is a combination of materials left at the site over a long time period, so perhaps some of them were Middle Palaeolithic and some were Upper Palaeolithic, but there really does not seem to be any way to know for sure, and arguably, no need to posit any Upper Palaeolithic contribution at all.

Table 11.3. Typologically identified pieces in layer 2. ***Pièces identifiées typologiquement de la couche 2***

Typology	English name	2 n	2 %	couche 4 %
Eclats levallois	Levallois flakes	7	2.9	11.0
Lames et lamelles	Blades and bladelets	80	33.2	23.2
Pointes levallois	Levallois points	4	1.7	9.2
Pointes moustériennes	Mousterian points	5	2.1	2.6
Racloirs simples	Single sidescrapers	37	15.4	14.3
Autres racloirs	Other sidescrapers	24	10.0	7.9
Grattoirs	Endscrapers	13	5.4	2.8
Couteaux à dos	Backed knives	2	0.8	1.2
Denticulés	Denticulates	12	5.0	2.3
Encoches	Notches	16	6.6	3.5
Bulbe enlevé	Bulb removed	11	4.6	5.2
Autres	Others	7	2.9	1.7
Nucléus levallois	Levallois cores	3	1.2	4.0
Nucléus non-levallois	Non-Levallois cores	20	8.3	10.9
Total		**241**	**100**	**100**

Figure 11.5. Piece number M2 2 8, an atypical endscraper and notch, of unidentified raw material (but judging by the picture, probably type Q, from Murs). ***Pièce numéro M2 2 8, un grattoir atypique et encoche, de matière première non-identifiée (mais à juger par la photo, probablement le type Q, de Murs).***

The raw material sample includes 2025 pieces, of which 1962 match the typological sample by catalogue number. Matching the pieces did necessitate moving some typologically identified pieces back into their original tamis groups, unfortunately, so tamis/collectif numbers are over-represented in Table 11.4, which gives the breakdown of those matched pieces by raw material source areas, and in Table 11.5, which gives the results as percentages.

The 47.0% "unidentifiable" pieces in this layer include 28.7% patinated, and the rest burned. In reality, 20.8% of pieces were classified into raw material types which are burned, but some of these were in a type which is still identifiable as to its source. These are low numbers: compared with other layers, Layer 2 is not very patinated and not very burned, leaving the majority of pieces identifiable. Of those, an unusually high number (27 pieces, or 1.4% of the matched assemblage) are in raw materials for which the source has not been identified. The remainder of the pieces come from 13 different source areas, making this the second most diverse assemblage in the Bau, after the couche 4, which had 16 source areas. This diversity is even more impressive when we consider that the assemblage from layer 2 is much smaller than that of the couche 4 (roughly one-tenth of the size in the raw material sample).

The pieces of unknown source include pieces in all 6 of my unknown types (potentially adding up to six more source areas!), and they appear in various typological categories. In the Liste Bordes set, the piece is a multiple side scraper (one edge straight, one edge bifacially retouched) in my type AB. The two cores are in the same raw material as each other (type M) and are prismatic (non-Levallois); one weighs 41 g and retains some cortex, the other weighs only 15 g and has little or no cortex left.

The assemblage includes pieces from relatively distant source areas in all directions, from Mormoiron in the west to the Rocher de Guérin area in the far south-east and Roussillon in the SSW. The three pieces in the distinctive

Table 11.4. Number of pieces by typology and source area in layer 2. ***Nombre de pièces par catégories typologiques et groupes de sources des matières premières dans la couche 2***

2 n	Liste Bordes	Debris	Eclats	Galets	Lames	TamisColl	Nucléus	Total
Unidentifiable	14	1	36	0	3	867	1	922
Unknown	1	0	5	0	4	15	2	27
Faraud etc.	2	1	10	0	1	183	0	197
Méthamis etc.	0	0	4	0	0	28	0	32
Les Sautarels etc.	0	0	1	0	0	9	0	10
Sault	13	0	9	0	6	246	2	276
Murs	19	2	22	0	6	242	3	294
Local	8	0	7	1	4	132	1	153
St. Jean de Sault	0	0	0	1	0	0	0	1
Roussillon	0	0	1	0	0	0	0	1
Murs-Bezaure Olig.	2	0	3	0	0	33	0	38
Ravin de la Treille	0	0	1	0	0	4	0	5
St. Trinit	0	0	1	0	0	1	0	2
Durance	0	0	0	0	0	0	0	0
Tertiary Calavon	0	0	0	0	0	0	0	0
Tertiary Murs VdeV	0	0	0	0	0	0	0	0
Mormoiron	0	0	0	0	1	2	0	3
R de Guérin	0	0	0	0	0	1	0	1
N.Aurel	0	0	0	0	0	0	0	0
Total	59	4	100	2	25	1763	9	1962

raw material type from Mormoiron (Figure 11.6) are all from zone K4: two in the tamis, and one small blade or bladelet (number 151). "Three" pieces does not sound like very much, but only four pieces of this type have been identified from the entire site (the other is in the couche 4), so this seems noteworthy to me. The one piece from the Rocher de Guérin, of type AP, was in a tamis from zone J5, while the single piece of type V, from Roussillon, was a flake in zone L3. The most distant source to the north, Ravin de la Treille, is much closer to the site than these other sources, as it is only a few km north of the gorge of the Nesque, but it was also used, as is shown by one flake and four pieces in the tamis.

In addition to these few pieces from distant source areas, there is an unusually high proportion of local material, not just from the local source area, but also from nearby St. Jean de Sault, source of a single cobble of flint of type AM. It was found in zone N8 and weighs 735 g; it has little or no cortex, but no sign of having been flaked, so it is a naturally-worn cobble, not a core. It might have served as a hammerstone. The really local flint, type S from Saume Morte, a few hundred metres from the Bau, is very well represented in the layer 2 assemblage, appearing as retouched tools, flakes, blades, one core, and pieces in the tamis. The one *galet* (cobble) is in the local limestone (my type AF), and was probably brought up from the river below the site. It is worth mentioning that there is little or no flint in the bed of the Nesque, and the type AM cobble can not have come from there.

Finally, the usually most-important source areas were also used, with the largest number of pieces from Murs, but that area does not dominate the assemblage, as it does in some other layers. It is closely followed by Sault, then the "Faraud etc." source area, and the local source is fourth in terms of abundance. Méthamis is farther down the list than usual, with its 32 pieces (all but 4 in the tamis) representing only 1.6% of the total matched assemblage. Only three source areas (Murs, Sault, and Local) have pieces representing all parts of the *chaîne opératoire*; Faraud comes close but has no cores.

Figure 11.7 shows the distribution of raw material source areas used for the layer 2 assemblage. Table 11.6 recalculates the percentages of pieces from sources in different geographic areas, based on the entire raw material sample (not just the pieces matched to typology),

Table 11.5. Percentages of pieces by typology and source area in layer 2. ***Pourcentages de pièces par catégories typologiques et groupes de sources des matières premières dans la couche 2***

2 %	Liste Bordes	Debris	Eclats	Galets	Lames	TamisColl	Nucléus	Total
Unidentifiable	23.7	25.0	36.0	0	12.0	49.2	11.1	47.0
Unknown	1.7	0	5.0	0	16.0	0.9	22.2	1.4
Faraud etc.	3.4	25.0	10.0	0	4.0	10.4	0	10.0
Méthamis etc.	0	0	4.0	0	0	1.6	0	1.6
Les Sautarels etc.	0	0	1.0	0	0	0.5	0	0.5
Sault	22.0	0	9.0	0	24.0	14.0	22.2	14.1
Murs	32.2	50.0	22.0	0	24.0	13.7	33.3	15.0
Local	13.6	0	7.0	50.0	16.0	7.5	11.1	7.8
St. Jean de Sault	0	0	0	50.0	0	0	0	0.1
Roussillon	0	0	1.0	0	0	0	0	0.1
Murs-Bezaure Olig.	3.4	0	3.0	0	0	1.9	0	1.9
Ravin de la Treille	0	0	1.0	0	0	0.2	0	0.3
St. Trinit	0	0	1.0	0	0	0.1	0	0.1
Durance	0	0	0	0	0	0	0	0
Tertiary Calavon	0	0	0	0	0	0	0	0
Tertiary Murs VdeV	0	0	0	0	0	0	0	0
Mormoiron	0	0	0	0	4.0	0.1	0	0.2
R de Guérin	0	0	0	0	0	0.1	0	0.1
N.Aurel	0	0	0	0	0	0	0	0
Total	**100**	**100**	**100**	**100**	**100**	**100**	**100**	**100**

Figure 11.6. A raw material sample of the distinctive flint type AO, from Mormoiron. ***Un échantillon du type de matière première AO, un silex particulier provenant de Mormoiron.***

and excluding those that are unidentifiable or of unknown source. These values have been added to the table which has been growing throughout this book, so that layer 2 can be compared to the other layers.

Based on these numbers, layer 2 is unlike any other layer, but perhaps closest to the couche 4. Layer 2's assemblage, first of all, has a truly striking proportion of local material, compared to all those layers that have little or none of it. It has one of the higher values for material from the NE, but nothing unusual; it is also midrange for the SW and the Near SW. It has the second-lowest percentage of material from the WSW, after layer G (which was one of the smaller assemblages); layer 3 is similar in this respect, but different in terms of the Local and NE material. The layer 2 assemblage was obviously created by people using the same lithic resource landscape as the people that created every other layer assemblage, but, once again, using it in their own way.

In Wilson & Browne (2014), the GLM for layer 2 provided very good results, with eight of the ten factors showing individual statistical significance (Table 11.7). Quality of the raw material was most heavily weighted,

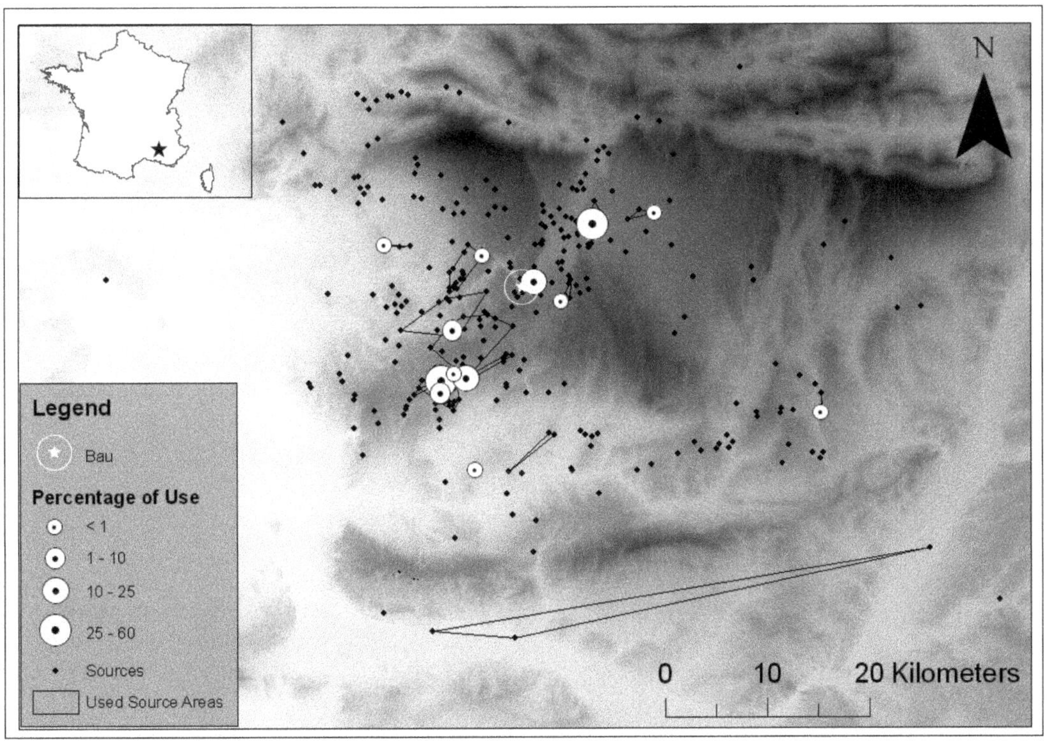

Figure 11.7. Map of the raw material source areas used for layer 2. ***Carte des aires de sources de matières premières utilisées pour la couche 2.***

Table 11.6. Percentages of identified pieces (excluding unidentifiable and unknown source pieces) from each geographic region for layers K2, K1-J, I, H, G, E, C, 5, 4, 3 and 2. ***Pourcentages des pièces identifiées (excluant les pièces indéterminables ou de sources inconnues) de chaque région géographique pour les couches K2, K1-J, I, H, G, E, C, 5, 4, 3 et 2***

Geographic area	K2 %	K1-J %	I-I4 %	H-H2 %	G %	E %	C %	5 %	4 %	3 %	2 %
Local	0.2	3.4	1.3	0.7	0	0	0.5	1.4	6.94	0.5	15.0
NE	19.4	20.4	12.3	15.7	41.5	31.0	39.5	5.7	22.95	37.0	27.7
E	0	0.3	0.1	0	0	0	0	0	0.09	0	0.1
Far SE	0	0	0	<0.1	0	0	0	0	0.05	0	0.1
Far S	0	0	0.1	0	0	0	0	0	0	0	0
S	0	0	0	0	0	0	0.5	0	0.01	0	0
SSW	0	0	0.1	0	0	0	0	0	0.04	0	0.1
SW	44.3	54.1	35.0	30.0	48.4	25.4	17.6	59.0	32.18	34.7	33.5
Near SW	23.0	15.0	29.1	34.8	9.6	29.9	30.7	10.5	24.68	22.7	19.3
WSW	12.4	6.6	21.9	18.6	0.5	12.7	11.2	22.9	12.61	4.6	3.4
W	0	0	0	0	0	0	0	0	0.01	0	0.3
NW	0.7	0.1	0.1	0.2	0	1.0	0	0.5	0.42	0.5	0.5
Total	**100**	**100**	**100**	**100**	**100**	**100**	**100**	**100**	**100**	**100**	**100**

followed by three of the rock sizes, Small and Medium, which were negative (i.e. sources with more small and medium rocks were less likely to be used), and Large rocks, which was positive. When the Wald Chi-Square results were converted to percentages (Table 11.8), the five raw material factors added up to a total value of 69.5%, which is the highest value of any of the layers we tested (Wilson & Browne, 2014). The Terrain variables added up to 29.8%, the second lowest value for any layer, after layer 3 (the relatively small Intercept value in layer 2 accounts for the difference). In fact, the difference between the Raw Material total and the Terrain total was greater (39.7%) than in any other layer. This assemblage seems to have been made of rocks primarily selected on

Table 11.7. Generalised linear model for raw material source choices in 2. P-values in bold font are statistically significant. Sample: 1057 pieces from 13 source areas

Overall:
Omnibus test: likelihood ratio chi-square = 566.524, df = 10, p = <0.001

Variable	Beta	S.E.	Wald Chi-Square	P-value	AIC
Quality	2.001	0.3967	25.445	**<0.001**	184.342
Extent	1.127	0.4919	5.251	**0.022**	
Difficulty (Cal/km)	-0.006	0.0190	0.085	0.770	
Small rocks	-10.140	2.6402	14.750	**<0.001**	
Medium rocks	-5.193	1.4135	13.495	**<0.001**	
Large rocks	1.843	0.5152	12.793	**<0.001**	
Very large rocks	1.207	0.8187	2.173	0.140	
Calories	-0.002	0.0006	10.586	**0.001**	
AOSA	-2.1 e-7	9.2 e-8	5.047	**0.025**	
AOSISA	5.5 e-5	1.9 e-5	8.485	**0.004**	
(Intercept)	3.174	4.0589	0.612	0.434	

Table 11.8. Wald Chi-Square values as percentages, assemblage from layer 2

Variable	Wald Chi-Square	Wald Chi-Square as %
Quality	25.445	25.8
Extent	5.251	5.3
Difficulty (Cal/km)	0.085	0.1
Small rocks	14.750	14.9
Medium rocks	13.495	13.7
Large rocks	12.793	12.9
Very large rocks	2.173	2.2
Calories	10.586	10.7
AOSA	5.047	5.1
AOSISA	8.485	8.6
(Intercept)	0.612	0.6
Total	**98.722**	**100**

the basis of their raw material characteristics, rather than on their locations or ease of access (although that always matters, but less in this case). The most heavily weighted Terrain factor is Calories (10.7%, Table 11.8), which may be due to the abundance of Local raw material in this unusual assemblage: obtaining flint from the local source requires very few calories, compared to the other sources. This shows up in the model as a negative Beta value for Calories, because the "low-calorie" local source tended to be used preferentially, rather than the "higher-calorie" sources farther away – unless they had good quality, large rocks available.

11.7. Conclusion: Layer 2

In layer 2, as usual, we have a palimpsest of materials left by an unknown number of people, during an unknown number of visits to the site, spread over an unknown length of time. The climate was cold, at least most of the time, and continued to be cold after the bulk of the deposits had been laid down, allowing signs of cryoturbation to develop. It is possible that the top of layer 2 includes Holocene sediments, but there was very little sign of archaeological content in those topmost sediments, so it is safe to say that most of layer 2 accumulated during the Late Pleistocene.

During the time of accumulation of this layer, the hominins that came to the Bau brought with them diverse animals, of a variety of sizes (but mainly medium to large), and from a variety of preferred habitats. They also brought raw materials obtained at distant sources in various directions, and left an assemblage which is not dominated by any one source area, and which also includes the highest proportion in this site of local raw material. The variety of raw materials suggests that highly mobile groups were passing through, bringing their personal toolkits, but the amount of local material might seem to contradict that picture. If the local material consisted only of cores and knapping by-products, we might think that it had been brought in to be worked in order to make new tools, during refurbishment of the toolkits. The newly made tools would then have been carried away. However, the local material is also present in the retouched tools, so the occupations seem to have lasted long enough that local raw materials were collected, and tools were made, used, and discarded. . . which does not sound like groups passing through. Were the tools really used? Although no use-wear studies have been done, perhaps we can use breakage as a proxy for use, since if a tool was being used and it broke, it would make sense for it to be discarded. When we described the typology of the pieces in 2013-14, we generally made a note if a tool, flake or blade was broken. In layer 2, a total of 17 tools, flakes and blades in identified raw materials were described as broken: 5 of them are in the local type, S. That is actually the highest total of broken pieces for any raw material source (the others are 4 from Sault, 4 from Murs, 3 from Faraud etc., and 1 unknown). Put another way, of the 19 known tools, flakes and blades made in local raw material, 5 are broken. Either this raw material is particularly fragile, or these tools were seeing hard use.

The use of local material is only surprising because so little of it was used in the lower layers. Once again I have to ask why that is. Layer 2 is less patinated than other layers, so is the local rock just showing up better? Probably not: the next highest proportion of local material is in the couche 4, and that material is often patinated. Was the local material not geologically exposed at the source earlier on, so that previous hominins did not have as much access to it? Perhaps, except that although layers 2 and 4 have respectable proportions of it, layer 3 has almost none. Or was it just a tactical choice that these hominins made, but that earlier ones did not? We may never know the true reason. In fact, if all we knew of the Bau were layer 2, it would be the diversity of their lithics which would be interesting, since so many different source areas are represented; the use of local material would be entirely unremarkable. In this layer highly mobile groups of people, using a large territory and carrying their toolkits, occasionally passed through the site. While they were at the Bau, they hunted local animals, such as the ibex and chamois that may have come from the gorge itself, and they used local lithic resources to make tools to process the carcasses, and perhaps for other purposes. Then they moved on.

11.8. Layer 1

There really is not much to say about layer 1, but it does deserve to be included. It was a modern layer, dusty grey and fine-grained, with remnants of rootlets here and there, and occasional specks of charcoal. In the early days of our excavations it was sometimes as much as a few cm thick, but given our habit of cleaning the site every year before starting to excavate, there is nothing left of it now. Or rather, there may be a new layer 1 forming, now that the site has been left unexcavated and uncleaned for many years.

We had heard reports that earlier excavators found a few potsherds in the uppermost layer, and in the early days of our dig we did find a few ourselves (along with other, even more modern, garbage). We then went several years without finding anything apparently post-Palaeolithic, until 1999, when we extended our excavations to the east of the tranchée L, into zones O2 and P2, and uncovered the shelf-like naturally hollowed out area in the rear wall (see Figure 11.3). At the top of those sediments, tucked away into smaller hollows in the rock shelter wall, we found some bones of small animals (possibly dragged in by some animal that had a den in there), and a few potsherds. One of those is unforgettable to me. It was a large piece of an earthenware bowl, with a bit of the base on one side, but otherwise just the right size and shape to be a dirt-stained, rather thick, human skull fragment. I got a bit of a shock when I saw it, until I turned it over and saw the base and realised what it really was. I was having quick visions of there being a burial back there, but no, phew! It's a bowl. The diggers and I had a laugh over that. Then Serge came along and I said "Look what they found, Serge", holding it out the way I had first seen it. His reaction was the same as mine (I believe the phrase "*tu me donnes une crise cardiaque!*" may have been involved), but I quickly turned it over and reassured him. I still laugh about it, though: it looked so much like a skull fragment.

These few finds from layer 1 are evidence that humans (and smaller animals) continued to use the site even after the Palaeolithic, but once the Neolithic started there would have been less use for a rock shelter, at least until archaeologists started hanging around. The few remains left behind over that rather short time period do not amount to much more than evidence of occasional passage. Still, the Bau has been known – or rediscovered many times – and used for a total of well over 200,000 years. The next chapter will try to sum up all of that use.

11.9. Sommaire du chapitre

Ce chapitre concerne le reste des dépôts: les sédiments cryoclastiques meubles de la couche 2 qui se sont accumulés au-dessus du dôme de brèche dans la tranchée L, ainsi que les sédiments modernes de la couche 1 dont nous avons parfois trouvé les traces en surface. La couche 2 constitue une couche archéologique importante, plus épaisse au fond du site, près de la paroi, probablement parce que les sédiments ont pu s'accumuler là où la surface de la brèche est en pente vers cette même paroi. On voit en effet que les cailloux et les os sont souvent roulés, et qu'ils ont été transportés par de l'eau avant leur mise en place. Des phénomènes de cryoturbation sont également visibles avec des petits cailloux qui ont été déplacés par les effets du gel et du dégel. Ces sédiments cryoclastiques ont été formés pendant une période froide du Pléistocène supérieur, et la couche doit être beaucoup plus jeune que les couches 3 et 4. En effet, la formation et l'érosion de la brèche et des autres dépôts représentent très certainement un laps de temps important. Néanmoins nous n'avons pas assez de matériel diagnostique pour déterminer si la couche 2 comprend des éléments du Paléolithique supérieur.

Malgré le passage de l'eau qui a roulé les objets dans les sédiments, on ne voit pas de cimentation dans ces dépôts, et les pièces lithiques sont moins patinées que d'habitude. Ces dernières peuvent montrer des croûtes calcaires incrustées sur le bord des pièces, mais la plupart n'ont qu'un début de patine. Peut-être parce qu'elles sont plus jeunes, ou alors parce qu'elles ont été soumises à des conditions géochimiques différentes en comparaison avec les pièces provenant des autres couches. Les restes sont également moins brûlés que dans certaines couches de la séquence.

Les pollens ne sont pas abondants et sont dominés par le pin; les quelques autres espèces traduisent un climat assez doux, contredisant ainsi la nature des sédiments. Il est possible que ces pollens aient été amenés postérieurement dans les sédiments par les eaux d'infiltration. La faune n'est pas bien préservée mais néanmoins diversifiée et en relation avec des habitats différents correspondant à des environnements boisés, ouverts ou escarpés.

Une dent humaine est attribuée à la couche 2, mais les circonstances de sa découverte et sa nature nous permettent de dire qu'elle est hors contexte. En effet, elle semble assez moderne (Pléistocène supérieur terminal ou Holocène), mais ceci ne veut pas dire que la couche 2 est aussi jeune.

La couche 2 a livré le troisième assemblage lithique le plus abondant, après ceux des couches 4 et H1, en dépit de l'absence de réel sol d'occupation, avec seulement quelques concentrations plus ou moins lenticulaires et des vestiges dispersés dans les dépôts. Comparé à l'assemblage de la couche 4, celui de la couche 2 offre plus d'éléments du Paléolithique supérieur (plus de lames, moins de technologie levallois), mais une attribution au "Moustérien progressif" serait quand-même plus convaincante. Par ailleurs, n'oublions pas que les couches inférieures de ce même site, beaucoup plus anciennes, renfermaient également beaucoup de lames. Par exemple, la couche J4-K1 en avait plus que la couche 2. De fait, on ne peut plus vraiment considérer la présence de lames comme un fossile directeur du Paléolithique supérieur.

Les matières premières de la couche 2 sont très diverses, presqu'autant que dans l'assemblage de la couche 4, qui est dix fois plus grande. Il y a quelques pièces dont la source n'a pas encore été identifiée. Les autres proviennent de gisements localisés dans toutes les directions, parfois éloignées, bien qu'en faibles quantités dans ce cas. Néanmoins, dans le même temps, parmi toutes les couches du Bau, la couche 2 contient la proportion la plus élevée de silex provenant du gisement près de l'abri, de même qu'une bonne quantité de matériel des sources habituelles (Murs, Sault, etc.). On voit que les hominidés de la couche 2 fréquentaient le même territoire lithique que ceux présents dans les autres couches, mais ils en exploitaient les ressources et se déplaçaient selon leurs propres stratégies.

Dans la couche 2 nous avons comme d'habitude un palimpseste de vestiges laissés par un nombre incalculable de préhistoriques lors d'un nombre inconnu de leur passages sur le site, représentant une durée de temps inconnue. Le climat était froid la plupart du temps, et ce froid a continué après l'abandon du site, comme en témoigne la cryoturbation. Les préhistoriques ont amenés des animaux de toutes sortes, souvent grands ou de taille moyenne, et des matières premières lithiques diversifiées issues de différentes origines et régions d'approvisionnement. Par ailleurs, ils ont aussi utilisé le silex local. La présence de matériaux exotiques peut suggérer qu'il s'agissait de gens très mobiles, de passage, mais la quantité de matériel local sert à nuancer cette idée. L'assemblage ne contient pas seulement les nucléus et le débitage qui suggéreraient une remise à neuf de la trousse à outils des préhistoriques; il contient également bon nombre d'outils en silex local. Ces derniers sont en effet souvent cassés, comme s'ils avaient été utilisés de façon intense. Ainsi, ces groupes étaient sans doute très mobiles et souvent de passage, mais lors de leurs visites au Bau ils exploitaient bien les ressources locales: le silex, ainsi que le bouquetin et le chamois qu'ils chassaient dans les gorges.

La couche 1 mérite d'être mentionnée, mais il n'y a pas beaucoup à dire. Elle est composée de sédiments modernes, grisâtres et poussiéreux. Au début de nos fouilles elle se composait de quelques centimètres d'épaisseur au sommet du remplissage, mais elle n'existe plus désormais. Quelques tessons de poteries avaient été trouvés par nos prédécesseurs et nous en avons trouvés quelques-uns aussi (un d'entre eux ressemblait d'ailleurs beaucoup à un morceau de crâne humain avant que l'on identifie le pied du bol). Ces morceaux témoignent d'une utilisation périodique du site même après la Paléolithique, mais à partir du Néolithique cet abri-sous-roche difficile d'accès ne semble pas avoir été très fréquenté, du moins pas avant que les archéologues commencent à venir y traîner.

12

"Sounds of Laughter, Shades of Life"

12.1. Introduction

This book is about people: two hundred thousand years' worth (give or take a few millennia) of hominins whose lives intersected with the rock shelter we know as the Bau de l'Aubesier. It is also about the rock shelter, which protected the people from the weather, served as a locus of their activities, preserved the traces of their actions, and challenged and informed the archaeologists who eventually came to dig up those traces. The Bau is a special place, and not just for its contents. It is a place of life. It glows in the evening sun reflected off of the nearby cliffs. People who come to visit for the first time almost always say "Oh! It's bigger than it looks in the pictures!" (That warns me that I will not be able to put enough pictures in this book to get across how big it is.) Then, after being shown the different parts of the site and having had aspects of the excavation explained to them, many such visitors pause, stand quietly looking around, give a bit of a sigh, nod, and thank me for sharing this place with them. Most people are reluctant to leave. The Bau has a spirit, if I can say that in an otherwise strictly scientific book. It is old, but not elderly; benevolent, but detached. Go there and open yourself to it, and you can feel it, commune with it, maybe even communicate – but it will not just tell you its stories. You have to work for that.

The work that archaeologists did starting in 1901 provides the evidence for the interpretations presented in this book and allows us to tell the Bau's story. In the first eleven chapters we saw the site's context and the layers within it. This final chapter attempts to tie all of the results together, without being just a summary of what went before. Some repetition is necessary, but only in order to address two final questions: first, how do the layers in the tranchée L match up to the layers in the Tranchée Moulin? And after that, the ultimate question: what does the Bau tell us about the hominins of the Middle Palaeolithic?

12.2. Tranchée L vs. Tranchée Moulin

The tranchée L and the Tranchée Moulin are both located in the upper part of the deposits, and both exposed a thick mass of breccia on top of a black archaeological layer on top of a fine-grained, very hard dalle (which can be traced from one trench to the other), with other archaeological layers underneath that. The layers in each trench were given different labels in order to eliminate any chance of presupposing relationships which would later turn out to be inaccurate. Can we now finally determine what those relationships really are? Probably yes, and maybe not.

Figure 12.1 gives a simplified sketch of the stratigraphy in each trench, placed in their approximately correct relative elevations by z value. Dashed lines show the potential linkages between layers. Layer 2 consists of loose sediments on top of breccia, and may correlate to a now missing thickness of sediments on top of the breccia in the Tranchée Moulin, or perhaps to part of the breccia. Layer 3 was found in the breccia (and associated sediments), so might be expected to match up with layer C. Layer 4 is on top of the dalle so it "should" match layer D, while layer 5 was under the dalle so it should match layer E. . . or maybe F. How well do these suppositions match up to what we know of the layers?

We can compare layers by their sedimentological characteristics, by their faunal content (Table 12.1), the typological characteristics of their lithics (Table 12.2), the source areas exploited for their lithic raw materials (Table 12.3), and the use of geographic areas shown when source areas are grouped into regions (Table 12.4 repeats Table 3.5, and shows how the source areas were combined by geographic region, while Table 12.5 repeats the results, as previously given in Table 11.6). Finally, Table 12.6 shows the Wald Chi-Square results, converted to percentages, obtained in the GLMs for 11 layers (from Wilson & Browne, 2014).

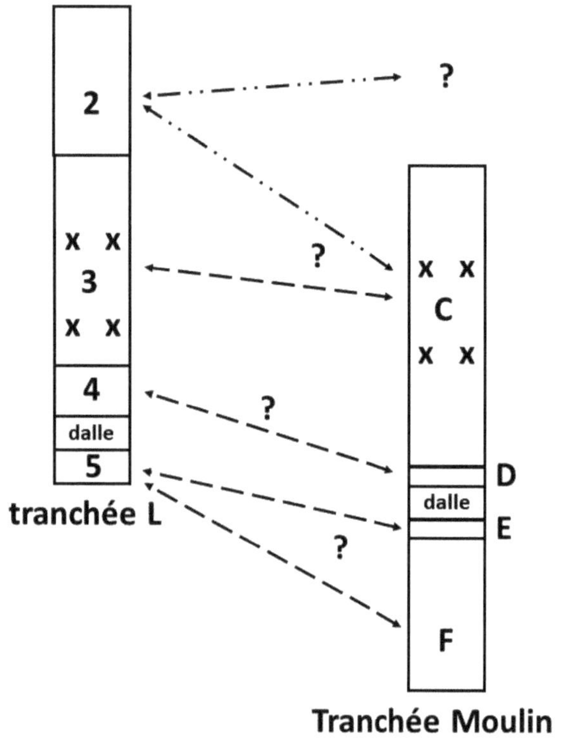

Figure 12.1. Possible correlations between layers in the two upper trenches. See text for discussion. Xs mark layers of breccia. ***Corrélations possibles entre les couches dans les deux tranchées. Voir texte pour explications. Les Xs indiquent couches de brèche.***

Starting at the bottom, layer 5's stratigraphic position is similar to that of layer E, but it is coarse-grained and essentially unburned, so sedimentologically it resembles layer F more than the very fine-grained, very burned layer E. In terms of fauna, there are some similarities between layer 5 and both layers E and F, so that is inconclusive. For source areas and geographic regions, there are no results for layer F, but layer E does not seem to match layer 5 (especially when the areas are grouped into regions). On the other hand, the basic typological breakdown of layer 5 is closer to that of layer E than that of layer F. Finally, the GLMs could not be run on layer F, so there are no Wald Chi-Square results for that layer, but layer 5 and layer E do not resemble each other. With these conflicting signals, we will have to continue to consider layers F, E and 5 as three separate layers.

A simple consideration of the sedimentary appearance of layers D and 4 (both fine-grained and black) lends credence to the idea that they might match up. This is complicated by the fact that as we dug layer D, it pinched out towards the east, so it might not link up to the couche 4 at all. We have no faunal data for layer D, and the lithic data is sparse, because the assemblage is highly patinated (so we can not tell where the rocks came from, or run a GLM) and some of the pieces seem to have been mislaid, so we have no typological results for them – and that includes any retouched items that might have been found. There is therefore no firm reason to come to any conclusion. The couche 4 might be the equivalent of layer D, or not.

Layer 3 is a hodgepodge of sub-layers found with the breccia and associated loose sediments above the black layer, and so is layer C. Their faunal results do not match at all: only two species were identified in layer 3, and neither appears in layer C. On the other hand, although none of the lithic results (typology, source areas, GLMs) are exact matches to each other, they are fairly close. I am comfortable drawing tentative equivalences between the two trenches at this level.

Finally, layer 2 was found in loose sediments above the breccia in the tranchée L. In the Tranchée Moulin there was very little loose sediment available: only a bit of infilling between the back wall and the breccia in bands F and G. As chapter 7 describes, it is unclear how any archaeological material in those sediments was labelled: it may have been C4, or some of C1, or both. It does seem likely that some material existed, however, because the reports mention some material in layer C being rounded by water transport, and/or showing signs of mechanical breakage (Lebel, 1988), and we found the same thing in layer 2. To that extent, it is possible that layer 2 (or part of it) matches part of layer C. For interest's sake, C4 included 74 lithic pieces, all but one in the tamis (and very small); the one remaining piece is a core, but it is too patinated to identify its raw material type. Of the other pieces, only one was identified to raw material type, and it also was patinated. That is all rather uninformative, but by virtue of being nondescript, could fit in to layer 2. Overall, however, layer 2 does not match well to the complete layer C. Less than half of the faunal species match, and the lithic data match either poorly or not at all, depending on the category considered. The best conclusion seems to be that the equivalent of layer 2 is mainly missing from the Tranchée Moulin.

Should we expect the layers in both trenches to line up neatly? No, of course not. Different parts of the site might have been used at different times, leaving deposits in those different places: some people camping near the side wall, for example, and then other people a couple of centuries later camping near the back wall. By now, those two sets of deposits could easily seem to be in equivalent stratigraphic positions, but they would really have no relationship to each other. Alternatively, even if, say, layer 5 and layer E *were* formed at the same time by the same people, why would we expect them to have the same contents? They could represent different activity areas: butchering and processing, versus cooking and sleeping, for instance. It is therefore only natural that drawing correlations between layers in separate trenches is difficult. It is worth the effort, however, because if we *can* do it, we can learn more about the broader range of activities represented by both "layers" combined into one.

12.3. What the Bau teaches us about the Middle Palaeolithic

The Bau is a very rich site, and our years of excavation there have produced an abundant and very detailed

Table 12.1. Faunal results, from Faure et al. (1996), Fernandez (2001, 2006, **pers. comm.) for the ungulates, A. Argant (2006) for carnivores, and Crégut-Bonnoure et al. (2011) for *Ursus arctos*: Minimum Number of Individuals (MNI) for each species or % of assemblage, as available. N/a = species present but not quantified. ***Liste faunique, d'après Faure et al. (1996), Fernandez (2001, 2006, **comm. pers.) pour les ongulés, A. Argant (2006) pour les carnivores, et Crégut-Bonnoure et al. (2011) pour*** **Ursus arctos*****: Nombre Minimal d'Individus (MNI) de chaque espèce ou % de l'assemblage, selon la disponibilité des données. N/a = espèce présente mais non-quantifiée***

Taxon	common name	% K1-J	% I-I4	% H-H2	MNI G	MNI F	MNI E	MNI C	MNI 5	MNI 4	MNI 3	MNI 2
*Capra caucasica praepyrenaica***	ibex	-	-	-	1	1	-	1	1	6	-	3
Rupicapra rupicapra	chamois	0.6	0.6	1.3	-	-	-	1	-	1	-	2
Hemitragus cedrensis	tahr	2.3	5,4	34.6	-	-	-	-	-	-	-	-
Bos primigenius	aurochs	53.2	43.2	15.7	-	-	1	-	1	7	1	1
Equus sp.	horse	-	-	-	-	1	1	-	2	7	-	1
Equus cf. *taubachensis*	horse	-	-	18.0	-	-	-	-	-	-	-	-
Equus mosbachensis	horse	35.3	31.0	-	-	-	-	-	-	-	-	-
Capreolus capreolus	roe deer	1.1	0.6	15.7	1	-	-	-	-	2	1	3
Cervus elaphus	red deer	2.3	10.3	12.0	-	1	1	-	1	10	-	1
Dama dama	fallow deer	1.1	1.9	1.5	-	-	-	1	1	3	-	-
Megaloceros giganteus	Irish elk	2.3	4.0	2.2	-	-	-	-	1	1	-	-
Rangifer tarandus	reindeer	-	-	0.1	-	-	-	-	-	-	-	-
Dicerorhinus hemitoechus	rhinoceros	1.7	0.8	-	-	-	-	-	-	-	-	-
Vulpes vulpes	red fox	-	-	n/a	-	-	-	-	-	-	-	-
Sus scrofa	boar	-	2.3	n/a	-	-	-	-	-	1	-	-
Lynx spelaea	cave lynx	-	-	-	-	-	-	-	1	-	-	1
Panthera spelaea	cave lion	-	-	-	-	-	-	-	1	-	-	-
Ursus arctos	brown bear	n/a	n/a	n/a	-	-	-	-	1	4	-	2

Table 12.2. Lithic assemblages, summarised by typological category. ***Assemblages lithiques sommaires, par catégorie typologique***

Category	K2 %	K1-J %	I-I4 %	H-H2 %	G %	F %	E %	D %	C %	5 %	4 %	3 %	2 %
Liste Bordes	1.6	4.5	4.9	3.2	4.9	8.5	8.2	0	5.1	13.4	2.6	1.0	4.0
Debris	0.6	0.4	0.5	0.6	0	1.7	0	0	0	0	0.4	0.4	0.3
Eclats	22.9	21.4	17.6	14.4	7.3	20.3	7.6	19.5	1.8	5.7	13.2	2.9	4.8
Lames	3.2	3.5	1.9	1.0	2.4	3.4	4.7	1.3	1.1	4.5	1.2	0.5	1.7
Tamis/Collectif	70.2	67.3	72.8	78.0	80.5	62.7	75.4	79.2	91.2	74.3	81.6	94.6	88.3
Nucléus	1.5	2.9	2.3	2.7	4.9	3.4	4.1	0	0.7	2.1	1.0	0.5	0.8
Galets	0	0	<0.1	0	0	0	0	0	0	0	0	0	0.1
Total	**100**	**100**	**100**	**100**	**100**	**100**	**100**	**100**	**100**	**100**	**100**	**100**	**100**

record of fauna and lithics; accompanying analyses have placed these in their environmental context, and in their raw material source landscape. The dozen or more layers within the site span a couple of hundred thousand years of prehistory, from the ancestors of the Neandertals to modern humans (counting the few Neolithic deposits of layer 1, as well as the traces of 20th century archaeologists). Most of the deposits are Middle Palaeolithic, however, and that is the period that the Bau can tell us the most about.

The Middle Palaeolithic is defined on the basis of stone tool industries: a flake-based industry with frequent use of prepared cores, generally with a large proportion of sidescrapers (see for example Bordes, 1961), but it is also to some extent a chronological indicator (later Middle to Late Pleistocene) and is most commonly associated with the hominin type we call Neandertals, as well as their immediate ancestors. The Bau fits that description, but also stretches it and enriches our understanding of the period.

The lithic assemblages in most layers include flakes, blades, cores and retouched tools in varying proportions. The Levallois technique is present from the lowest levels on up, and sidescrapers are common. The blades (see chapters 2, 3 and 11) are more common in the early layers than they are "supposed" to be, and become very common again in the uppermost layer. Figure 12.2 shows some of these blades, laid out on a table for study. Blades have sometimes been considered to be an indicator of Upper Palaeolithic industries, but this can no longer be accepted: deliberate production of blades greatly pre-dates the Upper Palaeolithic, as Carmignani (2017, see also Carmignani et al., 2017, and references therein) and others have demonstrated.

The Bau is also remarkable for the intense use of fire displayed in some layers. Based on traces of fire having affected lithic pieces, every layer must have had fire somewhere in its vicinity, even if the bones and the sediments themselves are not burned. The least-burned layers are J4/K1, with 49 out of 1735 flint pieces (2.8%) categorised as raw materials which by definition are burned, and G, with 8 of 225 pieces (3.6%) in the burned categories. In some other layers, however, or at least in some zones within them, the majority of lithic pieces are burned, as are the bones, and the sediments are a black, greasy, mass of tiny charcoal fragments (see chapters 5 and 9, for instance). There can be no doubt that fire was repeatedly used in layers H1, E, D, and 4.

There has been a great deal of work recently on prehistoric use of fire (see for instance Sandgathe, 2017, and other papers in that supplemental journal issue). One question that has been raised concerns determining whether fire was deliberately made to be used, or just fortuitously happened upon, and curated to be used elsewhere (Sandgathe, 2017). This is difficult to resolve. There are a variety of ways to make fire, and even if some of them are difficult and time-consuming, they can be made to work. The benefits of fire mean that people will use such methods, as has been demonstrated ethnographically. The trouble is that some techniques result in the fire-making apparatus being burned, and if not, the apparatus can be kept and taken away for future use elsewhere. The chance of finding the piece of pyrite used as a strike-a-light, or the stick twirled between the hands to create friction, is probably no greater than finding the remnants of a match in the ashes in my wood stove.

Sandgathe (2017) suggested that if hominins depended on finding naturally-made fire, rather than making it themselves, there would be a correlation between fire use and climate periods when natural fires would be more common, i.e., when lightning strikes would occur to start them (Sandgathe, 2017). This does not seem to be the case at the Bau, however. The combustion level of layer H1 seems likely to date to a warm time period (MIS 5e), bracketed by two colder periods, and the fire level seems to have been used (based on the time of death of the animals) during the winter, whereas the deposits just above and below it seem to have accumulated more during the summer. H1 therefore reflects use of fire during the winter of a warm period. The couche 4, on the other hand, seems to have formed mainly from occupations during the spring

Table 12.3. Percentages of pieces (with both typological and raw material identification) by source area; layer G % recalculated from chapter 6 excluding two pieces whose source was not identified. ***Pourcentages de pièces (identifiées typologiquement et par matière première) par groupes de sources de matières premières; % pour la couche G recalculé par rapport au chapitre 6 pour exclure deux pièces dont l'origine n'a pas été identifiée***

% by layer	K2 %	G23/G24 %	J-K1 %	I-I4 %	H-H2 %	G %	E %	D %	C %	5 %	4 %	3 %	2 %
Unidentifiable	53.0	24.3	41.0	58.4	58.2	14.4	90.1	92.5	75.8	61.5	63.10	66.7	47.0
Unknown	3.8	0	0.5	0.2	0.4	0.9	0.1	0	0.3	0	0.19	0.3	1.4
Faraud etc.	9.9	10.0	8.8	12.0	14.4	8.1	3.0	2.0	7.3	4.0	9.00	7.3	10.0
Méthamis etc.	5.3	7.5	3.8	9.1	7.7	0.5	1.3	0.7	2.7	8.8	4.61	1.4	1.6
Les Sautarels etc.	0.3	0.8	0.2	0.1	0.2	0	0	0	0.3	0.4	0.53	0.8	0.5
Sault	8.1	16.7	10.9	4.6	6.4	35.1	3.0	4.8	9.4	2.2	8.27	11.9	14.1
Murs	17.7	31.4	27.9	13.1	11.1	39.2	2.1	0	3.3	22.0	11.03	8.2	15.0
Local	0.1	3.8	2.0	0.5	0.3	0	0	0	0.1	0.5	2.62	0	7.8
St. Jean de Sault	0	0	0.2	<0.1	0	0	0	0	0	0	0.04	0	0.1
Roussillon	0	0	0	<0.1	0	0	0	0	0	0	0.01	0	0.1
Murs-Bezaure Olig.	1.2	5.4	3.4	1.2	1.1	1.8	0.4	0	0.6	0.4	0.28	2.0	1.9
Ravin de la Treille	0.3	0	0	<0.1	0.1	0	0.1	0	0	0.2	0.14	0.2	0.3
St. Trinit	0.2	0	1.0	0.5	0.1	0	0.1	0	0	0	0.12	0.2	0.1
Durance	0	0	0	<0.1	0	0	0	0	0	0	0.00	0	0
Tertiary Calavon	0	0	0	0	0	0	0	0	0.1	0	0.01	0	0
Tertiary Murs VdeV	0	0	0.1	<0.1	<0.1	0	0	0	0	0	0.02	0.2	0
Mormoiron	0	0	0	0	0	0	0	0	0	0	0.01	0	0.2
R de Guérin	0	0	0	0	<0.1	0	0	0	0	0	0.02	0	0.1
N.Aurel	0	0	0	0	0	0	0	0	0	0	0.01	0	0
Total	100	100	100	100	100	100	100	100	100	100	100	100	100

Table 12.4. Raw material source areas grouped by geographic region. ***Aires de sources de matière première regroupées par région géographique***

Geographic area	Approximate distance from the Bau	Source areas included
Local	1 km	Local
North-East	5 – 15 km	Sault, St. Trinit, N. Aurel
East	5 km	St. Jean de Sault
Far South-East	30 – 35 km	Rocher de Guérin
South	15 – 20 km	Tertiary Calavon
Far South	35 km	Durance
South-South-West	18 km	Roussillon
South-West	10 – 15 km	Murs, Les Sautarels etc., Murs-Bezaure Olig., Tertiary Murs V de V
Near South-West	3 – 10 km	Faraud etc.
West-South-West	7 – 15 km	Méthamis etc.
West	12 km	Mormoiron
North-West	5 km	Ravin de la Treille

Table 12.5. Percentages of identified pieces (excluding unidentifiable and unknown source pieces) from each geographic region for layers K2, K1-J, I, H, G, E, C, 5, 4, 3 and 2. ***Pourcentages des pièces identifiées (excluant les pièces indéterminables ou de sources inconnues) de chaque région géographique pour les couches K2, K1-J, I, H, G, E, C, 5, 4, 3 et 2***

Geographic area	K2 %	K1-J %	I-I4 %	H-H2 %	G %	E %	C %	5 %	4 %	3 %	2 %
Local	0.2	3.4	1.3	0.7	0	0	0.5	1.4	6.94	0.5	15.0
NE	19.4	20.4	12.3	15.7	41.5	31.0	39.5	5.7	22.95	37.0	27.7
E	0	0.3	0.1	0	0	0	0	0	0.09	0	0.1
Far SE	0	0	0	<0.1	0	0	0	0	0.05	0	0.1
Far S	0	0	0.1	0	0	0	0	0	0	0	0
S	0	0	0	0	0	0	0.5	0	0.01	0	0
SSW	0	0	0.1	0	0	0	0	0	0.04	0	0.1
SW	44.3	54.1	35.0	30.0	48.4	25.4	17.6	59.0	32.18	34.7	33.5
Near SW	23.0	15.0	29.1	34.8	9.6	29.9	30.7	10.5	24.68	22.7	19.3
WSW	12.4	6.6	21.9	18.6	0.5	12.7	11.2	22.9	12.61	4.6	3.4
W	0	0	0	0	0	0	0	0	0.01	0	0.3
NW	0.7	0.1	0.1	0.2	0	1.0	0	0.5	0.42	0.5	0.5
Total	**100**	**100**	**100**	**100**	**100**	**100**	**100**	**100**	**100**	**100**	**100**

to fall period, i.e., summer, and probably during a cold climatic period. That makes the couche 4 match the levels above and below H1, rather than matching H1 itself. Even if we postulate that the winter of a warm period might not be very different from the summer of a cold period, that does not explain why there was so much less use of fire above and below H1. And in fact it does not seem probable that these two different climate and seasonal periods were equally likely to have been subject to lightning storms, so nature was probably not providing equal chances at finding natural fire. How, then, was fire used so much in those levels (H1 and 4) and not at other times?

With no direct evidence of the origin of the fires, we have to fall back on indirect evidence. At the Bau, at least in layers H1 and 4 (less is known about layers E and D), fire was repeatedly used, and maintained for long periods of time. A large amount of bone was burned as fuel. The fires were not confined into constructed and confined hearth areas, but they did tend to be placed repeatedly in

Table 12.6. Wald Chi-Square values as percentages. *Valeurs de Wald Chi-carré convertis en pourcentages*

Variable	K2	K1-J	I-I4	H-H2	G	E	C	5	4	3	2
Quality	16.1	16.8	20.8	14.5	8.5	7.4	15.6	12.3	22.7	22.8	25.8
Extent	12.9	15.3	19.5	10.8	10.2	14.5	5.9	7.2	5.2	9.5	5.3
Difficulty (Cal/km)	8.4	13.7	13.7	14.1	15.0	8.8	4.4	0.2	1.8	4.0	0.1
Small rocks	5.7	9.9	5.3	7.2	-	-	8.2	7.0	13.1	9.2	14.9
Medium rocks	6.4	4.6	4.1	5.9	6.3	1.4	12.1	6.5	13.1	12.5	13.7
Large rocks	16.2	11.0	14.8	11.6	10.2	24.2	27.0	25.6	5.7	23.8	12.9
Very large rocks	0.2	0.1	<0.1	0.5	3.2	0.5	0.8	3.9	1.3	0.5	2.2
Calories	4.4	16.6	6.8	1.6	9.6	2.1	0.3	<0.1	12.7	2.5	10.7
AOSA	5.5	3.0	0.1	6.5	0.2	5.6	6.8	7.6	6.0	4.1	5.1
AOSISA	22.6	8.2	14.5	27.3	27.9	25.1	17.0	18.2	13.4	9.3	8.6
(Intercept)	1.6	0.8	0.4	0.1	8.9	10.3	1.9	11.4	4.9	1.8	0.6
Total	**100**	**100**	**100**	**100**	**100**	**100**	**100**	**100**	**100**	**100**	**100**

Figure 12.2. Blades and other tools from the lower layers, spread out to be studied. *Lames et autres outils des couches inférieures, étendus pour l'étude.*

approximately the same places. Such intensive use of fire suggests that it was an important, even necessary, part of these hominins' lives. If they depended upon it, it seems likely that they knew how to make it at will. However, that leaves the question of why fire is not so intensely used in other layers. If the hominins had mastered fire-making and used it so much that they seem to have depended on it, why did they not always use it? Were the missing fires somewhere else (off stage left)? Have the traces not been preserved? Were the occupation periods too short to have any great amount of fire residue build up, compared with the subsequent sedimentation? It is true that layers H1 and 4 are also the two richest in terms of faunal remains and lithics, and therefore probably represent the two layers when hominins were spending the most time at the Bau. Or maybe fire was required for some activities (e.g. smoking and drying meat), but the hominins were not always doing that at the Bau. . . There is no clear answer, so more research is required!

The Bau also shows us some slight glimpses of social relationships between the hominins. The couche 4 hominins were a family group (or groups), which we know because they had children who lost deciduous teeth. Some of those children died, but we do not know the circumstances of that. An adult hominin in layer I also died, after a lengthy period of periodontal illness (see chapter 4). That hominin would have been unable to eat properly for weeks or months, and with the combination of illness and hunger would have been unable to function well for a considerable time – and yet was kept alive by his or her group.

The dental problems in the layer I hominin and the carious teeth in the couche 4 (chapter 9) are among only a few examples of such problems from any Palaeolithic site (Lebel & Trinkaus, 2002b). The tooth groove on the isolated tooth from layer I (Lebel & Trinkaus, 2002a), on the other hand, is an example of a hominin taking care of his or her teeth: it is too bad we can not know whether that hominin learned from the example of the one with the terrible tooth problems. In any case, these specimens tell us that people cared for each other, and that (at least in the couche 4), the Neandertal diet included more than just meat, since carbohydrates are necessary to feed the bacteria that cause dental caries (Trinkaus et al., 2000b).

Meat was definitely a major contributor to the diet of all of these hominins, however. Every layer contained bone fragments, and some contained large numbers of identified individual animals: see chapters 3, 4 and 5, for instance, about layers J through H, and Table 12.1. Most of these animals are medium to large in size, with "large" loosely defined as meaning at least a couple of hundred kg (aurochs, horse, Irish elk, red deer, rhinoceros, brown bear and cave lion), "medium" meaning body mass in the several tens of kg (ibex, chamois, tahr, reindeer, roe deer, fallow deer, and boar), and "small" meaning less than about 20 kg (fox and lynx). (Reindeer could be considered large, not medium, but this would make very little difference to the results.) Table 12.7 shows the proportions of these size classes in each layer, based on the values given in Table 12.1, with the MNI numbers converted to percentages, for easier comparison. Large animals are preponderant, especially in layers with more than a few individuals (see Table 12.1). Only layers H and 2, of the layers with fairly abundant fauna, show a majority of medium-sized animals. In layers with only a few animals, they can be large or medium or some of each.

Many of these animals were killed as prime-aged adults, but there are other scenarios, too: in some cases animals were taken as juveniles, for instance (see individual chapters for details, and Fernandez, 2001, as well as Fernandez & Legendre, 2003; Fernandez et al., 2006 for horses). In almost every case, the age distributions show that they were deliberately hunted, not opportunistically encountered or scavenged. Given the variety of sizes and behaviours of the animals, this translates into the fact that these hominins knew and used several different hunting techniques, which they adopted as necessary for each species – even within the same layer and therefore time period. They were versatile.

Choice of prey seems to have varied through time and also by season of the year, as we have seen in several layers. This seasonality of the fauna shows us seasonality in the use of the site, as was mentioned above. In many cases it seems that the site was used in the summer during cold periods, and in layer H1 it was used in the winter during a warmer climatic period. That seasonal use is not necessarily always true, especially for the layers where we have evidence only of very short, "passing through" types of visits. Still, if the tendency is for summer use of the Bau during cold periods, perhaps the hominins were at lower elevations in the winter and moved up to the Bau in the spring. That might have been so that they could avoid the colder winter temperatures at higher elevations, but it

Table 12.7. Faunal data grouped by animal size. ***Données fauniques par taille des animaux***

Fauna size	% K1-J	% I-I4	% H-H2	% MNI G	% MNI F	% MNI E	% MNI C	% MNI 5	% MNI 4	% MNI 3	% MNI 2
large	94.8	89.3	47.9	-	66.7	100	-	70.0	69.0	50.0	35.7
medium	5.1	10.8	53.2	100	33.3	-	100	20.0	31.0	50.0	57.1
small	-	-	n/a	-	-	-	-	10.0	-	-	7.1

might also have been that they wanted to avoid the heat of summer lower down – an early form of the transhumance, bringing flocks of sheep up to the summer pastures, that still exists in this area. Not that they had flocks to drive up there, but they may have been following herds of prey animals, or they may just not have liked to get too hot themselves. Neandertal bodies are adapted for life in a cold climate, after all. During the warm climatic period they used the Bau in the winter, possibly because winter conditions there may now have been more tolerable, but the real question is: where were they in the summer? Did the milder climate expand their potential winter range into higher elevations, or did it shift their entire range upwards? If we could ever find summertime sites dating to this time period higher up on the Mont Ventoux, rather than down in the valleys, that would suggest the latter notion.

At the Bau, we have the seasonal data from the fauna, and we also have territorial data from the lithics: where the raw materials come from. We can never know the exact path that a rock took to get from its source to the Bau, but we do know that someone had to pick it up at the source, that someone had to carry it, and that someone (not necessarily the same person) left it at the Bau. That gives us a chance to see the movements of people across the landscape. The combination of an unusually detailed record of lithic raw material source use, the characteristics of the sources and the landscape, and the seasonal use of the site as shown by the fauna, gives us a glimpse of the strategies the hominins used to keep themselves supplied with the things they needed.

It was shown in chapter 5, for instance, that within layer H we can detect three strategies of use of the territory: (1) concentrate on the SW and the Near SW, (2) use the SW with more WSW and less Near SW, and (3) move the centre of gravity of the territory to take in more NE and Near SW material, while reducing use of the SW and especially of the WSW. None of these strategies was exclusively used, but there are tendencies to favour one more than others, and strategy number 3 seems to match up with winter-time use of the site, reducing the need for difficult travel in snowy conditions. At that time, too, the fauna are more commonly the medium-sized animals (especially tahr) which would have been available in the gorge itself. The GLM reported in Wilson & Browne (2014) showed (Table 12.6) that although the quality of raw material at the source mattered (as did the availability of a large size of nodule), it was above all the ease of access to raw materials that was driving source area choice in layer H. It also seems that layer H is a case of provisioning of place (Kuhn, 1995), based on the abundance and size of the cores (244 cores in the typological sample, or 2.7%, with an average weight of 55.8 g).

The other layer that shows a majority of medium-sized animals (Table 12.7) is layer 2, and it presents an even more extreme case of this reduced centre of gravity, since not only does it have a relatively high proportion of material from the NE (Table 12.5) and a very low proportion from the WSW, it has a remarkably high proportion (for the Bau) of local raw material. On the other hand, layer 2 also has a very diverse lithic assemblage, with material from many different sources in many different directions, in small quantities. These were a highly mobile people, carrying their toolkits with them as they moved around in their large territory, but when they came to the Bau they took advantage of local resources, both lithic and faunal. The GLM (Table 12.6) shows that raw material characteristics (quality and rock sizes available) were much more important in influencing their choice of sources of raw material than the ease of obtaining it (although that did still matter), which makes some sense when we consider that they seemed to have provisioned the individual (Kuhn, 1995) with their personal toolkits, rather than the place. This is corroborated by the relative scarcity and small size of cores: there are 23 in the typological sample, or 0.8%, with an average weight of 29.9 g – a bit over half the size of the cores in layer H.

Thus we see that the hominins that used the Bau also used entire territories, and they used them in versatile ways. There are no changes through time that we can (falsely) point to as "improvements" or "progress". Instead, they seem to have had a large bag of tricks to delve into, and they varied their behaviour through time as they saw fit and as the environment around them changed. We have to remember that we are seeing only a very small part of what went on in the region over a very long time. The unusual raw materials show us where their territory extended to (it may have stretched farther – we cannot know), but we do not have information on what they did while in those other areas, nor at what time of year they went there. Other sites in the Vaucluse exist (La Combette, Bérigoule, the Baume des Peyrards, etc.) but they are not contemporaneous, so although they can add more pieces to the jigsaw puzzle, there are still not enough to create a full picture at any one time. The pieces we do have at the Bau, however, give us better than usual views of how those hominins lived.

12.4. Words are flowing out . . .

It often seems to me that people imagine the lives of prehistoric hominins to have been unrelentingly bleak. Cold, wet, and hungry, they huddled together for protection against marauding predators . . . Really? They mastered fire. They had sophisticated toolkits and mobility strategies that allowed them to exploit comfortably large territories. They hunted large, dangerous animals: *they* were the marauding predators! And they, like us, were primates that lived in social groups, cooperating and interacting with each other.

The archaeologists that excavated the Bau de l'Aubesier were, and still are, social primates. Our excavation activities included a wide range of human behaviours. We worked hard (some shirked), we sweated, we bled, we argued, we cried, but we also loved, and we laughed. The sounds of jackhammers, of hammers and chisels, of "oof!" and "gro-o-an!", were punctuated by the sounds of laughter.

The hominins that came before us also had emotions and interpersonal interactions. Why not? They lived and worked together, as we do. Why would they not take pleasure in a job well done? We can imagine them sheltering from a wild storm, but we can also imagine them enjoying a warm sunny day. Watching the bats leave the cave in the evening, and the birds soaring in the updrafts of the gorge in the afternoon. Stepping outside of the rock shelter to see where the sun was going during a solar eclipse – as we did; and watching it come back again. Taking care of family members, and grieving, in their way, when someone died. Expressing the equivalent of "Oh, *man* that's good!" while eating a particularly well cooked piece of aurochs. And laughing. Laughing when someone did something silly. Laughing when someone hit a flint core just right – and doesn't that sound good? Laughing and proud when the little one walked for the first time.

They have not left us the sounds of their laughter, but they have left us the shades of their lives, inciting and inviting us to learn more about them. They hunted, they ate, they made tools . . . they lived. They teach us that although our world does change, there are universals, too: across space – and across time.

12.5. Sommaire du chapitre

Ce livre est consacré au site archéologique du Bau de l'Aubesier, aux gens qui l'ont visité pendant environ 200 000 ans, et à l'abri-sous-roche lui-même. Il a abrité des générations de groupes humains en leur offrant un lieu d'activité et de repos, préservé les vestiges de leurs travaux, et il a défié et informé les archéologues qui sont venus fouiller les traces du passé. Le Bau est spécial: c'est un lieu de vie, brillant le soir dans la lumière qui se reflète dans les falaises avoisinantes. Il a un esprit, ancien mais pas vieux, bienveillant mais détaché. Si l'on prend le temps de s'ouvrir calmement à lui, on peut communier, peut-être même communiquer avec lui – mais il ne va pas simplement d'emblée vous raconter ses histoires. Il faut travailler pour cela.

Le travail des archéologues, commencé en 1901, nous permet de retracer l'histoire du Bau. Dans les onze premiers chapitres de ce livre nous avons décrit ses dépôts, ses vestiges et leur contexte. Dans ce douzième chapitre nous abordons deux questions finales. La première, peut-on trouver des liens ou correspondances entre les couches de la Tranchée Moulin et celles de la tranchée L? La seconde constitue en fait la question ultime de cet ouvrage: qu'est-ce que le Bau nous apprend des hominidés du Paléolithique moyen?

Les deux tranchées sont localisées dans les dépôts supérieurs et comprennent une masse de brèche très dense avec, plus bas, une dalle de brèche à grain très fin. Entre les deux se trouve une couche archéologique noire: la couche 4 dans la tranchée L, et la couche D dans la Tranchée Moulin. Tout semble simple: D devrait être l'équivalent de 4. Cependant, ce n'est pas le cas. La Figure 12.1 montre la situation schématiquement. On peut comparer les couches archéologiques par leurs positions stratigraphiques, leur contenu sédimentologique, leurs restes fauniques, leurs matières premières lithiques et leurs typologies, etc. . . . mais très souvent on trouve des problèmes. Ainsi, la couche D s'efface juste à l'est de la Tranchée Moulin, et on ne sait pas si elle continue sous la brèche. A l'exception de sa carbonisation, son contenu n'est pas du tout comme celui de la couche 4. Au final on manque d'indices et la question d'équivalence reste ouverte.

Autre exemple, sous la dalle, dans la tranchée L nous avons la couche 5, tandis que dans la Tranchée Moulin se trouve la couche E, qui surmonte la couche F. Selon les critères choisis la couche 5 ressemble, soit à la couche E, soit à la couche F, aux deux ou alors ni à l'une ni à l'autre. Là encore, pas de conclusion possible.

Plus haut, à l'intérieur de la brèche dans la tranchée L se trouve la couche 3. D'après sa nature sédimentologique, et par la nature dispersée de ses sous-couches, cette couche ressemble à la couche C de la Tranchée Moulin. Nous manquons d'informations sur la faune, mais les données sur le lithique sont similaires, même si elles ne sont pas identiques, et on peut dire qu'il est probable que les couches 3 et C sont équivalentes.

Enfin, au dessus de la brèche dans la tranchée L, la couche 2 se caractérise par des sédiments meubles. Dans la Tranchée Moulin il n'y avait que très peu de sédiment de cette nature, seulement des dépôts dans l'espace au fond entre la brèche et la paroi rocheuse. Lorsque nous avons trouvé du matériel archéologique dans ces sédiments, nous l'avons regroupé dans la couche C, qui peut alors correspondre partiellement à la couche 2, ainsi qu'à la couche 3. En revanche, il est également possible que la couche 2 soit tout simplement absente de la Tranchée Moulin.

Dans tous les cas, il n'y a pas de raison de croire que les couches des deux tranchées doivent obligatoirement correspondre à des dépôts équivalents. Les préhistoriques ont pu utiliser différentes parties du site lors de visites bien espacées dans le temps. Pendant une seule occupation, par contre, ils ont pu entreprendre des activités distinctes dans les deux parties du site, laissant des vestiges différents (restes de cuisine versus lieu de repos ou taille de silex, par exemple). Cependant, cela vaut la peine de chercher des liens entre les tranchées, pour essayer d'avoir une vision plus globale du mode de vie des préhistoriques.

Qu'est-ce que le Bau peut nous apprendre sur le Paléolithique moyen? D'abord, il nous oblige à élargir notre conception de l'industrie lithique de cette période. Il nous livre des éléments typiquement moustériens: beaucoup de racloirs, utilisation de la technologie levallois, etc. Dans plusieurs couches, cependant, y compris les plus anciennes, on trouve beaucoup plus de lames que les industries du Moustérien n'en contiennent habituellement. La lame a auparavant été considérée comme une indication d'industries de type Paléolithique supérieur. Il est maintenant évident que cette interprétation ne peut plus être retenue.

Le Bau est également remarquable pour l'utilisation intense du feu dans certaines couches. Il y a une sorte de bruit de fond du feu dans ce site, avec quelques vestiges brûlés dans chaque couche, mais les couches H1, E, D et 4 sont complètement carbonisées: sédiments, os et lithiques. Est-ce que les hominidés de ces couches maitrisaient le feu? Sandgathe (2017) a suggéré que les Néandertaliens devaient chercher et sauvegarder des feux naturels, et qu'il y aurait donc une corrélation entre l'utilisation du feu et la période ou saison quand ces derniers sont plus abondants (i.e., quand la foudre est la plus fréquente). Au Bau, cette corrélation ne semble pas exister. L'aire de combustion dans la couche H1 semble avoir été très active en majorité pendant l'hiver, tandis que la couche 4 reflète des occupations plutôt estivales. Ce qui est certain, c'est que les préhistoriques du Bau ont utilisé le feu de manière récurrente et habituelle sur de longues périodes. Ils ont utilisé les os comme combustible, et s'ils n'avaient pas de foyers aménagés, ils avaient au moins l'habitude de placer leurs feux à peu près aux mêmes endroits. Ainsi, au moins dans les couches noires, le feu était un outil important qu'ils savaient très bien utiliser.

Le Bau nous donne également quelques informations sur les relations sociales des hominidés. Dans la couche 4, il y avait un ou plusieurs groupes familiaux comme en témoignent les dents de lactéales retrouvées. L'adulte de la couche I est mort après une maladie de longue durée; l'aide de ses proches a sans doute prolongé sa vie. Les problèmes dentaires de cet adulte, ainsi que les caries trouvées sur deux dents de la couche 4, placent ces dernières parmi seulement quelques exemplaires de ces problèmes chez les hominidés. En revanche, nous avons une dent de la couche I qui montre des "soins dentaires", avec une rainure creusée par l'utilisation fréquente d'un cure-dent.

Les caries dentaires prouvent que ces gens devaient manger des glucides, mais la quantité d'os impressionnante dans ce site témoigne du rôle important de la viande dans leurs vies. Dans la plupart des couches, les restes fauniques des animaux de grande taille (aurochs, cheval, etc.) dominent; les couches H et 2 sont les seules (avec un bon échantillon faunique) où la majorité des animaux serait de taille moyenne (bouquetin, chamois, tahr, etc.). La distribution des âges des animaux suggère très souvent une chasse sélective, adaptée à chaque espèce en fonction de la saison de l'année, et ce quelle que soit l'éthologie ou les comportements des espèces chassées. Ces groupes étaient polyvalents dans leurs stratégies d'acquisition.

L'utilisation saisonnière du site n'est pas systématique, mais probablement orientée vers des visites estivales pendant les périodes froides, et plutôt hivernales en période tempérée. Il se peut que les préhistoriques suivaient les troupeaux d'animaux, en une sorte de transhumance à l'ancienne. Il est également possible que les Néandertaliens préféraient éviter la chaleur à plus basse altitude pendant la période estivale des périodes tempérées. Il serait intéressant de trouver d'autres sites avec des indications de fréquentations saisonnières, en particulier plus bas dans les plaines, ou plus haut sur les flancs du Mont Ventoux.

Les matières premières représentées dans les assemblages lithiques nous donnent des informations précieuses concernant les déplacements territoriaux des groupes (mais malheureusement pas d'indications saisonnières!). Ainsi, nous avons identifié plusieurs stratégies d'approvisionnement et de mobilité à travers le territoire dans les différentes couches. Les hominidés ont pu choisir parmi ces stratégies pour varier leur mode de vie, tout en utilisant très souvent les mêmes ressources minérales disponibles dans le même paysage lithique. Encore une fois, ils ont fait preuve de comportements polyvalents.

Flots de paroles. . . Il me semble souvent que les gens ont une idée des vies des préhistoriques comme étant implacablement morne. Frigorifiés, mouillés, affamés, ils se sont blottis ensemble, cherchant un peu de protection contre les prédateurs en maraude. . . Sérieusement? Ils ont maitrisé le feu. Ils avaient des trousses à outils bien adaptés et des stratégies de déplacements qui leur permettaient l'exploitation de grands territoires. Ils chassaient de grands animaux souvent dangereux: c'était *eux*, les prédateurs en maraude! Et comme nous, ils étaient des primates, vivant en groupes sociaux, coopérant et interagissant avec leurs proches.

Les archéologues qui ont fouillé au Bau de l'Aubesier étaient et sont toujours des primates sociaux. Pendant nos fouilles nous nous sommes comportés comme les humains font toujours: nous avons travaillé dur (quelques-uns ont esquivé), transpiré, saigné, pleuré, et nous nous sommes parfois disputés. Mais nous avons aussi aimé, et nous avons ri. Les sons des marteaux-piqueurs, des ciseaux, des marteaux, les gémissements et les "ouf!" étaient souvent ponctués par des rires.

Nos prédécesseurs préhistoriques n'avaient-ils pas eux aussi des émotions? Ils vivaient et travaillaient ensemble, comme nous. Ils ont dû apprécier le bel ouvrage ou une tâche bien faite. Si l'on peut facilement les imaginer s'abritant pendant une tempête, pourquoi ne pas les envisager profitant d'une belle journée? Ils ont pu observer le vol des chauve-souris qui abandonnent l'abri au crépuscule, ou des oiseaux montant par les courants ascendants dans les gorges l'après-midi. Quitter l'abri, comme nous l'avons fait, pour voir où le soleil se cache pendant une éclipse, et scruter son retour. Prendre soin d'un membre de la famille, et pleurer sa mort. Dire l'équivalent de "Oh là là que c'est bon!" en mangeant un morceau particulièrement bien cuit d'aurochs. Et rire. Rire quand quelqu'un faisait quelque chose de drôle. Rire au son si satisfaisant de la frappe parfaite et précise sur un nucléus en silex. Rire et être fier quand le petit marche pour la première fois.

Ils ne nous ont pas laissé les sons de leurs rires, mais plutôt les ombres de leurs vies, nous incitant et nous invitant à en apprendre plus. Ils chassaient, ils mangeaient, ils fabriquaient des outils . . . ils vivaient. Ils nous apprennent que, bien que notre monde change, il existe de façon immuable des caractéristiques humaines universelles: à travers l'espace – et le temps.

Bibliography

Agam, A., Azuri, I., Pinkas, I., Gopher, A., & Natalio, F. 2020 Estimating temperatures of heated Lower Palaeolithic flint artefacts. Nature Human Behaviour, https://doi.org/10.1038/s41562-020-00955-z.

Argant, A. 2006 Les Carnivores du Bau de l'Aubesier (Vaucluse): Rapport d'étude préliminaire. In S. Lebel (Ed.) Le Bau de l'Aubesier: Abri sous roche moustérien des Gorges de la Nesque (Commune de Monieux, Vaucluse), rapport de mission 2006. Rapport soumis au Ministère de la Culture et de la Communication de France, DRAC-PACA. Unpublished.

Argant, J. 2000 Le Bau de l'Aubesier (Vaucluse) – Prélèvements 1999 pour l'analyse pollinique et premiers résultats. In S. Lebel (Ed.) Le Bau de l'Aubesier (Vaucluse, Monieux): Rapport des activités de fouilles 1998-1999-2000. Manuscrit déposé auprès du Ministère de la Culture et de Communication de France; pp. 71-71g. Unpublished.

Argant, J. 2006 Analyse pollinique de la coupe C15 et d'un spéléothème. In S. Lebel (Ed.) Le Bau de l'Aubesier: Abri sous roche moustérien des Gorges de la Nesque (Commune de Monieux, Vaucluse), rapport de mission 2006. Rapport soumis au Ministère de la Culture et de la Communication de France, DRAC-PACA. Unpublished.

Ben-Dor, M., & Barkai, R. 2020 The importance of large prey animals during the Pleistocene and the implications of their extinction on the use of dietary ethnographic analogies. Journal of Anthropological Archaeology 59, 101192; https://doi.org/10.1016/j.jaa.2020.101192.

Blackwell, B., Schwarcz, H.P., Lebel, S., Porat, N., & Wilson, L. 1991 $^{230}Th/^{234}U$ dates for stalagmites and Electron Spin Resonance (ESR) dates for enamel from the Pleistocene site, Bau de l'Aubesier (Provence), France. Abstract submitted to the annual meeting of the Geological Association of Canada & Mineralogical Association of Canada.

Blaser, F., Bourguignon, L., Sellami, F., Rios Garaizar, J. 2012 Une série lithique à composante laminaire dans le Paléolithique moyen du Sud-Ouest de la France: le site de Cantalouette 4 (Creysse, Dordogne, France). Bulletin de la Société préhistorique française 109(1), 5-33.

Bordes, F. 1961 Typologie du Paléolithique ancien et moyen. Delmas, Publications de l'Institut de Préhistoire de l'Université de Bordeaux, Mémoire n° 1.

Browne, C.L., & Wilson, L. 2011 Resource selection of lithic raw materials in the Middle Palaeolithic in southern France. Journal of Human Evolution 61, 597-608; doi: 10.1016/j.jhevol.2011.08.004.

Buisson-Catil, J. (Ed.) 1994 A la rencontre des chasseurs néandertaliens de Provence nord-occidentale. Notices d'Archéologie vauclusienne 3: Le Paléolithique moyen en Vaucluse. Avignon: Service d'Archéologie de Vaucluse.

Carmignani, L. 2017 From the flake to the blade: Technological evolution of the Middle Paleolithic blade phenomenon. IDQP "International Doctorate in Quaternary and Prehistory" Ph.D. Thesis, Universitat Rovira i Virgili, Tarragona Spain.

Carmignani, L., Moncel, M.-H., Fernandes, P., & Wilson, L. 2017 Technological variability during the Early Middle Palaeolithic in Western Europe. Reduction systems and predetermined products at the Bau de l'Aubesier and Payre (South-East France). PLoS ONE 12(6): e0178550. https://doi.org/10.1371/journal.pone.0178550

Courtin, J. 1974 Le Néolithique de la Provence. Société préhistorique française, Paris, Mémoire 11.

Courty, M.-A. 2006 Rapport sur la micromorphologie. In S. Lebel (Ed.) Le Bau de l'Aubesier: Abri sous roche moustérien des Gorges de la Nesque (Commune de Monieux, Vaucluse), rapport de mission 2006. Rapport soumis au Ministère de la Culture et de la Communication de France, DRAC-PACA. Unpublished.

Crégut-Bonnoure, E. 2011 Sur la présence du lion des cavernes *Panthera leo spelaea* Goldfuss en Provence (sud-est de la France). Quaternaire Hors-série 4, 237-244.

Crégut-Bonnoure, E., & Guérin, C. 1996 Famille des Bovidés. In C. Guérin & M. Patou-Mathis (Eds.) Les grands mammifères plio-pléistocènes d'Europe, pp. 62-106. Paris: Masson, Collection Préhistoire.

Crégut-Bonnoure, E., Slimak, L., Lewis, J.E., & Brochier, J.E. 2011 Nouvelles données sur les sites pléistocènes et holocènes à *Ursus arctos* du Vaucluse (sud-est de la France). Quaternaire Hors-série 4, 147-183.

de Lumley, H. 1971 Le Paléolithique inférieur et moyen du midi méditerranéen dans son cadre géologique. Vième supplément à Gallia Préhistoire, tome I, Ligurie-Provence.

de Lumley, M. A. 1973 Anténéandertaliens et Néandertaliens du bassin méditerranéen occidental européen. Etudes Quaternaires 2, 1–626.

du Merle, P., & Guende, G. 1987 Présentation du Mont Ventoux. In P. du Merle (Ed.), Ecologie du Mont Ventoux. La Terre et La Vie, suppl. 1 (pp. 11-20). Paris: Société nationale de Protection de la Nature.

Faure, M., Guérin, C. & Lebel, S. 1996 Les modes de subsistance des Néanderthaliens dans le Paléolithique moyen du sud-est de la France: l'abri du Bau de l'Aubesier à Monieux (Vaucluse). In La vie préhistorique (pp. 160-165). Dijon: Faton.

Fernandez, P. 2001 Etude archéozoologique et paléontologique des niveaux d'occupations moustériens au Bau de l'Aubesier (Vaucluse): implications palethnologiques et biochronologiques. Thèse de Doctorat, Université Claude-Bernard, Lyon I.

Fernandez, P. 2006 Etude paléontologique des ongulés du gisement moustérien du Bau de l'Aubesier (Vaucluse, France): Morphométrie et contexte biochronologique. Documents du Laboratoire de Géologie de Lyon, no. 161.

Fernandez, P., & Legendre, S. 2003 Mortality curves for horses from the Middle Palaeolithic site of Bau de l'Aubesier (Vaucluse, France): methodological, palaeo-ethnological and palaeo-ecological approaches. Journal of Archaeological Science 30, 1577-1598.

Fernandez, P., Faure, M., Guérin, C., & Lebel, S. 1998 Stratégie de chasse des Néandertaliens du Bau de l'Aubesier: choix et opportunisme. In XVIIIè Rencontres Internationales d'Archéologie et d'Histoire d'Antibes, Economie préhistorique: les comportements de subsistance au Paléolithique. Sophia Antipolis, France: Editions APDCA, 309-323.

Fernandez, P., Guadelli, J.L., & Fosse, P. 2006 Applying dynamics and comparing life tables for Pleistocene Equidae in anthropic contexts (Bau de l'Aubesier, Combe-Grenal) and carnivore (Fouvent) contexts with modern feral horse populations (Akagera, Pryor Mountain). Journal of Archaeological Science 33, 176-184.

Gagnepain, J., Gaillard, C. 2005 La grotte de la Baume Bonne (Quinson, Alpes de Haute-Provence): synthèse chronostratigraphique et séquence culturelle d'après les fouilles récentes (1988-1997). In N. Molines, M.-H. Moncel, and J.-L. Monnier (Eds.), Données récentes sur les modalités de peuplement et sur le cadre chronostratigraphique, géologique et paléo-géographique des industries du Paléolithique inférieur et moyen en Europe, BAR Publishing International Series 1364, Actes du Colloque International de Rennes, 22-25 septembre 2003, 73-85.

Gagnepain, J., Gaillard, C., Notter, O. 2004 La composante laminaire dans les industries lithiques du Paléolithique moyen du Verdon (sud-est de la France). In Le Paléolithique moyen. [The Middle Palaeolithic], Actes du colloque UISPP, Université de Liège, 2001, Section 5, Bar Publishing International Series, 1239, 57-65.

Jaouich, A. and Paquin, D. 1996 Diffractométrie aux rayons-X et caractérisation de la couche H-1. In S. Lebel (Ed.) Le Bau de l'Aubesier (Vaucluse, Monieux): Numéro de site 84 07 900 2AP. Rapport de synthèse des activités de fouille 1993-1994-1995. Manuscrit déposé auprès du Ministère de la Culture de France et du Conseil de Recherches en Sciences humaines du Canada; section 2.6. Unpublished.

Kennedy, S. 2002. Hemitragus jemlahicus. (On-line), Animal Diversity Web. Accessed July 07, 2020 at https://animaldiversity.org/accounts/Hemitragus_jemlahicus/.

Kuhn, S.L. 1995 Mousterian Lithic Technology: An Ecological Perspective. Princeton: Princeton University Press.

Kuhn, S.L. 2004 Upper Palaeolithic raw material economies at Uçagizli cave, Turkey. Journal of Anthropological Archaeology 23, 431-448.

Lebel, S. 1987 Le Bau de l'Aubesier, Vaucluse, Monieux: Rapport des activités de fouille 1987. Avec la collaboration de R. Brandi. Manuscrit déposé auprès du Ministère de la Culture de France et du Conseil de Recherches en Sciences humaines du Canada. Unpublished.

Lebel, S. 1988 Le Bau de l'Aubesier, Vaucluse, Monieux: Rapport des activités de fouille 1988. Avec la collaboration de R. Brandi. Manuscrit déposé auprès du Ministère de la Culture de France et du Conseil de Recherches en Sciences humaines du Canada. Unpublished.

Lebel, S. 1990 Le Bau de l'Aubesier, Vaucluse, Monieux: Rapport intermédiaire des activités de fouille 1990. Manuscrit déposé auprès du Ministère de la Culture de France et du Conseil de Recherches en Sciences humaines du Canada. Unpublished.

Lebel, S. 1993 Le Bau de l'Aubesier, Vaucluse, Monieux: Rapport intermédiaire des activités de fouille 1993. Manuscrit déposé auprès du Ministère de la Culture de France et du Conseil de Recherches en Sciences humaines du Canada. Unpublished.

Lebel, S. 1994 Le Bau de l'Aubesier, Vaucluse, Monieux: Rapport intermédiaire des activités de fouille 1994. Manuscrit déposé auprès du Ministère de la Culture de France et du Conseil de Recherches en Sciences humaines du Canada. Unpublished.

Lebel, S. 1996 Le Bau de l'Aubesier (Vaucluse, Monieux): Numéro de site 84 07 900 2AP. Rapport de synthèse des activités de fouille 1993-1994-1995. Manuscrit déposé auprès du Ministère de la Culture de France et du Conseil de Recherches en Sciences humaines du Canada. Unpublished.

Lebel, S. 2000 Le Bau de l'Aubesier, Gorges de la Nesque (Vaucluse, Monieux): Rapport des activités de fouille 1998-1999-2000. Numéro de site 84 07 900 2AP. Manuscrit déposé auprès du Ministère de la Culture et Communication de France et du Conseil de Recherches en Sciences humaines du Canada. Unpublished.

Lebel, S. 2006 Le Bau de l'Aubesier: Abri sous roche moustérien des Gorges de la Nesque (Commune de Monieux, Vaucluse), rapport de mission 2006. Rapport soumis au Ministère de la Culture et de la Communication de France, DRAC-PACA. Unpublished.

Lebel, S. & Trinkaus, E. 2001 New discoveries of Middle Palaeolithic human remains from the "Bau de l'Aubésier (Vaucluse, France)", Nouvelles découvertes de restes humains du Paléolithique moyen du Bau de l'Aubésier (Vaucluse, France). Bulletin et Mémoires de la Société d'Anthropologie de Paris, n.s., 13(1-2), 15-21.

Lebel, S. & Trinkaus, E. 2002a Middle Pleistocene human remains from the Bau de l'Aubesier. Journal of Human Evolution 43, 659-685.

Lebel, S. & Trinkaus, E. 2002b Short Note: A Carious Neandertal Molar from the Bau de l'Aubesier, Vaucluse, France. Journal of Archaeological Science 28, 555–557. doi:10.1006/jasc.2001.0689.

Lebel, S., Trinkaus, E., Faure, M., Fernandez, P., Guérin, C., Richter, D., Mercier, N., Valladas, H., & Wagner, G.A. 2001 Comparative morphology and paleobiology of Middle Pleistocene human remains from the Bau de l'Aubesier, Vaucluse, France. Proceedings of the National Academy of Science, USA 98, 11097-11102.

Legal, S. 2007 Histoire géologique du Lubéron. Bulletin de la Société des Naturalistes du Pays d'Apt 21, 63-66.

Masse, J.-P., 1993 Valanginian to Early Aptian carbonate platforms from Provence (S.E. France). In T.A. Simo, R.W. Scott, & J.-P. Masse, (Eds.), Cretaceous Carbonate Platforms (pp. 363-374). Am. Assoc. Petrol. Geol. Mem. 56.

McNabb, J. 2011 The British Lower Palaeolithic: Stones in Contention. Routledge; Taylor & Francis e-Library, 419 p.

Moncel, M.-H. 2005 Baume Flandin et Abri du Maras: deux exemples de débitage laminaire du début du Pléistocène supérieur dans la Vallée du Rhône (sud-est, France). L'Anthropologie 109 (3), 451-480.

Moulin, F. 1904 L'Abri moustérien du Bau de l'Aubesier (Vaucluse). Bulletin de la Société Préhistorique de France 1, 14–20.

Parush, Y., Assaf, E., Slon, V., Gopher, A., Barkai, R. 2015 Looking for sharp edges: Modes of flint recycling at Middle Pleistocene Qesem Cave, Israel. Quaternary International 361, 61-87. https://doi.org/10.1016/j.quaint.2014.07.057

Pelletier, M., Desclaux, E., Mallye, J.-B., & Crégut, E. 2020 Identifying the accidental-natural mortality of leporids in the archaeological record: insights from a taphonomical analysis of a pitfall without evidence of human presence. Journal of Quaternary Science 35, 677-694. 10.1002/jqs.3203.

Sandgathe, D. 2017 Identifying and Describing Pattern and Process in the Evolution of Hominin Use of Fire. Current Anthropology 58, supplement 16, S360-S370.

Sauzade, G. 1983 Les sépultures du Vaucluse du Néolithique à l'Age de Bronze. Etudes quaternaires 6.

Triat, J.-M. 2015 Pierres de Provence: Richesse et originalité du patrimoine géologique provençal. Cavaillon: Imprimérie Rimbaud.

Trinkaus, E., Lebel, S. & Bailey, S.E. 2000a Middle Paleolithic and Recent Human Dental Remains from the Bau de l'Aubesier, Monieux (Vaucluse). Bulletin et Mémoires de la Société d'Anthropologie de Paris, 12(3-4), 207-226.

Trinkaus, E., Smith, R.J. & Lebel, S. 2000b Dental Caries in the Aubesier 5 Neandertal Primary Molar. Journal of Archaeological Science 27, 1017–1021. doi:10.1006/jasc.1999.0512.

Truc, G. 2012 Histoire géologique. In G. Truc, G. Sauzade & J.-M. Hermant (Eds.), Découverte géologique du Bassin de Mormoiron (pp. 10-43). Communauté des Communes des Terrasses du Ventoux: création denis plat.com.

Ughetto, G., & Ughetto, A.-L. 2007 Aperçu de l'histoire géologique du Mont Ventoux. In F. Melki, & M. Briola (Eds.), Ventoux, géant de nature: Guide des richesses biologiques du mont Ventoux (pp. 30-39). Mèze, France: Collection Parthénope, éditions Biotope.

van Vuure, T. 2002 History, morphology and ecology of the aurochs (*Bos primigenius*). http://citeseerx.ist.psu.edu/viewdoc/summary?doi=10.1.1.534.6285.

Wilson, L. 1988 Petrography of the Lower Palaeolithic Assemblage at the Caune de l'Arago, France. World Archaeology 19(3), 376-387.

Wilson, L. 2007 The Vaucluse raw material project: artifact provenance and landscape context in the Middle Palaeolithic of southern France. In L. Wilson, P. Dickinson, & J. Jeandron (Eds.), Reconstructing Human-Landscape Interactions: Papers of the Developing International Geoarchaeology 2005 conference. Newcastle-upon-Tyne: Cambridge Scholars Publishing, 234-251.

Wilson, L., & Browne, C.L. 2014 Change in Raw Material Selection Criteria and Subsistence Behaviour Through Time at a Middle Palaeolithic Site in Southern France. Journal of Human Evolution 75, 28-39; http://dx.doi.org/10.1016/j.jhevol.2013.12.018.

Wilson, L., Webb, H., & Baker, C., 2010 Inventaire des silex du Vaucluse disponible sur site web et utilisation de fossiles dans la détermination de l'origine du silex vauclusien. Proceedings of the conference "Silex et territoires préhistoriques. Avancées des recherches dans le midi de la France", C@hiers de Géopré 1, 38-50.

Wilson, L., Browne, C.L., & Texier, P.-J. 2018 Provisioning on the Periphery: Middle Palaeolithic raw material supply strategies on the outer edge of a territory. Journal of Archaeological Science: Reports 21, 87-98.

Wilson, M. 1997 Etude sédimentologique et micromorphologique du Bau de l'Aubésier, France: site archéologique néandertalien (Würm moyen). Mémoire présenté comme exigence partielle de la Maîtrise en Sciences de la Terre, Université du Québec à Montréal.

www.ingramcontent.com/pod-product-compliance
Lightning Source LLC
LaVergne TN
LVHW071208210926
848482LV00015B/748
9781407358383